ASTROLOGY, SCIENCE AND SOCIETY
HISTORICAL ESSAYS

ASTROLOGY
SCIENCE AND SOCIETY
HISTORICAL ESSAYS

Edited by Patrick Curry

THE BOYDELL PRESS

First published 1987 by The Boydell Press
an imprint of Boydell & Brewer Ltd
PO Box 9, Woodbridge, Suffolk IP12 3DF
and Wolfeboro, New Hampshire 03894-2069, USA

ISBN 0 85115 459 X

British Library Cataloguing in Publication Data

Astrology, science and society: historical essays.
1. Astrology – Europe – History
I. Curry, Patrick
133.5′094 BF1676
ISBN 0-85115-459-X

Library of Congress Cataloging-in-Publication Data

Astrology, science, and society.
'Evolved out of a conference held at the Warburg Institute, London, in March 1984, on the history of medieval and Renaissance astrology in Europe' – Acknowledgments.
Bibliography: p.
1. Astrology – History – Congresses.
2. Europe – Intellectual life – Congresses.
I. Curry, Patrick. II. Warburg Institute.
BF1676.A28 1987 133.5′094 86-23304
ISBN 0-85115-459-X

Printed in Great Britain by
St Edmundsbury Press, Bury St Edmunds, Suffolk

CONTENTS

LIST OF ILLUSTRATIONS

ACKNOWLEDGEMENTS

These essays evolved out of a conference held at the Warburg Institute, London, in March 1984, on the history of medieval and Renaissance astrology in Europe. I would like to record my thanks to several people, for their help both then and since: to Jacques Halbronn, who originally suggested holding the conference; to the Warburg Institute, for permitting their conference facilities to be used, and assisting with the arrangements; and of the many others I could name, to two in particular: Angus Clarke, and my wife, Suzanna, for their constant encouragement and advice. I am also grateful to Dr Clarke for translating a paper, along with those who translated their own papers and one anonymous translator. Above all, I would like to thank the contributors to this volume, who patiently stayed with me during the long and often trying process of bringing it to fruition. (In this connection, I am also grateful to Professors E. S. Kennedy and R. Klibansky, for giving papers at the Warburg conference which I regret being unable to include here.)

The publication of this book has been assisted by a grant from the *Twenty-Seven Foundation*. The Twenty-Seven Foundation is a registered private charity which gives grants for research and publishing in history, c/o the Institute of Historical Research, Senate House, Malet Street, London WC1 7HU.

Publication was also assisted by a grant from the *Urania Trust*. The Urania Trust is a registered educational charity, dedicated to the study of the relationship between man's knowledge and beliefs about the heavens and every aspect of his art, science, philosophy and religion. It welcomes inquiries c/o 49 Castle Street, Frome, Somerset BA11 3BW.

Together with the publisher, I would like to express my thanks to both bodies for their generous help.

BIBLIOGRAPHY

This is a short and highly selective bibliography for the history of astrology. It consists almost entirely of leading secondary sources in English, and mainly with reference to the period covered by the essays in this volume. It includes some historiographical writings of interest. Many other references of interest will be listed in this volume's individual papers.

Allen, D. C. *The Star-Crossed Renaissance* (1941; reprinted N.Y.: 1966, 1973).

Capp, B. *Astrology and the Popular Press. English Almanacs 1500–1800* (London: 1979).

Clarke, A. 'Giovanni Antonio Magini (1555–1617) and Late Renaissance Astrology' (Ph.D. thesis, University of London: 1985).

Curry, P. 'Revisions of Science and Magic', *History of Science* 23 (1985), 299–325.

Eade, J. C. *The Forgotten Sky: A Guide to Astrology in English Literature* (Oxford: 1984).

Field, J. 'A Lutheran Astrologer: Johannes Kepler', *Archive for History of Exact Sciences* 21 (1984), 189–272.

Garin, E. *Astrology in the Renaissance: The Zodiac of Life* (1976; translated London: 1983).

Klibansky, R., Panofsky, E. and Saxl F. *Saturn and Melancholy: Studies in the History of Natural Philosophy, Religion and Art* (London: 1964).

Lilly, W. *Christian Astrology* (1947; reprinted London: 1985).

Lindberg, D. (ed.) *Science in the Middle Ages* (Chicago: 1978).

Neugebauer, O. 'The Study of Wretched Subjects', *Isis* 42 (1951), 111.

Pagel, W. 'The Vindication of Rubbish', *Middlesex Hospital Journal* 42 (1945), 42–45.

Ptolemy, C. *Tetrabiblos*, edited and translated by F. E. Robbins (London and Cambridge, MA: 1940).

Saxl, F. *Lectures* (London: 1957).

Tester, J. *A History of Western Astrology* (forthcoming, Woodbridge, Suffolk: 1987).

Thomas, K. *Religion and the Decline of Magic* (London: 1971 and Harmondsworth: 1973).

Thorndike, L. *A History of Magic and Experimental Science* (N.Y.: 1923–58; 8 vols).

Thorndike, L. 'The True Place of Astrology in the History of Science', *Isis* 46 (1955), 273–78.

INTRODUCTION

Patrick Curry

The essays in this volume extend from late thirteenth-century Italy to late seventeenth-century England, and readers looking for a close unity of immediate subject-matter and approach are forewarned – they stand rather as a demonstration of the scope and fruitfulness of the subject as a whole. Its relative neglect alone is adequate justification, I think, for bringing together such widely-ranging papers, and I have no intention of attempting to generate a spurious unity beyond that. Equally, this is not the place for a detailed historiographical survey; besides amounting to a book in itself, it would not introduce the subject so much as throw the reader into its deep end. What I want to do here is simply to place these essays in overall context, with some remarks on the history of astrology as such, and as an intellectual discipline. In the first place, it will be necessary to examine its travails; then, the signs of hope.

Over half a century ago, Fritz Saxl remarked that 'The history of astrology is still regarded as an obscure and unimportant subject.' With some exceptions, that remains sadly true today. Certainly the subject has attracted some first-rate scholars, who have produced lasting work. Even leaving aside the history of non-European astrology, there are the names of Bouché-Leclercq, Boll, Neugebauer, Thorndike, Cumont, Garin and Thomas – without by any means exhausting the list. Nonetheless, such figures stand out as relatively isolated figures in the ranks of past and present historians. And that would seem to be something of a puzzle. For the fact is – and here it is only possible to state this, with extreme compression – that even without looking any earlier in its millenial history, astrology between (say) 1300 and 1700 was neither obscure nor implausible. It was, rather, an integral part of European life and culture. Astrological ideas and practices were vitally involved in (and, of course, influenced by) everything from philosophy, the arts and sciences to politics, medicine, agriculture and navigation. In various forms and degrees of complexity, astrology was present throughout society, from the peasantry to the clergy and aristocracy.

With the Reformation and Counter-Reformation, however, astrology came under increasing attack. Both movements tended to reject it, as part of their compaign to 'reform' and control popular culture. Furthermore, astrology's hitherto secure (if slightly cramped) place in the dominant Thomist synthesis, which sanctioned the influence of the stars, had crumbled. Early modern science, too, dissociated itself from astrological ideas, often while covertly courting them. But most importantly, drastically changing social circumstances permitted the moral and intellectual criticism –

itself nothing new – to take hold. By about 1700, astrology had lost its footing in elite and educated European culture. It survived among the lower and labouring classes, where it came under renewed attack from scientific popularizers in the nineteenth century; but this campaign, too, met with only limited success, since an urban, middle-class astrology appeared late in the century which is still with us today.

At the least, therefore, astrology was a significant part of the cultures and times studied by historians of medieval, Renaissance and early modern Europe. Even afterwards, the story of its fall from grace (and survival) is fascinating. Why, then, is it neglected? Appropriately, I think most of the answer can be deduced from its history. One consequence of the development of our modern intellectual disciplines was the dismemberment of the terrain that astrology, in its far-flung ambit, had united. One part went to astronomy, and much of the remainder – the human pole – became psychology and sociology. It is understandable that with its apparent intellectual extinction, some people doubted whether such an exotic beast had ever lived. Even more to the point, however, are the intellectual lineages of those whose profession is the past: historians, and historians of science and medicine. (There is a persistent division of labour between these two which further complicates the picture.)

Astrology was customarily an inseparable mixture of what is now distinguished between as 'science' and 'mathematics' on the one hand, and 'magic' on the other. The former element makes it difficult for historians of science to avoid completely; but the latter, equally, makes it (along with alchemy) uniquely irritating. This reaction undoubtedly stems from an often uncritical loyalty of historians of science to modern science. To put it another way, the efforts of early modern science to define itself against magic and neo-Aristotelianism has rubbed off on many of its historians – an attitude aggravated by the continued existence of astrology, in defiance of scientific enlightenment. This is quite evident in the literature, beginning with the doyen of history of science, George Sarton, who was unable to mention astrology (in relation to early Greek science) without descending into abusive caricature, explicitly related to his feelings about modern astrologers. For this he was reproved by Otto Neugebauer, in a short but powerful paper entitled 'The Study of Wretched Subjects', which defended against such destructiveness 'the very foundations of our studies: the recovery and study of the texts, regardless of our own tastes and prejudices.'

Also common has been the strategy of simply ignoring astrology, along with other disreputable relatives. Very often, that may be entirely in the order – to pick one example, to the extent that the agenda is restricted to the purely intellectual and 'internalist' history of science after Newton. But the ban is usually much wider; and the justification, when given, is that for reasons that are unstated, 'it is impossible to write in the same sentence of the victors' and losers' view of a battle'. Obviously, a history of science that views its job in this way will soon partake of hagiography. It also follows that the continued presence of astrology, which has been defined as a superstitious anachronism, will act as an additional spur to suppress such historical traces in the lineage and relations of science. It is true that this

'Whiggish' approach has waned in recent years, but it is far from over. With the 'occult' in history, unfortunately, it still seems necessary to fight for basic historiographical rights for the subject, and for its human subjects!

The discipline of history as such is perhaps less problematic in these respects. But here too, one must contend with inimical priorities and archaisms. Too much historical writing seems still to be based on Carlyle's view of history as 'but the biography of great men', and the progressive unfolding of (their) great ideas. Such elitism (often reflected in the institutional positions of its leading adherents) is quite happy to concern itself with the powerful and successful, at others' expense. It sits comfortably with the traditional dominance of political history, as it is commonly practised. It also resonates with much of ecclesiastical history, top-heavy with major figures and victorious ideas. Other branches of history, such as economic, can be more open-minded but are limited by the constraints of isolated specialization. The picture is not entirely dark, however. To the extent that it successfully resists such narrowness, the recent healthy revival of social history is a very promising development, not least for the history of astrology. This revival is itself beset, of course, by contemporary considerations – none the less exigent for being pragmatic and political – affecting all the historical disciplines. In common with the other humanities and human sciences, they are faced with the effects of continued recession. The criteria for financial support are increasingly dominated by technological and ideological usefulness. And this situation is presently exacerbated, in the Anglo-Saxon world at least, by governments' openly anti-intellectual and even anti-educational philistinism.

Fortunately, many individual historians, including those of science, continue to produce excellent and exciting work. More specifically for our concerns here, this book is itself a testament to the vitality of its subject, and the quality of work it contains can only assist in bringing the history of astrology into its rightful standing and relationship with other matters of historical importance. Furthermore, it appears amid some other encouraging signs. In the history of science, the Whiggish mould was at least cracked by pioneering studies in the 1960s, which considered metaphysical influences in early modern science – especially those of Frances Yates, though such was not her primary intention. These were followed by a reinvigorated sociological history of science in the 1970s and '80s, whose wide and contextual view refuses anachronistic judgements of winners and losers. In both areas there have been some subsequent retrenchments of the older views (and some sharpening of ideas), but the overall effect has been liberating.

Similiarly, in history; for example, the early Annales school initiated a vigorous interest in 'collective mentalities', that in recent years has attracted a host of historians working to deepen and refine our picture of the often unorthodox and synchretistic, but vital, religious beliefs of the common people. In a closely related development, the work of Carlo Ginzburg and others has stimulated healthy controversy over the differences between elite and popular culture. And last but not least, English Marxist historians such as Christopher Hill and E. P. Thompson have pioneered an impressive body of work on the social and political history of the labouring and middling classes, starting 'from below'. Journals like *Past and Present* and *History Workshop Journal* continue to extend this work. All these developments often have different starting points and complex strands of variance, but the net intellectual effect has been to let in fresh air and new light – an encouraging context in which to approach hitherto 'obscure' subjects like astrology. Such gains are certainly not imperishable, but they are durable. They raise the question, how may they be built upon, and – which is the same question – how can the history of astrology now best be approached?

As the papers here demonstrate, the very term 'astrology' – like 'science' or 'religion' – conceals a challenging multiplicity of ideas and activities. Very often these serve quite different purposes in different circumstances. It follows from this fundamental point that astrology should be studied in (as far as practically possible) its fullest social and intellectual contexts. Often the historian will only be interested in astrological material in relation to other concerns. But even the historian of astrology as such must keep firmly in mind (especially when generalizing) this kind of question: what did the astrologies of, say, a Babylonian high priest, a Renaissance magus, and an English Interregnum almanac-writer have in common? The shared conceptual constraints, when closely examined, are unlikely to extend beyond a nominal committment to the importance of the stars; and other circumstances will differ radically. Any tendency to over-isolate and reify an apparently continuous thousand-year-long tradition must therefore be resisted. For one thing, it is almost bound to be bad history; for another, and closely related reason, it plays directly into the hands of those whose primary interest is to protect the supposed purity of either astrology, or elements of its historical context (usually science). Traditions only exist in so far as they are continually re-invented and reconstructed by the historical participants; and that process should be a central part of the historian's data, not an unexamined assumption in crude form on his or her part.

There are more practical implications of this position, too. For example, a journal devoted to the history of astrology would undoubtedly help a great deal to fill the present gap, and stimulate more studies. But it would run the risk of isolation by the wider intellectual community. Astrology's polymorphism, combined with limited resources, suggests the necessity for at least an equal and parallel effort to put astrology onto the overall historical agenda – through books, and journals already extant, and in departments of history and history of science and medicine.

I hope it is clearer now why the history of astrology is a fascinating and potentially fruitful enterprise, worthy of our attention and respect. As these essays show, it springs from a scholarly committment to fidelity and critical open-mindedness, combined with a determination to consolidate and extend recent gains in making accessible, as far as possible, our whole past and the past of us all. It promises new lessons about our historical self-constitution, and new light from an unaccustomed source: in short, a triumph of reason, not a failure.

MEDIEVAL CONCEPTS OF CELESTIAL INFLUENCE A SURVEY

J. D. North

Although I use the word 'influence' in a modern and rather general sense, it is worth remembering that for many centuries it – and other words cognate with the Latin *influere* – would have carried connotations of an inflowing of something or other. Such a model of the ways in which the planets produce their effects is not by any means without alternatives. It was not the model at the back of Aristotle's mind, for instance, when he provided an explanation of how the Sun produces changes on the Earth. This was the starting point of his doctrine of efficient causation in *De generatione et corruptione* II.10, a fundamental text for those who tried to justify astrology in the scholastic tradition.

Aristotle had, in the *Physics*, shown to his own satisfaction that motion is the primary cause of change. What, then, do we need to explain a *de*generation? If generation needs motion, corruption requires a retreating motion in some sense – but a backwards motion is just a motion. Aristotle's answer was that motion along the *ecliptic* is what is needed, for this implies a certain 'advancing and retreating'.[1] He had to face up to another problem. In his universe, bodies fall, smoke rises, water falls, and so on, according to circumstances. Should not all the element fire have been lost to the Earth's surface? Again the Sun is the key to the solution of the paradox: it produces a cyclical interchange between elements. There is no mention here of the planets, although some later writers managed to substitute the planets for the Sun, in explaining the passage in question.

Another important passage from Aristotle is that section of the *Meteorologica* in which he discusses phenomena in the upper reaches of the sublunar region. Not only did they include wind, rain, thunder, lightning, and so on, but also comets and the Milky Way. Again, the Sun is given pride of place in the explanations offered, and the Moon is pushed into the background, even in connection with tidal phenomena.[2] Others in the ancient world appreciated the Moon's importance here, and medieval commentaries on the *Meteorologica* tended to introduce the lunar theories. They were, as is well known, often linked with the theory of critical days in the course of human disease, thanks to the supposed quasi-tidal behaviour of the humours. It is easy to

[1] There are problems with the text of II.10. Does it imply, for instance, that Aristotle thought there was a variation in the Sun's distance from the Earth? (Remember that his cosmos was supposedly homocentric.) These are not problems for a brief survey.

[2] See Heath 1913, p. 307.

distinguish three quite distinct medieval traditions concerning the tides. One was a philosophical tradition, in which the tides were fitted into the general Aristotelian picture. A second was the empirical tradition, in which the monthly movements of the Moon were rather carefully correlated with the tides at particular ports. The third was an astrological tradition, mingling Aristotelian philosophy and astrology from wheresoever.[3]

In addition to the Aristotelian element in the medieval discussion there is much with Stoic roots. Note, for example, the largely biological implications of their philosophical principles. Note the analogy drawn by Cleanthes between the renewal of the human soul by exhalation from the blood, and the sustaining of heavenly bodies by exhalation from the oceans and other liquids of the world's body.[4] He thought of the World Soul as a heat permeating the cosmos, just as Chrysippus took it to be a permeating *pneuma* – which is also responsible for making man an organic, living, whole. The language of both has connections with the language of influence in astrology. The same is true of the language of a related work *De mundo*, that for long went under Aristotle's name, but that probably derives from the Stoic Posidonius (d. ca.51 BC).[5] In his work there is a general Aristotelian patterning of the spheres and elements, and a lunar tidal theory. Everything in the universe is seen as 'ordered by a single power extending throughout, which has created the whole universe out of separate elements . . . *forcing* the most contrary natures to live in agreement with one another'. There is, in this work so often translated in the high and late middle ages, the notion that proximity to God determines the degree of orderliness – a sort of 'law of force'. God is a principle of organisation with an irresistible binding of action. Astrologers liked the linking of God's cosmic action with a philosophy of destiny – a blend of Platonic imagery and sentiment which was all the better for having, as it was thought, the authority of Aristotle behind it.

As seen through the eyes of the medieval astronomer, the most respectable apologist for astrology in ancient times was of course Ptolemy, who fell into both the Aristotelian and Stoic traditions. (Boll thought that Posidonius was an important influence on him.)[6] According to Ptolemy, a certain power (δύναμις) emanates from the aether, causing changes in the sublunar elements, and in plants and animals. Effluence from the Sun and Moon – especially important was the Moon, by virtue of her proximity – affects things, animate and inanimate, while the planets and stars also have their own effects. Given the key to calculating these effects – and that is what the *Tetrabiblos* is all about – one ought to be able to work out the weather and human character in advance. From a scientific point of view these sentiments are perfectly laudable. Optimism only becomes a vice when it closes its eyes to counter-evidence.

[3] For examples, see North 1976, vol. 3, pp. 206–8.

[4] Cf. D. E. Hahm 1977, pp. 136 ff. The heat of the cosmos is not moved by another force, for there is none stronger. It is self-moved, and thus of the nature of soul. The heavenly bodies are moved 'by their sensation and divinity', not by nature nor by force, but by will. There are many medieval echoes of the idea.

[5] See the introduction to the Oxford translation, E. S. Forster (1914) 1931.

[6] Boll 1894, pp. 131ff.

There are three ideas more or less explicit in Ptolemy that seem to have been taken as almost intuitively evident. Celestial force was taken to depend on (a) *position* pure and simple, of a planet in the zodiac; (b) the relative *separation* of a planet and the subject it is supposed to affect; and (c) the *rising and setting*, ascending and declining, of a planet in relation to its subjects.[7] Such ideas were not at all obvious to Plotinus.[8] We, with our consciousness of Newtonian and other forces, and how they operate, find difficulty perhaps only with (a), and the idea that *absolute* position can be of any influence. For Plotinus the stars were taken to have influence (they constantly pour forth warmth, for instance), but taken individually they were seen only as signs of a harmonious whole, a sovereign unity under the control of the One. Divination was supposed to be possible from observations made on them, as on birds and other animals; but this was no astral physics of the sort I am seeking. Notice that Plotinus was strongly influenced by Stoic thought, but that his influence on the middle ages was – at least as far as Europe is concerned – indirect.

The number of classical references to the 'meteorological' tradition, especially starting from solar and lunar radiation and the tides, is potentially very large. If I fail to do justice to them, or indeed to intervening Indian and Islamic civilizations, this is because I wish to concentrate on developments in the Latin West. There, without any doubt, the most influential Islamic writer was Abū Ma'shar (786–866), who was read and quoted constantly from the time of the translations of John of Seville and Hermann of Carinthia in the twelfth century to the decline of the subject in the seventeenth.[9] The work in question was his *Introductorium*, to give it its Latin title; and since I am discussing the Latin tradition I shall refer to its author as Albumasar, the name by which he was known in the West.

Albumasar's astrology was an amalgam of Hellenistic and Indian ideas, the former with both neo-Platonic and Aristotelian accretions. The planets were the dispensers of power. In Hermetic philosophy they tended to be taken to *mediate* between the sublunar world and the One, whereas in Indian astrology they were supposed to dispense *their own* power, and that according to very complicated mathematical rules. These rules linked planet with man; but they also linked acts of a former existence with the present – an idea that did not gain currency in the West.

In giving an account of change in the world, Aristotle had said virtually nothing about the *planets* (as opposed to the luminaries, the Sun and Moon). Albumasar thought that they, the planets, offered the key to the enormous *diversity* of change in our world. Otherwise, he summarized Aristotle's views on generation and corruption, emphasizing a factor I have not yet mentioned – a *receptivity* to planetary influence, on the part of terrestrial things. He acknowledged two sorts of action, (a) by contact, and (b) through a medium. In the second case, one might have (1) voluntary action bringing contact

[7] See especially *Tetrabiblos* I.2 and (for examples of his application of the principles) III.11.

[8] *Enneads* II.3.i.

[9] For a comparison of the translations and an assessment of the Aristotelian content of the *Introductorium*, see Lemay 1962, passim.

about, through a medium; and (2) natural action, as when heat is transmitted through a vessel. More interesting is (3) natural action transmitted through an invisible intermediary, as when a magnet attracts iron through invisible air. This magnetic analogy later became very popular. Note here the idea of the attracted thing, the iron, being suited to receive the attraction.

The analogy might well be seen as the reduction of one mystery to another. Actually Albumasar fell back on a neo-Platonic explanation when he tried to go deeper, one in terms of emanation rather than of mechanism. He passed on to the twelfth century a doctrine of the formation of individual beings in terms of the properties of the elements of which they are formed, their forms, and 'all else that is of consequence'! The last category is where the heavenly bodies entered the picture. They were supposed to produce the *variety* of life here below. Albumasar also helped to fix the old idea that the planets are intelligences of some sort: for him, since their effects are (in part) rational in nature, so must they be rational. Others deduced the rationality of the planets from the fact that they were, or so it was supposed, *self*-moving.

When Ptolemy had been considering the form and character of the human body he had correlated different physiques with different planetary positions (stations) at the birth of the person. The idea here was that the planets' virtue depends on motion with respect to the fixed stars, which in turn depends on position on the epicycle. Ptolemy also wanted to take into account their positions with respect to the *horizon*, making them most powerful when in mid-heaven (i.e. on the meridian).[10] So far so good. No doubt we may see an analogy with the case of the Sun; but alas, the parallel with thermal radiation breaks down when Ptolemy for some reason said that they are second most powerful when on the *horizon* (or in the succedent place). Here we see an example of how Aristotelian physics could be mingled with an incompatible horoscopic astrology where horizon effects, having nothing to do with heat or growth in the world, counted for most.

As in Aristotle, and in Albumasar, it was movements that produced effects. The more rapid the movement, the bigger the effect. In Albumasar, though, we get that other idea of receptivity of effects, or connaturality, as it was sometimes called: the Moon, as it repeatedly rises and sets over the sea, is supposed to *draw* the water by a *cognate force*. The sea, that is, the tide, follows this tractive force *of itself*. In this connection, the *Introductorium* contained one particularly influential chapter in which were listed eight causes of tidal inequality.[11] In brief, lunar radiation is a vital factor, and this depends, for example, on the separation of the Sun and Moon. Certain humid stars are also taken to have an effect; and likewise the position of the Moon on its eccentric; and so one could go on. There is a measure of good scientific sense in all this, and it is noteworthy that Albumasar refrains from spending much time in the discussion of lunar influence on phenomena that are not quantifiable.

The influence of this work of Albumasar was perhaps nowhere more apparent than in the *De essentiis* of one of its translators, Hermann of

[10] *Tetrabiblos* I.24.
[11] *Introductorium* III.5.

Carinthia (1143). Hermann referred to Albumasar by name more often than to any other author, apart from the arch-astrologer Ptolemy. Hermann did not so much account for the bond between the heavens and the sublunar regions in terms of motion, radiation, and a fitting receptivity, as in terms of musical proportions – all very neo-Platonic.[12] Through the vagueness of Hermann's prose it is possible to detect such ideas as that '*the Same*', the celestial sphere destined to carry the stars with a steady motion, was needed by God to provide what was *stable* in the world, while the complex of spheres having to do with the Sun's oblique motion ('*the Different*') provides variety. Whether his readers knew what he was talking about, it is very hard to say; but we, of course, recognize here Aristotle's *De generatione et corruptione*, hidden under a layer of eastern dust. It is when Hermann comes to the *medium* between the Same and the Different, the medium being in the form of the planetary spheres, that he finds for us a strong bond, filled, as he says, 'with every musical proportion'.[13] This mediating bond reconciles extremes by means of love.

This material might seem astrologically unpromising, but if it does nothing else, it does at least show that any typology of celestial influence must be drawn generously.

Hermann went on to say how, in secondary generation and corruption, two principles were needed for each; and he related these to the planets, albeit in a very obscure way, in an attempt to rationalize the favourable and unfavourable, cooperative and opposing, planetary characters. At the end of his first book he comes out with a 'physical' explanation of influence, as we might describe it: *sympathetic vibration in the sublunar world.* This makes it rather difficult for him in the second book, when he takes over material from Albumasar on the tides. For all Albumasar's inconsistencies, his explanations do fit reasonably well with our ideas of directed forces. For Hermann, this simply will not do. For him the Sun could be at a particular place in the zodiac on two different occasions and produce different effects, by virtue of the different sorts of planetary configuration (collaboration). Again following Albumasar, he ascribes *singular* guidance to the stars and *universal* guidance to the planets. He has the image of a bond (nexus) between a dragging Essence (the heavens) and a dragged Substance.

Daniel of Morley is an example of one who at this time tended to reject the Latin Platonic tradition in cosmology, and in drawing heavily on William of Conches to introduce his readers to the astrologized Aristotle of Albumasar. He took over the tidal material, for example, and was probably responsible for passing it on to some thirteenth and fourteenth century writers – in particular to Richard of Wallingford in his *Exafrenon*, a work of astrological meteorology.[14] There you will find, for instance, an ancient doctrine to the effect that there is a factor that may diminish the efficacy of planets, namely their proximity to others. This is the doctrine of *rays.* Each planet is supposed to extend rays along the zodiac for a short distance – nine degrees

[12] Burnett 1982, pp. 41ff.
[13] Burnett 1982, 63vH.
[14] For the text see North 1976, vol. 1, and for commentary vol. 2, especially pp. 113ff.

for Jupiter, eight for Mars, fifteen for the Sun, and so forth. A planet falling under the Sun's rays would be burned (*combustus*). This is not to be connected with my theme in a significant way. The rays (*radii*) are more like a glory round the planets than a *lumen*, a light they pass to other objects to illuminate them, or than a radiation poured down on the Earth to change the elements.

By the early thirteenth century, scholars were wont to examine their textual sources more critically than had earlier been the case. Robert Grosseteste, around 1209 – say roughly half way through a long life, ending in 1253 – wrote an interesting little tract on astrological meteorology with the title *De impressionibus aeris* [etc.].[15] Here he placed great emphasis on motions, whether of the luminaries or the planets, but whereas his predecessors had left this as a rather vague concept, or spoken just of annual or monthly motions, for Grosseteste it was the *daily* motion across our sky that was supposed to determine planetary influence on the Earth. When the Moon is *furthest* from us it is moving *fastest* across our sky with the daily motion, and then its effect is greatest, thought Grosseteste. This is perhaps counter-intuitive, even though he seems to be following the Albumasar rules. The same sort of argument applies to the Sun, and its effects. The planets are rather more complicated. As for the Moon, Grosseteste is guilty of a certain sleight of hand, having to do with the difficult problem of explaining why there are two ebbs and two flows every day; but I will not expose the trick to public view on this occasion.[16]

Grosseteste at this time had not yet developed his light metaphysics, the arrival of which can be roughly associated with his *De luce* (c.1225–30). He had earlier taken what might be called the 'common astrologer's view', that the planets and fixed stars act on the lower world in virtue of qualities they share with it (hot, cold, wet, and dry). This of course is a basic premise of much alchemy, and was not in keeping with orthodox Aristotelianism. With the more mature, post *De luce*, writings, his position on this point changed. The centrality of the Earth in the cosmos was in order that it might be, so to say, a focus for the light from the spheres. He had now become convinced that every natural body has in it a celestial, luminous, nature, and a fiery luminosity, at that. Light had stepped into a role played earlier by the qualities of the elements. One might even see traces of the new idea in his *De cometis*, written, surprisingly enough, in ignorance of the *Meteorologica*.[17] According to Grosseteste in the *De cometis*, the power of the stars and planets sublimates from the Earth spirits of the nature of the Sun. This power (*virtus*) is yet another to add to our list of celestial influences.

The advantage of the metaphysics of light was that it carried with it a body of physical theory – a geometrical optics, and even a sort of photometry, based on the notion of a pyramid of light propagation. It is all rather

[15] Baur (ed.) 1912, pp. 41–50.

[16] For a fuller account of the contents of the present lecture, and more details of Grosseteste's argument, see J. D. North (1986).

[17] The opuscula by Grosseteste mentioned here are in Baur 1912, supplemented by later versions. See Thomson 1940 for bibliography, and McEvoy 1983 for further bibliography (and comment).

impressionistic, and I shall omit the details, but notice something here of no less interest than his crude geometry of forces: he has found an explanation, as he believes, for the piling up of water on the side of the Earth away from the Moon. The Moon's rays are supposed to be reflected by the heavens back to the other side of the Earth. We cannot see the luminosity, for it is indeed very tenuous, but the waters of the Earth feel its effect. Those who feel inclined to smile at this explanation should ask themselves whether they can feel the gravitational force of the Moon.

In the *Hexaëmeron*, Grosseteste's final position on astrological belief is stated at some length.[18] Superficially it is hostile – astrology books are written at the dictation of the devil, and should be burned – but his hostility has to do with the issue of determinism, free will, and theological values. His belief in celestial influence was as strong as ever. He thought that the *science* of the astrologers must fail because the influences they sought are so precisely focussed in accordance with the momentary stellar configuration, that even the most accurate astronomer would not find them. They were *real* enough, in Grosseteste's view.

There is another work of relevance to my theme that has often been taken as Grosseteste's, namely one under the title *De fluxu et refluxu maris*.[19] Although by someone intellectually close to him, it does not seem to me to be by Grosseteste. It is worth mentioning, though, for the fact that the verbalization of the problem of the Sun's assistance to the Moon in causing the tides is beyond the author's powers – hardly surprising, of course.

One way or another, Grosseteste contributed much to the subject of celestial influence, however inadvertently, through his writings on light and his new paradigm – explaining light propagation and vision in terms of species. Species were likenesses of objects, given off regardless of the perceiving being; but the word could also mean the force or power by which an object acts on its surroundings. For Roger Bacon, *all* natural causation was attributed to the multiplication of species, and it was natural for him to couch reference to celestial influence in terms of it. We have seen how it was traditionally supposed that the heavens and the world must *share* something, for the heavens to influence the world. First – and for the early Grosseteste – it was a sharing of elemental qualities. Later, for Grosseteste, it was a binding of light in matter. For Bacon, going back in his *De multiplicatione specierum* to Aristotle's *Metaphysics*, it was that they shared the same *matter*. Later he settled on power (*virtus*). It is not that he was writing astrology here, but that he took the basic fact of astrology – the existence of influence – for granted, and tried desperately to find for it a theoretical basis. No doubt he would have said, had we asked him, that he was doing what Aristotle would have done in *De generatione et corruptione*, had he been privileged to live in the thirteenth century.[20]

Bacon's theory is of considerably greater ingenuity than anything to be found in straight astrological texts, but it rarely connects with experience. It

18 Edited in Dales and Gieben 1982.

19 There is a text with translation in Dales 1966.

20 See Lindberg 1983, pp. liv–lv, 71, 89, 209–11, for these views held by Bacon.

has some curious points. The law of variation of force (we may think of it as light-force) with distance was not universal, but applied only to forces along radial directions in the universe. Sad to say, the whole theory of species was ill suited to a theory of forces and photometric laws. Bacon knew the power of geometrical methods, though, and one should not forget that the praises he so often heaped on them were largely due to their value in 'speculative and practical' astrology and astronomy, by which we learn, as he said, 'the celestial causes of the generation and corruption of all lower things'.[21] At another place he tried to explain the miracles wrought by words, in terms of species. The species or power of the heavenly configuration at the time the words are uttered should, he thought, be taken into account here. It all connects nicely, of course, with the doctrine of charms and their magical properties. Bacon was an eclectic. His remarks on the tides, for instance, pick up much from Albumasar, but also from Grosseteste. What distinguished him from most of his contemporaries was that his was a reasoned eclecticism, selective and systematic, up to a point.

When considering Albumasar we saw that he likened celestial influence to magnetic powers, acting – as they do – without contact. That is not to say that celestial influences are by nature magnetic; it is simply that they are *like* magnetic influence. The famous letter of Petrus Pergrinus on the magnet (dated 1269), which hints at earlier discussions not now extant, settles however on a celestial solution to the problem of the direction seeking properties of the lodestone.[22] Some writers thought that the correct explanation was rather the existence of deposits of the stone at the Earth's poles. Bacon opted for the celestial alternative. Neither he nor Petrus was prepared to ascribe the power to the Pole Star, as was commonly done. It was for them *the heavens as a whole* that matter, each part of the heavens (Petrus) corresponding to a different part of the magnet. They both thought that if one could only pivot a perfectly balanced sphere of the stone smoothly, it would turn with the turning of the heavens, serving incidentally as a timepiece. William Gilbert and Galileo had an enjoyable time rejecting the idea as illogical – but they were wrong, from the point of view of geocentricism.

The 'astrological optics' of Grosseteste, Bacon, and John Pecham (who had interesting things to say in the same tradition, as regards the correlation existing between points of the Earth and sky)[23] had one property that contemporaries must have found attractive: it explained how twins could be born more or less simultaneously and yet have different biographies. The reason was that the aggregates of radiant pyramids from the heavens falling upon the children at birth were subtly different. This was also said to be why

[21] *Opus tertium*, Brewer (ed.) 1869, p. 107.

[22] There is a convenient translation, with bibliography of relevant editions and commentary, in Grant 1974, pp. 368–76.

[23] Since I have mentioned Pecham, let me here observe that his solution to the problem of the tidal waters away from the Moon was different from Grosseteste's, but still given in terms of light. He supposed the Earth to be slightly transparent, so that the Earth acted as a less, rather than the universe an optically dense medium. See Lindberg 1970, p. 99.

plants are so diverse, a problem that others (e.g. Maimonides) had solved by supposing that each plant corresponded to an individual star. The new explanation was somewhat more logical.

Grosseteste, as I have said, changed the ways in which scholars spoke of celestial influence, but he did not displace the old Aristotelian idea that *motion* was at the root of the problem. The motion hypothesis had its difficulties. How was the motion transmitted inwards, if the spheres are perfectly frictionless? There were fashionable models still around which lent themselves to the idea of an inward transmission of motion along Aristotelian lines – for example al-Biṭrūjī's – but their thirteenth century popularity was for philosophical rather than astrological reasons, and by the fourteenth century their popularity was fading fast. Aquinas is a notable example of a philosopher who hesitated as between al-Biṭrūjī and Ptolemy, and who indeed changed his mind like a retrograde planet on this very significant point.[24]

In view of the multiplicity of belief as to the *character* of the fundamental motions in the universe, the planetary motions, it is rather surprising that there was not greater variety in the theories of causation by them on the world below. Were they intelligences? Were they bound, or not, to their planetary bodies? Of what substances were those bodies? Were they given an initial impulse at the creation of the world? Many were the opinions on these points, but few were the philosophers who tried to develop a systematic vision of planetary causality. Aquinas saw the need. He tried his hand at many explanations. He started, for example, with an analogy, an absolute monarchy and its ruler being made analogous with the universe and its Prime Mover. In *De Trinitate* he tried out a doctrine according to which the planets control the adjustment of the balance between the elements in mixtures. Throughout his many writings on these topics (Litt gives more than a hundred and thirty passages on celestial influence alone) his angelology is there, waiting in the wings, directing his thoughts, as it seems to me. It is out of the question to try here to do more than scratch the surface of his far-reaching schemes. Let me note his idea in *De veritate* that bodies get some, at least, of their properties from the elements of which they are constituted, and other properties from the 'impression' of the heavenly bodies. Here he introduced the magnet, but now no longer on an analogy. Magnetism is an *effect* of heavenly influence, the impress of celestial bodies. He does not tell us how to draw the line between explanations in terms of primary elemental qualities and celestial explanations. The latter have all the appearances of a back-stop, to catch whatever the first cannot. Certain stones and herbs have occult properties, from the celestial sort of cause. It is not enough to distinguish between Aquinas' view of *natural* powers as God's vicarious actions and *supernatural* actions as God's direct intervention in the world. There is a large area of middle ground, and this is by no means without its own subdivisions.

It could be said that Aquinas did his best to determine what sorts of concepts should be used in what situations. The causality of celestial bodies

[24] See Litt 1963, ch. 18.

was for him *irreducible* to the causality we know in our study of the physics of the elements. That is what it is not. We know where we stand; but the sense of mystery surrounding the idea of celestial influence is heightened. Aquinas was rational and systematic but not – in the eyes of those looking for a physics of celestial influence – in the best tradition of natural philosophy. But this is just another way of saying that one prefers light rays to angels, in one's physics.

It is out of the question, here, to pursue every minor variation in medieval belief on the subject of celestial influence, but there are a number of fourteenth century writers who added significantly to the material they inherited. Dante, on the threshold of the century, made excellent use of the ideas I have been discussing. In the *Convivio* he used the notion that *stars* influence *things* to explain by analogy the passing of varying degrees of perfection from the *sciences* to *us*. Putting the analogy this way round shows to what extent the belief in the power of the stars was an axiom for Dante. I believe it was so for everybody, and I have certainly never come across any text in which the idea is openly challenged.

Like Dante, most scholars had a rather vague notion of the mechanism of influence. For them it was usually taken to be just a question of 'aspects'. There were scholars, though, with more carefully worked out schemes, rivalling those we have already examined, and one quintet deserving special mention were the Parisians John Buridan, Nicole Oresme, Marsilius of Inghen, John of Saxony and Henry of Langenstein. Buridan is the key figure, and may be seen as mentor, or grand-mentor, of the others. He introduced as an important factor the *heating* effect of light, taken along with motion as a candidate for what brings about the heat.[25] He would not accept that light carries the powers of the stars, and he went so far as to argue that there are more fundamental qualities than heat, cold, moistness and dryness, in the celestial influx. He was perceptive enough to see that there are limits to what one should try to explain: it is not necessary, he thought, to say *why* light warms – it just *does*. And he introduced the notion of energy (if that is not too grand a way of expressing it) when he said that the actions of the heavens do not fatigue them. These actions were thus somehow disengaged from physical forces here on Earth. That is of course utterly compatible with the Aristotelian line of thought.

Buridan investigated a curiously 'Aristotelian' problem. Why, if motion is nothing other than the mobile thing, should the heavens need to move to get their effects? Averroes had looked at the question, and had suggested that the *aggregate* of spheres was what counted, and that they were like an animal. This analogy is obviously in the macrocosm/microcosm tradition. Buridan offered another solution, taken – as he thought – from Peter of Auvergne. (He was modest enough to suggest that we make our own choice between alternatives.) It was suggested that the Sun's sphere causes a quality in the lower spheres other than heat, and that this, when reaching heatable things, generates heat in them. The explanation might be seen as highly sophisticated

[25] Moody (1942) 1970, quaestio 15 of Book II, concerning *De caelo* II.7, is the source of most of the following account.

and thoroughly scientific, by those attracted by the introduction of theoretical concepts with no immediately visible (or otherwise sensible) counterpart; but there will no doubt be others who regard it as a typically medieval introduction of the supernatural. It is hard to adjudicate between the alternatives, since Buridan did not take the example much further, scientifically speaking. He did, however, illustrate it by analogy with the electric torpedo fish, which shocks the fisherman holding a net, but which does not shock the medium, the net. The example goes back to Simplicius. There is a somewhat lengthy rationalization of the sort of explanation offered here, in Henry of Langenstein's *Tractatus contra astrologos*: primary qualities give rise to accidental and substantial change, and to secondary qualities, some of which may be occult and hidden from the senses.[26]

As an example of innovation by a natural philosopher of the group I mentioned, let me take Oresme, who is better known for his use of celestial causality to attack astrology than for his ideas on the former. His development of a system of graphing quantities of quality and motion is well known. One may think of what he calls a *configuration* in either of two ways: as a total graph of a thing as it changes; or as the internal ralations between the parts of the graph. Oresme's theory of configurations could, as he thought, explain just about every aspect of daily life, at least in principle: projectile motion, beauty, friendship, magnetism, hostility, . . . the list was potentially endless. It was, as he said, a question of working out the compatibility, or harmony, of the configurations.[27] Magnet and iron had a natural bond, according to him, because there was an accord between the configurations of their qualities. It is hardly surprising that he extended the principle to the classic case of harmony – 'the figuration of celestial velocity'. The odd thing about this is that Oresme took his inspiration from William of Auvergne's *De universo* (1231/6?), in which William had extended the classical astrological doctrine of aspects to other phenomena – medical, herbal, psychological, and so forth. And now we find Oresme simply accepting the extended principle, and turning it back on itself to explain astrological influence. Faced with two mysteries, who is to say where we should begin? We are reminded here of Dante's recourse to the concept of *love*, to explain celestial influence, something I did not have occasion to mention earlier.

Henry of Langenstein took a stand against occult qualities – an ambiguous term, which I shall not here try to make more precise. In effect, he told astrologers to change their ways, and to give their attention to the terrestrial end of the process, that is, to the terrestrial objects from which were elicited effects that were in some sense already present. He certainly believed in these effects, these influences, and had an interesting and curiously old-fashioned way of accounting for them in terms of seminal reasons, and activity down a chain of action, with God at the top. He was one of those, though, who in his youth had become a fan of Oresme's, with all that wonderful new jargon

[26] Edited in Pruckner 1933, in which edition see pp. 197–9.

[27] See especially Clagett 1968, particularly pp. 243–5, 299, 115–21, and 543, for the following ideas.

about the latitude of forms, which seemed capable of introducing precision into a very imprecise subject. But it never quite fulfilled its promise, as we know.

Natural philosophers of the middle ages tended to regard the celestial spheres as God's instruments. In a very broad sense we can see here the makings of a divide between medieval and post-Reformation protestant theology. Both Luther and Calvin, for example, played down the importance of natural powers, and put greater emphasis on immanent divine causation. (The very example Luther took to illustrate the point though, the example of the Sun's rays helping to procreate the chicken in the egg within the hen, could have come out of any neo-Aristotelian text on the celestial causation of generation.) The theologians were beginning to put more weight on the workings of God's word than on natural effects, and Luther's tactics, if carried out by every scholar, could well have thrown scientific practice into disarray, if not into the grave. Talk of celestial influence might not have paid very obvious dividends, but it did help to keep alive a very valuable frame of mind.

Such talk did not always have a strong bond with the natural sciences as we see them now. This was so when bandied around by such as Ficino, Pico and Campanella. One might ask why this should matter, since as a purely contingent matter, celestial influence was a concept leading nowhere. It would be very special pleading to hold that it contributed much to seventeenth century mathematical theories of gravitation, for example – although in their philosophical embellishments many seventeenth century theories were replete with medieval notions and terminology. There were attractions and subtle spirits, fluxes and emanations, and notions that most scholastics would have looked on with disdain. Hooke studied the brain of man (it could be seen through a hole in his head) that was said to wax and wane in turgidity with the Moon. Robert Boyle, patron of the astrologer John Bishop, hoped to identify subtle but corporeal emanations from the Sun, emanations distinct from light and heat. The old ideas survived, and if they were mingled with scepticism as to astrological practice, this was in itself nothing new. The beliefs might have led nowhere, in the way the corresponding medieval beliefs led nowhere, but they had an intellectual binding power within the cosmological systems that made use of them, and that is a good enough reason for valuing them, historically.

BIBLIOGRAPHY

BAUR, L., *Die philosophischen Werke des Robert Grosseteste, Bischofs von Lincoln*, BGPM, ix (Münster i. W., 1912)

BOLL, F., *Studien über Claudius Ptolemaeus* (Leipzig, 1894)

BREWER, J. S., *R. Bacon: Opus tertium* (London, Rolls Series, 1869)

BURNETT, C. (ed.), *Hermann of Carinthia: De essentiis* (Leiden, 1982)

CLAGETT, M., *Nicole Oresme and the Medieval Geometry of Qualities and Motions* (Madison, 1968)

DALES, R. C. and Servus Gieben (eds), *Hexaëmeron* (Oxford, 1983)

FORSTER, E. S. (trans. of *De mundo*), *The Works of Aristotle translated into English*, vol. 3, under the editorship of W. D. Ross (Oxford, 1931)

GRANT, E., *Source Book in Medieval Science* (Cambridge, Mass., 1974)

HAHM. D. E., *The Origins of Stoic Cosmology* (Athens, Ohio, 1977)

HEATH, Sir Thomas, *Aristarchus of Samos* (Oxford, 1913)

LEMAY, R., *Abu Ma'shar and Latin Aristotelianism in the Twelfth Century* (Beirut, 1962)

LINDBERG, D. C., *John Pecham and the Science of Optics: Perspectiva communis* (Madison, 1970)

LINDBERG, D. C., *Roger Bacon's Philosophy of Nature* (Oxford, 1983)

LITT, T., *Les Corps célestes dans l'univers de Thomas d'Aquin* (Louvain, 1963)

McENVOY, J., *The Philosophy of Robert Grosseteste* (Oxford, 1983)

MOODY. E. A., *J. Buridan: Quaestiones super libris quattuor de caelo et mundo* (Cambridge, Mass., 1942, repr. 1970)

NORTH, J. D., *Richard of Wallingford. An Edition of his Writings, with Introductions, English Translation and Commentary*, 3 vols (Oxford, 1976)

NORTH, J. D., 'Celestial influence – the major premiss of astrology', in P. Zambelli (ed.), *Astrologi Hallucinati* (Berlin, 1986)

PRUCKNER, H. (ed.), *Studien zu den astrologischen Schriften des Heinrich von Langenstein* (Leipzig, 1933)

THOMSON, S. H., *The Writings of Robert Grosseteste, Bishop of Lincoln, 1235–53* (Cambridge, Mass., 1940)

PETER OF ABANO AND ASTROLOGY

Graziella Federici Vescovini

Introduction

An interest in the stars and their relationships with the sublunary world pervades the writings of Peter of Abano (Petrus Aponensis, Paduanus), the Paduan physician, philosopher and astrologer who was active around the late thirteenth and early fourteenth centuries (d.ca.1315). He writes systematically about astrology in three important and still unpublished works: *Lucidator dubitabilium astronomiae, De motu octavae sphaerae* and *Imagines*, or *Astrolabium planum*. In addition, Peter collected and translated into Latin Abraham ibn Ezra's writings about judiciary astrology – these remained popular and influential until the Renaissance when they were first printed.[1]

There is no doubt about the attribution of the first two manuscript works: Peter refers to them several times in his *Conciliator*. He actually declares that he composed the *De motu octavae sphaerae* in 1303 and he mentions his *Lucidator*. The composition of these astronomical works is mentioned in a postilla at the end of a manuscript of his *Physonomia* (Oxford, Bodl., Canonici Misc. 46): 'composuit librum de motu octavae sphaerae; librum de discordantiis utriusque partis astrologiae'. Peter only gives information about the composition of his *De imaginibus* or *Astrolabium planum* in the *Lucidator* and in the *De motu octavae sphaerae*. We may therefore accept the *De imaginibus* as a genuine astrological-astronomical work of Peter's, although we only have a single, and late (fifteenth-century) manuscript.[2]

Earlier this century historians (like Sarton and Duhem, for example)

A research grant from the Ministero della Pubblica Istruzione, administered by the University of Sassari during the academic year 1982/83, enabled me to carry out the manuscript research on which this paper is based. This paper was translated by Angus Clarke.

[1] For preliminary information about manuscripts and rare editions see Favaro, A., 'Pietro d'Abano e il suo *Lucidator astrologiae*', *Atti del R. Istituto veneto*, 75, 1916, pp. 515–27; Thorndike, L., 'Peter of Abano', *A History of Magic and Experimental Science*, II, New York, 1923, pp. 917ff; idem, 'Manuscripts of the writings of Peter of Abano', *Bulletin of the History of Medicine*, 15, 1944, pp. 201–19. See idem, 'The Latin Translation of the Astrological Tracts of Abraham Avenezra', *Isis*, 25, 1944, pp. 293–302; Norpoth, L., 'Zur Bio-Bibliographie und Wissenschaftslehre des Pietro d'Abano', *Kyklos*, 3, 1930, pp. 292–353; Federici Vescovini, G., 'Un trattato di misura dei moti celesti: il *De motu octavae sphaerae* di Pietro d'Abano', *Mensura, Mass, Zahl, Zahlensymbolik im Mittelalter* (*Miscellanea medievalia* 16/2), Berlin, 1984, pp. 277–93, regarding Peter's translations of Abraham ibn Ezra ('Abraham Judei Avenare cuius libros etiam in latinum ordinavi'); idem, 'Le fonti greco-arabo-latine del *Lucidator*' in Atti del Convegno, 'Primo Umanesimo e Filosofia a Padova', Padova 1985, *Medioevo*, 1985.

[2] See note 41 below.

tended to regard astrology as a superstitious blot on the pages of certain medieval thinkers. Now, thanks to the work of Warburg, Saxl, Boll, von Bezold, Gundel, Thorndike, Carmody, Garin and many others,[3] the importance of medieval astronomical and astrological texts for the history of culture and the history of ideas is much more fully appreciated. Some insight into the premises of the study of the stars, and into the reasons why it was regarded, quite justifiably, as a science, has led us to acknowledge how important astrology is in reconstructing the scientific and philosophical activity of the Middle Ages, when it most flourished.

In this context Peter of Abano's astronomical and astrological works are especially significant. According to one popular traditional view Peter was an occultist and a necromancer; but on closer scrutiny it is apparent that his writings lie squarely in the rational scientific and philosophical mainstream of his day. He does not defend the theory of the stars and their influence on earthly events *tout court*, but argues that astrology is a rational and philosophical science and tries to give it with the clearly defined status that other theoretical and practical disciplines enjoyed.

1 *The philosophical basis of Peter's astrological theories*

Like the author of the *Speculum astronomiae*, which seems to have influenced many passages of the *Lucidator dubitabilium astronomiae*, Peter of Abano believed that God acts on the sublunary realm through the heavenly bodies. All movements lower in the cosmological hierarchy result from the motions of the stars. Since stellar motions are regular and any variations in them can be mathematically expressed, an intrinsic regularity, order and rationality will be detectable in all terrestrial phenomena. It is not the stars which cause motion: it is the motions of the heavenly spheres, with their different characteristics, natures and qualities – the objects of the astronomer's research – which cause the phenomena that depend on them. Stars, planets and stellar images are not, in Peter's opinion, secondary gods, or demons. They are, in Aristotelian terms, principles of motion, governed by precise rules or by 'natural' causes. Although, as will be seen, Peter describes beliefs in planetary intelligences and spirits (referring to Averroes' *De substantia orbis*[4]), such beliefs are controverted by his theory of 'median' causality arising from the

[3] The bibliography of the subject is enormous, but see especially: Boll, F., Bezold, C., and Gundel W., *Sternglaube, Sterndeutung, die Geschichte und das Wesen der Astrologie*, 4th ed., Berlin, 1931; and Thorndike, L., 'The true Place of Astrology in the History of Science', *Isis*, 45, 1954, pp. 273–78. On the same topic see Rossi, P., 'Considerazioni sul declino dell'astrologia agli inizi dell'eta moderna', *L'opera e il pensiero di Giovanni Pico della Mirandola nella storia dell'umanesimo*, Firenze, 1965, II, pp. 316–33. See Thomas, K., *Religion and the Decline of Magic*, London, 1971, chs 11–12, pp. 323–385. See also Garin, E., *Lo zodiaco della vita, la polemica sull'astrologia dal Trecento al Cinquecento*, Bari, 1976. For Duhem's attitudes towards medieval astrologers, see Lejbowicz, M., 'Pierre Duhem et l'histoire de l'astrologie', *104e Congrès National des Sociétés savantes, Bordeaux 1979*, Paris, 1979, IV, pp. 147–57.

[4] *Conciliator differentiarum*, diff.9, propter tertium, Venice, Gabriel Petrus Tarvisiensis, 1476.

complexions of the planets, and by his theoretical denial of the reality of demons and divine intelligences.

The development of Peter's theory of astrological causality, derived from a tradition based on Ptolemy and Abu Ma'shar, was deeply influenced by Aristotle's physical and cosmological writings. He found a middle way between the Aristotelian distinction between *casus*, and necessary cause *per se*:[5] either events happen because of necessary causes, *per se*, or they are indeterminate. Consequently chance and fortune cannot be necessary or *per se* causes, but can only be accidental and never definitive, causing only indeterminate events: *ergo non sunt*.

Though not a necessitarian religion, Islam does regard the deity as omnipotent will and the world as ruled by contingence. Muslim astrology, especially in the work of Abu Ma'shar who theorised a celestial causality *per influxum*,[6] furnished another principle by which contingent and accidental events might be rationally explained.

Peter of Abano felt impelled to found a science which would be capable of investigating that which lies between being and not-being, between the necessary and the impossible. As a physician he had to deal with illness – the corruption of the form, the *ens in fieri*. As an astrologer he studied and strove to predict that which could be generated and corrupted: the succession of the seasons, the months, the weather – in short everything whose generation and corruption depended on the heavens. He was trying to uncover the rational principles which would account for the causes of the *ens in fieri* and of the *ens fore* or *futurum*. Astrological theories of causality designed to explain contingent events, the *ens in fieri* or *ens fore*, furnished him with the tools to resolve the problem. He was thinking therefore in terms of possibility rather than in the necessitarian terms of the Aristotelian metaphysical *ens esse*. Peter's science was designed to give a rational account of the contingent. In the case of medicine, this meant studying the being of illness – which lies between the *ens* and the *non ens* – and trying to predict its future course. His theory presupposes the active potentiality of matter, the existence of hidden, contracted or occult seminal virtues which are activated by astral influences. These ideas are expounded in various places in his writings, and in particular in his *Conciliator* where one particular passage compares the philosophical theory of *latitatio formae* with the notion of creation *ex nihilo*, from without, and with the notion of an external generation through the *dator formae*, proposed by Avicenna and by Algazali.[7] In addition to astral causality which happens 'nutu dei' ('at God's command')[8] and which is subject to universal

[5] *Physics*, II, v, vi, vii (196 b22–23, 197 a35): *Metaphysics*, IX, iii (1047 a24–26), iv (1047 b6–14).

[6] For Abu Ma'shar's works see especially Lemay, R., *Abu Ma'shar and Latin Aristotelianism in the Twelfth Century*, Beirut, 1962.

[7] *Conciliator*, diffs. 101, 16; and in particular, Nardi, B., 'La teoria dell'anima e la generazione delle forme secondo Pietro d'Abano', *Saggi sull' aristotelismo padovano dal secolo XIV al XVI*, Firenze, 1958, 1–8.

[8] Peter of Abano, *De motu octavae sphaerae*, MS, Venice, Museo Correr, Provenienza Cicogna 2289, XVth, cap. 1, f. 111r–v: 'Volens utique causam transmutationum maximarum huius machinae mundialis operis causa et ipsius intentione sensibili indagari, motum octavi orbis cum stellis in ipso signatis, inspexi causantem praecipue transmutionem huius *nutu Dei*'. This concept

and mathematical laws, Peter, as one of the faithful, affirmed the interventions of God, the *agens supernaturale*, whose actions are obviously miraculous because they are immediate or at least 'absque motu et transmutatione' ('without motion and transmutation'), and this 'quidquid dicat peripateticus' ('whatever Aristotle's disciple may have said').[9]

Causality, or rather, the action of the stars, unfolds in a threefold fashion[10] wherein the universal cause can be distinguished from the particular and singular cause, and in between lies the 'middle' (*media*) cause. Instead of *causa* Peter prefers the terms *impressio, influxus* and *actio*. The *impressiones* and the *influxus* of the stars, planets and their images are in some way 'intermediate' (*mediane*) between the universal and particular actions and they too are referred to the *virtus contracta* of the 48 celestial images.

The universal cause operates only through movement and light ('corpora superiora his duobus modis universaliter in haec inferiora agunt', namely, 'motu et luce'). The intermediate cause arises from the complexions of the planets ('modus vero operandi medius est qui per eorum complexionem operatur, nam quaedam earum dicuntur frigidae et siccae, ut Saturnus . . .'). What precisely Peter means by complexions – another notion which can be traced to Abu Ma'shar[11] – is revealed presently when he states that they do not exist in a planet in a formal way, nor do they somehow constitute the form of a planet; rather a complexion is attributed to a planet as something 'effective', which must be understood in terms of the effects which the planet produces. This, Aristotelians might argue, is the fifth nature or quintessence which is divine and, by contrast with the natures of the four elements, ethereal: to postulate the ether is tantamount to ascribing a divine nature to the planets, but Peter seems not to share this belief.

In addition the planets send 'good fortune' and 'misfortune': this depends on people's individual differences which, in turn, depend on the configuration of the heavens at birth and how it develops ('de nativitate et revolutionibus'). A given planet does not necessarily send good or bad fortune, but gives only relative to the other planets and to each specific individual.[12]

As a matter of fact there is one further specific way in which the heavens may act. This 'influxus stellaris particularis' depends on the configuration of sky or on the revolution of the planets at the moment and exact location of the individual birth. It depends, therefore, on several conditions, on the parents and on the virtue of the place in which the influence is concentrated ('ideo omnis planeta secundum diversitatem nativitatis et revolutionis uni existit fortuna, aliis vero infortuna').

---

receives a certain amount of support throughout Peter's writings, and it had already been influentially espoused by Albert the Great: see especially *Speculum astronomiae*, ed. Zambelli, P., Caroti, S., Pereira, M. and Zamponi, S., Pisa, 1977, p. 5 – a work which is attributed to Albert.

9 *Conciliator*, diff. 156, propter tertium.

10 *Conciliator*, diff. 12, propter secundum; the notion is examined from different angles and further clarified in diffs. 9, 10, 16, 21 and 28.

11 Abu Ma'shar, *Introductorium maius*, Augsburg, 1489, IV, ch. 1–2.

12 *Conciliator*, diff. 10, propter tertium.

The planets, signs and the combinations of the images of the paranatellonta can also act in an intermediate mode, namely – and here Peter borrows a term of Abu Ma'shar's – the *virtus contracta* of the 48 celestial images which shape the species and the inferior forms in their own likeness by giving them this particular *virtus*.[13]

This observation from the *Conciliator* brings us back to the manuscript *De imaginibus*, and indeed, images are also discussed in the *Astrolabium planum* which was attributed to Peter of Abano by Johannes Engel who published it in Augsburg in 1488 (and again in Venice).[14] The *Astrolabium*, however, lists 360 depictions of the influences of the degree ascendant at birth. It is properly a calendar and connects with Peter's notion of the particular influence deriving from the ascendant degree in the natal horoscope. He explains in the *Conciliator* that we are not meant to suppose that these figures physically exist in the firmament, but that they should be understood allegorically.

More will be said about the theory of the astrological images; for the moment it is sufficient to stress the importance of the astrological notion of the *virtus contracta* of the 48 images, and to point out that it was to be developed much more fully in the very different context of the philosophy of Nicolaus of Cusa.

2 The *Lucidator dubitabilium astronomiae*

Peter of Abano's manuscript *Lucidator dubitabilium astronomiae* – of which I am currently preparing an edition – is a careful and scrupulous defence of astrology. It is also a manageable compendium of late thirteenth-century and early fourteenth-century astronomical and astrological knowledge, as well as an encyclopedia of the medieval mantic arts situated in their theoretical and cultural contexts. Peter intended this work to demonstrate that astrology and astronomy were the same thing[15] and to reconcile astronomers, astrologers and theologians. The opposition between the two had been stated most strongly by theologians and by strict Aristotelians. Averroes, for example,

[13] *Conciliator*, diff. 101, propter primum: *Problema 22, Particula 1* (Mantua, 1475); *Compilatio physonomiae*, III, cap. 1, decisio I (Mantua, 1474).

[14] This work was probably rearranged by its editor, Johannes Engel, and derives from an original MS text which I have identified as Monaco CLM Lat. 22048, XVth, ff. 158v–176v, which bears the more precise title, *De imaginibus*. The difference between this version and Engel's edition suggest that Engel made a material contribution to the work: see Federici Vescovini, G., 'La teoria delle immagini di Pietro d'Abano e gli affreschi astrologici del Palazzo della Ragione a Padova', *Naturwissenschaft und Naturbeobachtung – Wissenschaft und bildende Kunst vom 14. bis 16. Jahrhundert* (Frankfurt, 16–18 July 1984), Weinheim, Chemie Verlag (Acta humaniora, 1987, pp. 27). See also Haage, B. D., 'Ein Handschriftenfund zum *Astrolabium planum* des Petrus von Abano', *Litterae ignotae*, 50, 1977, Göppingen, pp. 95–108, and idem, 'Dekane und Paranatellonta des *Astrolabium planum* in einem Nürnberger Fragment', *Archiv für Kulturgeschichte*, 60, 1978, pp. 121–39, note 18 to p. 24.

[15] At the end of the Oxford Bodl. codex Canon. misc. lat. 46, XVth, is Peter's *De physognomia*; f. 30v contains a list of his works and the *Lucidator dubitabilium astrologiae* is mentioned as *Librum de discordantiis utriusque partis astrologiae*.

had tried to separate astronomy, as a physical-metaphysical science, from astrology, defined as the technique of making astrological judgements and calculating planetary transits. In his prologue Peter declares that he wrote his *Lucidator* in order to defend astrology from the attacks of those whose ignorance of philosophical ideas prevented them from understanding Ptolemy, and to justify astrology to those sciences that had misunderstood astrology's principles.[16] Above all, he wrote it in order to clarify the ambiguities and difficulties posed by astrology, on the one hand prolix, on the other constantly needing to be verified by observation, and always open to error. He begins, therefore, by establishing what astrology has in common with the other sciences, and speaks of motion and its rules. From this he moves to the science of the 'judgements', posing *sigillatim* the queries of both sides of the argument. Like his *Conciliator litium medicinalium*, the *Lucidator* is arranged in 10 *differentiae* in the form of questions, each divided into four *particulae*. Only the first of the 10 *differentia* is concerned with the scientific justification of astrology; the remaining nine are devoted to genuine astronomical problems such as those concerned with the uniqueness or plurality of motions, the number of spheres and epicycles, and the location of the sun in relation to the moon, and so on.[17]

Here we shall examine only the first question: 'an astrologia <astronomia> cum his quae ipsius, extet scientia'. The first defence of astrology's scientific credentials is directed against Averroes' distinction between astronomy which is physical-metaphysical and astrology which is merely concerned with mathematical calculation: 'astrologia huius temporis nihil est in esse, sed est conveniens computationi non esse'.[18] Against this, and against the more general notion that astrology is concerned only with that which may be generated and corrupted – and must therefore be equally imperfect and corruptable – Peter advances a traditional argument in defence of astrology. Astrology investigates 'de corporibus superioribus perpetuis causantibus et designantibus quae inferioribus generabilia et corruptibilia sunt'; astrology is

16 'Quoniam astrologicae considerationis ambiguitates et discoliae propter ipsius grandem difficultatem constant non minimum, tum quidem ob eius prolixitatem, tum quia in ipsam pervenire non possumus nisi sensuum unico – puta visu in quo non parum inspicientes incaute circumvenit fallacia instrumentorumque similiter errore, tum propter crebas, laboriosas ac sumptuosas observationes, necessarias in illa, tumque propter raros inspectores eius, quorum etiam plurimi qui doctrinis phylosophicis indocti sermones illustrissimi et praecipue Ptolomei, prave intellectos suscepere, ac etiam quia scientiarum aliquae detestantur ipsam quam prave maximeque in eius principiis quae de natura scientia –visum siquidem mihi Petro Paduanensi editionem in ipsa contexere, qua ipsius difficultates ac discoliae in quantum possibile, declarentur eique adversantibus, resistatur. Quare *Lucidatorem* ipsarum eam malui appellare' (Paris MS Bib. Nat. lat. 2598, f. 99ra).

17 'An astrologia (astronomia) cum hiis quae ipsius, extet scientia', also expressed as 'an astronomia sit scientia cum eius appenditiis; an motus unus caeli communis ›sit‹ vel plures; an sphaerae sint novem plures aut pauciores; an sit ponere eccentricos et epiciclos; an planeta moveantur per se in eccentrico vel epiciclo; an Sol situetur supra Lunam immediate vel planetarum medio; an Solis declinatio sit 24 graduum vel aliter; an Solis aux sit mobilis vel immobilis; an centrum eccentrici Solis distet a terra 2 gradibus et 23 minutis sive ipsius aequatio; an Sol peragrat eccentricum eius in 365 diebus et 4 vel aliter' (Paris MS Bib. Nat. lat. 2598, f. 99ra).

18 Averroes, *Metaphysica*, XII, comm. 45.

therefore just as noble as those same heavenly bodies. Furthermore it is a mathematical discipline, and therefore a science, in so far that it consists of understanding demonstrable truths:

> Scientia est comprehensio veritatis ex demonstratione aquisita . . . sed talis est astrologia, cum dictum sit: eam mathematicarum fore unam quarum demonstrationes sunt in primo certitudinis ordine.[19]

Clearly this notion of science – which is also defined in the *Conciliator* as 'true knowledge verified by reason and confirmed by demonstration' – presupposes a theory where the human intellect and knowledge seem not to be conceived as a substantial and perfect ontological form, but rather as an intellectual acquisition, carved out by demonstration from a perceptible substrate and from principles on the intellect which is defined as a 'habitus intellectualis per demonstrationem elicitus'. In Peter's doctrine both *intellectus* and science are in fact *habitus*, or acquired dispositions, even though they may differ from each other. *Intellectus* is the *habitus* which allows axioms, propositions and principles to be truly and immediately understood, as soon as the terms of the question have been noted through the senses. Science is also a *habitus aquisitus* and is true knowledge on the same level as the *intellectus*, with the difference that it is acquired gradually, through reason and demonstration. This is not the place for a detailed analysis[20] of Peter of Abano's concept of science and of the compositive and resolutive method which he developed[21] to defend his notion of medicine as a science, at once practical and theoretical, and which he extended to embrace astrology as well. Nor do we have space to analyse each of the 10 arguments by which Peter maintains astrology to be a 'scientia certa, recto subiecto determinata, una, coaffinis, naturalis, nobilis, difficilis, utilis, prior et licita' and not at all conjectural ('coniecturativa').

3 *Astrology and the other sciences*

Astrology stands in a very clearly defined theoretical and speculative relationship to the other disciplines. First, astrology has nothing whatsoever to do with the seven mechanical or secondary arts which were by tradition: 'agricultura, pastoralis, et medicina, lanificium, armatura, navigatio et theatrica'. Of course Peter excludes medicine from this list which is given by Isidore of Seville and Hugo of St-Victor. Nor does astrology belong to the 'scientias sermocinales', grammar, logic and rhetoric. Logic is defined as the distinguishing ('discretio') between truth and falsehood, and guides the intellect in the ways of truth. It is useful to all the other arts – 'cum ad

[19] *Lucidator*, diff. 1, f. 99rb.
[20] *Conciliator*, diff. 3, propter primum.
[21] In particular *Conciliator*, diff. 8.

omnium methodorum principia viam habeat' – and can be further subdivided.[22]

The moral disciplines are three: 'ethica vel moralis dicta, monastica, quasi solitaria', 'oeconomica quae domus dicitur dispensativa' and lastly politics or 'civilis'.

Finally there is astrology, which is a speculative, philosophical and theoretical science; it belongs to the three essential modes of philosophy – physical, mathematical and divine ('naturalis, mathematicus et divinus').

It should be stressed that in his classification of the speculative and theoretical sciences, Peter has expanded on the role of the mathematical disciplines in the traditional quadrivium given by Boethius. By reinterpreting the traditional Aristotelian distinction between 'continuous' and 'discrete' quantities in the objects of mathematics, Peter extends mathematics to include optics ('perspectiva') and stereometry in addition to the usual geometry, arithmetic, music and astrology.

4 *Astrology-astronomy and its subdivsions*

If astrology is that science which uses reason to study bodies in continuous motion (and in such a case it literally is astrology), then it can also be called astronomy. According to Peter the two terms mean the same etymologically, because astronomy comes from *aster* and *nomos*, or *aster* and *lex*, and astrology comes from the same or from *aster* and *logos*, that is reason or rational discourse.[23] Astrology measures celestial phenomena, their sizes, distances, conjuctions and figures; it speculates about their universal and specific motions and because of this Abu Ma'shar defined it as the 'scientia de motibus'. Astrology is theoretical in so far that it quantifies motions mathematically, but practical because it is concerned with the *passiones* which are the sublunary consequence of these motions; in which case 'scientia iudicialis practica nominatur ex causis <superioribus>'.

Astrology is a universal science because the motions, natures and influences of the heavens are not particular realities but are different aspects of a single reality which has to be investigated and understood as a whole. Not only does the study of an individual aspect presuppose a knowledge of the

[22] *Lucidator*, diff. 1; *Conciliator*, diff. 2, propter tertium. After defining grammar, the *Lucidator* gives the following subdivisions of logic: 'Secundum quod eo sit discretio veri a falsi et tunc est logica sive rationalis dirigit enim intellectum in viam veritatis ... Cuius quippe, secundum Avicennam, octo erant partes, scilicet predicamentalis, interpretativa, syllogistica simpliciter demonstrativa, topica seu topicales, sophistica temptativa et rhetorica' (*Lucidator*, diff. 1, f. 100ra).

[23] 'Propter primum sciendum quod quidam assignarunt differentiam inter astronomiam et astrologiam, dicentes astronomiam fore illam quae partem motus pertractat: astrologia autem quae iudicia instruit. Sed illud neque ratio construit aut multorum usus persuadet, cum astronomia dicatur ab *astro* et *nomos*, *lex*; astrologia vero a *logos* quod *ratio*, et *sermo* et *logia*, *locutio*. Hoc autem indifferentia, similiter alterutrumque invenio in alterutro eius partem utramque proferri' (*Lucidator*, diff. 1, f. 100ra; and cf. *Conciliator*, diff. 10, propter primum).

others, but precise knowledge of an individual aspect is only possible in relation to the other aspects; so 'astrologus non debet dicere rem specialiter, sed universaliter'.

Astrology should therefore be divided into two branches, each of which falls in turn into two sub-branches. The astrology which studies celestial motions comprises: 1a – demonstrative, and 1b – narrative (or recapitulatory). The first is fully expounded and demonstrated in Ptolemy's *Almagest*; the second is to be found, Peter declares, in the works of Alfargani, Azarquiel, Thabit ben Qurra, Alkindi and others.

The astrology which investigates the influences of the heavens, also known as 'judicial astrology', falls into: 2a – the part which is introductory to judgements, and 2b – the practical (*exercitativa*) part, the actual making of the judgements. The first sub-category is of a general character and the second goes into particular specific cases ('Per ipsam actualiter possumus iudicare et in particulares causus descendere').[24]

The practical (*exercitativa*) part of judicial astrology falls in turn into four sub-categories. In his exposition of these Peter reveals himself to be thoroughly acquainted with the available manuals of practical astrology, including that by the author of the *Speculum astronomiae*. He frequently refers to the works of Abraham ibn Ezra which he claims to have 'systematised in Latin'. The four sub-divisions of executive judicial astrology are: revolutions, genitures, interrogations and elections. Of course, in keeping with the doctrinal habit of the time, electional astrology covered the science of the astronomical, astrological and magic *immagines* 'as it has been described', Peter declares, 'by Ptolemy, Thabit ben Qurra and Zahel in the book of Seals'.[25] The magical tradition also included with the astronomical images the necromantic images, that is, those whose efficacy presupposed the invocation of demons or spirits. Like the author of the *Speculum*, Peter provides an excursus on necromancy and the magical arts and makes his position on the subject clear. It is helpful to read this section of the *Lucidator* together with certain passages in the *Conciliator* where Peter explicitly states that he does

24 When mathematics is directed 'ad corpus continue motum', then it is 'astrologia dicta litteralis, ratione utens magis astronomia absolute nominata duplex, ut tactum prius: (1) quaedam scientia totius vel de motibus dicta et (2) altera de iudiciis iudicativa. Prima quidem dupliciter traditur (a) aut demonstratione – ut in *Almagesti* ubi demonstrative traduntur omnia . . . (b) vel narratione colligendo brevius quod in Almagesti prolixius demonstratum, seu Ptolomaeus, in *Introductorio* sibi ad artem sphaericam comparato Ametus dictus Alfarganus, Arzachel, Haben Chore (Thabit ben Qurra), Jacob Alchindi et alii . . . Quae autem iudicativa, duplex extat: una quidem introductiva in iudicia, docet enim praecognoscere necessaria iudicanti, habens se ad consequentem quodammodo ut Porfirius in predicamentis. Altera quidem est exercitativa dicta, per ipsam enim actualiter possumus iudicare ac in particolares casus descendere, velut per canones sanativos tandem actualiter curare. Primam siquidem et secundam tetigit aforistice Ptolomeus in *Centiloquio*, Albumasar *in Sadan*, Hermes Almansor et Albategni in eorum *Aforismis*, utcumque traditur siquidem iudicativa a Ptolomaeo diffusius, *Quadripartiti* primo, Albumasar *Introductorio*, Alkabitio et Abraham iudei Avenare, cuius libros etiam in latinum ordinavi, et Heben bia Zekele (Zahel Benbriz) cum aliis. Exercitativa vero quadruplex estat: una de revolutionibus dicta, alia de nativitatibus, tertia de interrogationibus et reliqua de electionibus' (*Lucidator*, diff. 1, f. 100va).

25 *Lucidator*, diff. 1.

not perform demonic magic. Demons, incubi and succubi are natural beings like any other; if they are exceptional it is only because they are born under extraordinary conjunctions or unusual astronomical phenomena like eclipses. On occasion, and quite unrelatedly, Peter did carry out operations of natural magic[26] such as constructing figurines or incising metals in order to cure certain illnesses.

5 *The magical arts*

In Peter's opinion books of necromantic images are obscene, hateful, evil and 'depravativi intellectus'. Magic is knowledge whose ends are perverted and falls into five categories: divination, mathesis, sorcery, conjuring and the casting of spells.

Magical divination can be either proper or improper. The proper kind proceeds from intrinsic causes and inspires hermits, anchorites and those who dwell far from the world and its concerns; it belongs to weak and afflicted spirits like epileptics and melancholics. Improper divination, by contrast, uses external signs and can be read in the four elements. It thus includes geomancy, hydromancy, aeromancy and pyromancy. Much of the wherewithal for geomancy is borrowed from astrology.

Mathesis, the second kind of magical divination, includes 'horospicium, haruspicium, augurium et auspicium', that is, finding omens in, respectively, the hours of the day, sacrifices, and the sounds and flights of birds. Peter does not ascribe much importance to these and dwells instead on one particular curiosity – *sternutatio*, the ancient custom of making predictions from sneezes (*Lucidator*, ms cit., f. 101 vb and *Problemata* XXXIII, 7).

Sorcery is the ability to foresee chance events and to understand and predict contingent events. Conjuring is the illusion which limits the visual abilities, the deception which makes the truth impossible. *Garamantia*, casting spells, is the evil magic art which weakens the body with spells which need the attention of a skillful physician just like any other illness. Necromancy is also a kind of *fascinatio*, of spell-casting,[27] and is even more serious because it brings with it the danger of death to the body and to the soul. The necromancer puts spells and curses on people, causing them injuries and pain purely by the power of the imagination.

26 'Et ego quidem fui expertus figuram leonis impressam in auro sole existente in Medio caeli, cum corde Leonis aut Venere aspiciente'. It is these and similar passages where Peter talks about the figure of a scorpion that Marsilio Ficino refers to in his *de vita coelitus comparanda*, ch. 18 – 'quales caelestium figuras antiqui imaginibus imprimebant ac de usu imaginum' (Venice, Aldus Manutius, 1516, f. 162v and cf. ch. 15, f. 160r).

27 'Necromantia vero dicitur a *necros* quod est *mors*, et *mantia*, id est mortis indivinatio. In ea namque ponunt periculum corporis et mortem animae consistere. Ipsa enim garamantia extat qua invocantur spiritus et animae incantantur defunctorum in responsis dandis in quatriviis, triviis et cimiteriis maximae . . . Et necromanticus est qui facit populo fascinamenta et facturas, seu malefitia, quae maxime dico imprimere propter imaginationem impressam laesionis talium . . . Garamantia est maleficium quod est fascinatio animalis, occupans vires ut sui compos esse non valeat, actum venereum impediens proprius' (*Lucidator*, diff. 1, f. 101va).

It would seem from Peter's descriptions that these arts belong to demonic magic, which he comprehensively abominates. He tends to describe its effects as psychophysical disturbances, if not actually as physical illnesses, in weak individuals. He also seems to dislike alchemy. He defines it as a science of the 'fifth part' of nature; alchemy's pretences towards medicine resemble necromancy's abuse of its undeserved reputation.

Physiognomy and chiromancy, however, are two natural sciences of prediction. They are closely linked with medicine and astrology and are uncontaminated by evil. From an analysis of bodily features it is possible to reconstruct an individual's past, present and future. The word 'physiognomy' comes, Peter states, from 'physis' and 'nomos', and not from 'physis' and 'onoma', as was generally thought. Physiognomy thus investigates the harmonious natural laws which regulate the relationship between soul and body.[28] Peter's interest in physiognomy recurs in his *Conciliator* and, especially, in his *Compilatio physonomiae*, a work devoted exclusively to the topic. Chiromancy – or palmistry – assumes that features of the hands reflect the motions of the soul and Peter regarded it as important, though subordinate to physiognomy.

6 *The nobility and utility of astrology*

Theologians, philosophers and others who have attacked astrology did so from ignorance, incompetence and blindness. They are scandalised by the mathematical theory of eccentrics and epicycles espoused by some astrologers. They think that such notions controvert the perfection of the heavens. Such 'divini hypocrites' think this theory harmful because it attributes imperfection and irregularity to the prime cause. But this is not true, says Peter. It is the ignorance of the critics which is harmful: to allow a role to secondary causes, namely the planetary motions, whose mathematical expressions presuppose an apparent diversity of motions, is not to diminish the perfection of the prime cause – indeed, if anything it is a reverent recognition of the prime cause.

So, astrology can also be useful to theology because its object is divine and unambiguous, and because its demonstrations are based on the science of numbers and measure, it can yield certainties. Its object is divine because

28 'Physonomia sub naturali contenta scientia, ut etiam nomen denotat: a *physis*, natura dicta et *nomos, lex*. Ex quadam namque lege ac cohordinatione naturae inest quod talis corporis formae similis animae imprimatur potentia; quae siquidem scientia existit passionum animae naturalium corporisque accidentalium habitum vicissim permutantium utriusque, ut in *Physonomiae* declaravi *compilatione*. Huius siquidem pars est chiromantia, dicta a *chir* quod est manus et *mantia* quasi vaticinatio in lineis et proiectionibus' (*Lucidator*, diff. 1, f. 101va). In his classification of the sciences and disciplines Peter also refers to Afarabi's *De ortu scientiarum*. He devotes much more space to describing the divinatory arts – and astrology, naturally – that does, say, Robert Kilwardby whose commentary on the *De ortu scientiarum* barely mentions the magical arts; see the chapters on the mathematical sciences including astrology in Robert Kilwardby, *De ortu scientiarum*, ed. Judy, A. G. OP, Toronto, 1976, pp. 31ff.

from the universal celestial and planetary motions one passes to their cause – which is God. Indeed, that which is 'theologicum nunquam videtur neque comprenditur', if not from secondary causes which are the effects derived from God. Astrology thus shares the nobility of the highest disciplines on three grounds: its method, its object and the two together. It is not harmful to divine law as has been asserted by some hypocrites who pretend to honour God ('ypocrites deicoles se fingentes'): 'verum potius ›eam‹ vigorat et extollit'. These theologians detest and fight against judicial astrology in particular because the ignorant believe that on the one hand its proponents assert celestial necessity and deny man's free will, while, on the other hand, believing in miracles and portents. In reality it is not so because the stars only incline, they do not compel. Astrology is thus first among the sciences, including perhaps metaphysics.

In the *Lucidator* Peter does not bother to investigate the relationship between medicine and astrology – which is clearly fundamental to his thought – probably because he had considered the subject several times and from several different angles in his *Conciliator* and in his *Problemata*. Evidence of his interest in astrological medicine is his translation of the short pseudo-Hippocratic tract *Hippocratis libellus de medicorum astrologia* – a genuine manual for the subject.[29]

7 *Astrology and medicine*

Differentia 10 of the *Conciliator* is entirely devoted to astrological medicine. Peter refutes the argument that astrology is useless to medicine because the knowledge of the laws by which the light and motions of the heavens operate it has nothing to do with curing the sick. Peter's confutation rests strictly on medical grounds, stressing how the light and motions of the heavens provoke changes in the four fundamental qualities – heat, cold, dryness and moisture – and thus in the humours. Therefore it is ridiculous to say that the motions of the heavens do not influence the patient's physical state, when in fact they can modify the *crasi*, the correct humoral equilibrium of his organism, his complexion. The competent physician will therefore identify the influences of the stars on his patients and he will treat them accordingly. He will avoid crises by remedies which counter negative stellar influences and he will foster the positive state of *eucrasia* with remedies that are by nature similar to the beneficial influences.

Indeed, the complexions of the seven planets correspond to human complexions. Although the term 'complexion' is ambiguous when applied to the heavenly bodies, it is evident that for Peter each planet had a different nature, due to the predomination of one or other of the basic qualities. He

[29] Specifically about astrological medicine are the Ps-Hippocratic *Libellus de medicorum astrologia* which Peter translated, and numerous questions in the *Conciliator* and in the *Problemata*. Regarding Peter's translation of the former item, see Kibre, P., 'Astronomia' or 'Astrologia Yppocratis', *Science and History: Studies in Honor of Edward Rose*, Warsaw, 1978, pp. 133–56.

regarded the planets as warm, cold and so on, and according to their complexions, exerted differing influence on the sublunary realm in general and on man in particular. It follows that physicians must know the characteristics of the stars and bear them in mind both when applying the three systems of treatment – diet, medicine (*potio*) and surgery – to cure the sick and when maintaining the health of those who are not sick.

Equally, a knowledge of the complexions of the planets allows the physician to identify the rhythm of crises because, as Peter frequently observes, an illness is nothing more than the plethora of a particular humour and each humour, like each part of the human body, falls under the dominion of a planet whose conjunctions, oppositions and *triplicitates* can therefore bring the illness to a happy or a fatal outcome. Furthermore, by observing in the stars when periods favourable to contagions and epidemics are imminent, the astrologer physician can protect his patients, either by predisposing their bodies against the particular infection, or by fortifying them so that the illness only touches them lightly. So astrology's contribution to medicine is that it facilitates prognosis and can suggest the appropriate therapy. The skilled physician will know how to detect the influence of the sun in chronic conditions, and will watch the moon – which acts most intensely on acute illnesses – to know when he may proceed to bloodletting, lancing, cauterisation and surgery proper. The human body, like the heavens, is divided into 12 zones each of which is superintended by a sign of the zodiac: Aries has the head, Taurus the neck, Gemini the arms, Leo the chest and so on down to the feet under the dominion of Pisces. Surgery of any kind is contraindicated, if not downright fatal, says Peter, when the moon is in the sign which rules the part of the body that is to be operated on. No operations should be done on the head when the moon is in Aries; wounds to the arms received while the moon is passing through Gemini heal reluctantly, if at all. Peter, and also Roger Bacon, explained this extremely widespread belief by the fact that the excess of the moist humour engendered by the lunar transit delayed healing and encouraged suppuration.

The assumption behind this notion is that matter can only receive celestial rays if it is favourably predisposed to do so. The astrologer cannot prevent the occurrence of malign conjunctions – astrology can hardly interfere with the progress of the stars – but he can at least predict them and, forewarned, fortify the body against their dire effects.

It is not enough for the physician-astrologer to have an accurate knowledge of the motions and natures of the stars; he must be equally well acquainted with the patient's natal horoscope. Indeed, a celestial influence that is normally regarded as malignant can actually be favourable in relation to the zenith of a particular individual. It is extremely important for physicians too, therefore, to know the hour at which the patient was born. Knowing that, it is possible to calculate the *alcocoden* or *hyleg*, the planet which rules that moment and which will influence the native most directly, and it is possible then to calculate the native's lifespan.

In addition, Peter declares that the astrologer-physician can usefully exploit the *immagines* which, if correctly made, can attract beneficial celestial rays to the patient and shield him from harmful ones. The effectiveness of these images is not due to the intervention of ultra-terrestrial powers but is a consequence of the perfect, but hidden, correspondence of all parts of the universe. In any case, if the astrologer-physician wishes to help his patients he must be able to interpret the continuous correspondence between earth and heavens whereby the configurations of the heavens correspond perfectly with terrestrial phenomena. However, although every part of the body corresponds to a sign of the zodiac and every humour and bodily organ corresponds to a planet – to such an extent that man is a perfect representation of the vault of the heavens, a microcosmos (*Conciliator*, diff.47) – it does not follow that the individual contains within himself all the elements of the world, rather he contains them *in nuce*, potentially, and reflects the universe by analogy. The four humours of the medical tradition which Peter describes fulfil the same functions in the human body as the four elements in the universe; both may be expressed in terms of the four basic physical qualities: hot, cold, dry and wet. Peter insists on the functional similarity and not on the identity of the elements; he explains the microcosm/macrocosm relationship in analogical terms, that is, in terms of proportion. He does not postulate that the same elements are present in man and in the universe, but emphasises that man and world are analogous because their existences are orchestrated and conducted by the same proportions. Peter theorises not only that every part of the microcosm, whether bodily organ or individual person, is proportional to part of the universe, but also that just as the heavenly bodies are regulated by a precise mathematical law – that of the proportions and disproportions of the planetary aspects – so the parts of the microcosm are proportional to each other because they reflect the same relationships and aspects that are found amongst the stars in the firmament. Man therefore stands at the heart of a series of symmetries. He is proportional to the universe both because every part of him is proportional to a zone of the sky, and because the relationship between his organs reproduces and reflects the proportions which interconnect the heavenly bodies. This double analogy is clearly expressed in the example which Peter often repeats: the heart is to the body as the sun is to the universe: both, being sources of heat and being centrally located – a mark of their superior nobility – have the same vivifying function, and one superintends the individual's life by means of the arterial system just as the other, by illuminating them, rules the planets which surround it.

8 *The 'De motu octavae sphaerae' and an astrological philosophy of history*

The *De motu octavae sphaerae* is another of Peter's important astronomical works which circulated widely in manuscript and remains unpublished. It is cited by Cusanus in his treatise on calendar reform. Shorter and more schematic than the *Lucidator* it reflects prevailing astronomical wisdom about equinoctal precession (first discovered by Hipparchos, and well known to Greek, Arabic, Jewish and Medieval European astronomers) and Peter says that he composed it in 1303. It is an important work for several reasons, in particular for its critique of the theory of 'trepidation' (according to which the equinoctal points oscillate or alternately advance and recede in a millenial rhythm) advanced by Thabit ben Qurra (ninth century) in his treatise on the motion of the eighth sphere. This theory seems to have been widely known to the Latin Middle Ages, judging by the large number of manuscripts which contain ibn Qurra's and Peter's treatises on the motion of the eighth sphere.

From an astrological point of view measuring the motion of the eighth sphere – the sphere of the fixed stars, which is endowed with a triple motion – is extremely important. All earthly events depend on when this motion began and how it varies: all changes in the arts, in institutions, in peoples and all the ups and downs of history are accurately orchestrated by the universal and particular motions of the eighth sphere. Peter begins, therefore:

> Since I desired to investigate the cause of universal changes in the mechanism of the world, in its perceptible intention and in the cause of its acting, I decided to study the motion of the eighth sphere with the fixed stars that are in it, for this, according to God's will, is the proper cause of every change. The philosophers think that the world is eternal; faith and the scriptures which hold that the world was created, teach us otherwise. However, it is more correct to believe what the scriptures tell us about the beginning of human history. It is then important to know whether historical events began when the terrestrial sphere was moving in the same direction as all the other spheres, or was it during the first precession, or was it before the first precession, and so on *ad infinitum*.

How much time has passed since that moment, and how is that moment to be determined? Everything depends on determining the variations in the motion of the eighth sphere. Peter observes that there are extreme discrepancies between the sources; there are even different dates for the beginning of history. Instead of devaluing the study of the motions of the eighth sphere, this positively stimulates research, says Peter. Since the highest and most united virtue is transmitted to the lower realms when the spheres are moving in the same direction, there was a period when the whole world shone with wisdom and virtuousness. During that period it pleased God to reside here below with us as a man; there were illustrious rulers like Alexander, Darius, Julius Caesar and many others; the Stoics flourished in Rome, likewise the Peripatetics founded by Aristotle in Athens; medicine, astronomy and rhetoric bloomed with Galen, Ptolemy and Cicero; jurisprudence and the mathematics spread far and wide. In this period all parts of the universe were harmoniously in proportion and at one with each other. But with the first precession of the equinoxes, with the first shift of the eighth sphere – discovered by Hipparchos and known by Ptolemy – things began to move contrariwise and great historical changes came about.[30]

[30] See Federici-Vescovini, G., 'Un trattato di misura dei moti celesti: il *De motu octavae sphaerae* di Pietro d'Abano', *Mensura, Mass, Zahl, Zahlensymbolik im Mittelalter (Miscellanea medievalia* 16/2), Berlin, 1984, pp. 277–93.

Passages of astrologising historiography like these in the *De motu octavae sphaerae* (and also in the *Conciliator* where he speaks of the *De motu octavae sphaerae*)[31] are of great interest and significance for the history of medieval astronomy.

Peter gives a critical account of the variations in the motion of the eighth sphere and rejects ibn Qurra's theory of trepidation because it entirely fails to 'save the appearances'.

The small displacement in the annual revolution of the eighth sphere must be in one direction only, as Ptolemy maintained, although it represents a loss of movement (*incurtatio*) which is calculated at 'in 20 gradibus eiusdem cum 15 minutis fere'. Since I have presented a detailed analysis of this work from an astronomical point of view elsewhere, I shall not linger on it any longer.[32]

9 *The theory of the images*

Ancient tradition and the most modern historians credit Peter with a work entitled *De imaginibus* which is also the second section of the *Astrolabium planum* (Augsburg, 1488). Before going any further we should distinguish between the works in which Peter develops philosophically his theory of the astrological images, and the works in which he simply describes the individual figures of the images for each of the 360° of the zodiac, as in the *De imaginibus*. Peter's astrological-medical-philosophical theory of the images is mentioned by Agostino Ricci in his own treatise entitled *De motu octavae sphaerae*[33] and by Marsilio Ficino. The theory can be reconstructed from passages scattered throughout the *Conciliator*, the *Lucidator*, the *De motu* and the *Problemata*. From passages in the *Conciliator* we learn that Peter believed 'median' or complexional influences to come from the planets and the 48 images which accompany the signs of the zodiac and which are composed by the configurations of the 1022 or 1029 stars in the sky, are endowed with a virtue which can be contracted into images and the like which are proportional to them. Such images have an allegorical rather than a real value and in the *Lucidator* Peter explicitly and emphatically rules out any link between images and the invocation of demons and spirits, that is, any use of images for necromantic purposes. The 48 images of the so-called *sphaera graecanica* are nothing more than celestial configurations resulting from the

[31] 'Et ideo sive ponatur octavum caelum moveri in oppositum noni ab occidente in oriens in centum annis uno gradu secundum sententiam Ptolomei longitudine, latitudine vero minus, seu declaravi in editione huic consimili quam in astrologia composui, unde etiam specialem tractatum cum instrumento [i.e. the astrolabe] materiali ostensivo construxi anno gratiae, 1303 quo ego Petrus Paduanensis hunc librum construxi' (*Conciliator*, diff. 9, propter tertium).

[32] *De motu octavae sphaerae*, Venice MS cit., f. 118r–v. See note 30 above.

[33] Augustinus Ricius, *De motu octavae sphaerae*, Trento 1513 (Paris, Simon Colinaeus, 1521, f. 41).

appearances and disappearances of the paranatellontic constellations[34] in the ecliptic, or north or south of it. Descriptions of them were not unanimous and already in Abu Ma'shar's exposition of the subject they were confused with the figures for the decans – 10° segments into which each zodiacal sign was divided – and the images of the planets, thus giving rise to the *sphaera barbarica*. Moreover, in texts like the *Picatrix* these images are endowed with divine features, thus giving rise to astrolatrous cults.[35] Peter does refer in general terms to lengthy passages of Abu Ma'shar, probably from the Sadan 'excerpts'[36] where the planets seem to be presented as divinities to whom prayers may be addressed. However, he stresses that the celestial images have to be understood metaphorically and that figurines and drawings fabricated during particular stellar configurations have only a limited value in medical practice – so if he is involved in anything it is natural, rather than demonic, magic.

Several passages of the *Conciliator* explicitly describe special astrological methods for fabricating astronomical images which, in conjunction with *electiones* and *interrogationes*, can mitigate, modify or enhance astral influences. One such passage lays particular emphasis on the transit of the Caput Draconis over the mid-heaven.[37]

In summary, then, Peter upholds the doctrine of the *virtus contracta* which emanates from the 48 heavenly images, says that he was acquainted with Ptolemy's work on astronomical images, with Thabit ben Qurra's work on magical images and with the book of seals.[38] He also stresses the danger that this technique, when used in conjunction with *electiones* and *interrogationes*, can overlap with necromancy unless it is kept within the permitted bounds of the natural characteristics of the astronomical images.

34 Abu Ma'shar, *Introductorium maius*, II, ch. 1 and 2; for the paranatellontic constellations see Aurigemma, L., *Il segno dello Scorpione*, Torino, 1976, pp. 44ff, with bibliography, and Le Boeuffle, A., *Les noms latins d'astres et de constellations*, Paris, 1977. For the theory of the decans, see Gundel, W., *Dekane und Dekansternbilder*, Studien der Bibliothek Warburg, 19, 1936, and idem, 'Neue astrologische Texte des Hermes Trismegistos', *Abhandl. Bayer. Akad. Wissenschaften, Phil.-Hist. Kl*, new series, 12, 1936; *Astrologumena (Sudhoffs Archiv* 6), Wiesbaden, Steiner Verlag, 1966.

35 For the *Picatrix*, see Garin, E., 'Un manuale di magia', *L'età nuova*, Napoli, 1969, p. 406 with bibliography; Perrone Compagni, V., 'Picatrix latinus', *Medioevo*, 1, 1975, which includes a partial edition of the Latin text; and my 'L'astrologia tra magia, religione e scienza', *'Arti' e filosofia, studi sulla tradizione aristotelica e i 'moderni'*, Firenze, 1983, pp. 182ff.

36 I am currently researching this topic; see note 37.

37 *Conciliator*, diff. 156: 'Propter secundum sciendum quod experientia potest demonstrari et demum ratione persuaderi praecantationem conferre . . . Prima similiter astronomiae oratione placantur, et in subsidium concitantur nostrum, ut orationum epilogus insinuat planetarum. Unde Albumasar in Sadan: 'Reges Graecorum cum volebant obsecrare deum propter aliquod negotium, ponebant Caput Draconis in Medio caeli cum Iove . . .'. See my studies: 'L'astrologia tra magia, religione e scienza', pp. 184–7; and ' "Albumasar in Sadan" e Pietro d'Abano', *Le scienze islamiche nel Medioevo*, Roma, Accademia Nazionale dei Lincei, 1987, p. 25.

38 *Conciliator*, diff. 28; *Lucidator*, diff. 1.

10 *The 'De imaginibus'*

The fifteenth-century Monaco manuscript, already noted by Thorndike, is most interesting.[39] It describes the figures for each of the 360 degrees of the zodiac, that is each ascendant degree of the zodiacal year, according to an ancient reference to the astrological calendar of the meanings of the figures of the ascendants. An individual's nature and profession depended on the influences of these figures of the ascendant degrees. We can see why Peter emphasised the ascendant figure of the planets in the 360° of the year: in his classification of the three ways in which celestial causes operate – the universal influence of the heavens, the complexional influence of the planets and the contracted virtue of the 48 images – the third category is devoted to the particular virtue of the aspects of the planets to the ascendant at the moment of birth. A similar theory, known as *myriogenesis*, or the variety of births, was developed by Firmicus Maternus in his *Mathesis* (VIII, 1 & 18).

Broadly speaking the best known treatment of astronomical images in the Middle Ages was that given in the second book of Abu Ma'shar's *Introductorium maius*, along with various derivative compilations, such as the *De imaginibus* of Michael Scot. This tradition must be distinguished from that of the *De imaginibus* by Thabit ben Qurra – edited and studied by Carmody[40]

[39] Monaco manuscript (CLM lat 22048) f. 158v begins: 'Incipiunt jmagines signorum super triginta gradus quorum quilibet habet suam jmaginem propriam ut predicentur naturas et exercitia hominum . . .' (and explicit): 'Mulier tenens speculum se speculatur ut meretrix. Finiuntur gradus signorum Zodiaci cum vultibus celestibus iuxta mentem Petri de Abano' (f. 176v). For example, regarding the degrees of Aries: '(1) Ascendit et apparet vir tenens in dextra manu falcem et in sinistra manu balistam: aliquis laborat, aliquis bella exercet. (2) Homo cum capite canino cum dextra manu extensa et in sinistra habens baculum . . .: homo litigiosus, invidus ut canis. (3) Homo cum pileo vel gasione in collo, dextra manu ostendens varia mundi et in sinistra applicata cingulo: pacem diligens. (4) Homo crispus capillis in dextra tenens accipitrem, in sinistra flagellum: raro ditatur, illa que accumulaverit senex consumet. (5) Duo homines dextra secans lignum habens bipennem, alter tenens ceptrum in dextra manu, qui stat a sinistra pater familias appellabitur, in alio emisferio . . . stella. (6) Rex coronatus tenens pomum in dextra manu et in sinistra ceptrum; transcendit a medio consanguineos et vicinos. (7) Homo armatus totus in dextra habens sagittam in se custodiens. (8) Homo cum galea in capite et non alias armatus in dextra manu habens balistam: litigiosus et occisor. (9) Homo bene vestitus in dextra habens gladium, nudus capite: semper miratur, verbosus. (10) Homo capite nudus, alias vestitus transfigens ursum unum cum cuspide: venator.' [and so on for each of the 30 degrees of each of the 12 signs of the zodiac, closing with] Pisces (f. 176r): '(20) Luna lucens: vir instabilis. (21) Armatus perforat gladio inermem hominem captum. . . . (22) Mulier lacerata veste: impudica. (23) Mulier pulchra in navi: instabilis . . . (24) Jacentes vir et mulier in lecto: lascivia. (25) Vir proiciens lapidem cum funda: litigiosus. (26) Mulier amputat caput viri cum securi circa lectum in somniis, homicida. In alio emisferio . . . (27) Vir nudus urinat: impudicus. (28) Vir habet accipitrem una manu, serpentem in alia manu: fortis sensibus. In alio emisferio . . . (29) Magnus piscis: instabilis. (30) Mulier tenens speculum se speculatur: ut meretrix. Finiuntur gradus signorum zodiaci cum vultibus celestibus iuxta mentem Petri de Abano' (Monaco MS cit., f. 158v–176v). The text corresponds, with some variants, to the second part of the Augusta, 1488 edition of the *Astrolabium planum* prepared by Johannes Engel: under the 30th degree of Pisces, for example, is given: 'Mulier se in speculo cernens: homo superbus erit et incastus' (f. 92v).

[40] Carmody, F., *The Astronomical Works of Thabit ben Qurra*, Berkeley and Los Angeles, 1960, distinguishes the Ptolemaic astronomical images tradition from the magical hermetic current

– which only deals with the images of the planets and the construction of figurines corresponding to their influence. In addition, Latin commentaries on Sacrobosco's *Sphaera* – like the one by Blasius of Parma – often treat of the astronomical characteristics of the images, whether paranatellontic, planetary or zodiacal. Peter's text, if (as the Monaco manuscript suggests) it is his,[41] is part of this complex tradition. As early as the sixteenth and seventeenth centuries Scardeone, Gaffarel and others suggested that it might have provided the astrological imagery for the fresco cycle in the Palazzo della Ragione, Padua, which was begun early in the fourteenth century and continued after the fire in 1420.[42] This series of images for the 360° of the zodiac at a latitude of 45° – approximately that of Padua – is also found in the Augsburg, 1488 edition of the *Astrolabium planum* by Johannes Engel. The preface to the second book announces figures for the sky at the centre of the sixth *clima* – whose latitude is approximately 45° – for the 12 houses, as well as images for the aspects and degrees of each sign, and that these images were 'devised by the most excellent and skilful doctor of the medical faculty, Peter of Abano'.[43] Explanations are also given – probably by Johannes Engel – of

---

which, however, is founded on a number of composite works like the *Liber prestigiorum*, attributed to ibn Qurra or to Hermes and associated with the mythical figure of Elbedis. The Arabic Sabaean hermetic texts should be distinguished from the Greek hermetic texts edited by Scott. There are two Latin versions of ibn Qurra's *De imaginibus*, derived from different originals. According to Carmody, one or other – we cannot be sure which – is the work of John of Seville.

41 I assume that Peter of Abano composed an *Astrolabium planum* on the strength of: a reference by him in the *Lucidator* (ff. 101va and 109rb); the fact that he gives a practical example of the imaginary construction of the ninth and eighth spheres in the initial paragraphs of his treatise *De motu octavae sphaerae*; and an explicit reference in the *Conciliator* (diff. 9, propter tertium). However, I have not found any MS which predates the Monaco MS which is fifteenth-century and very late, so some doubts must remain about the authenticity of the work in the Monaco codex, *De imaginibus* or *Astrolabium planum*. According to Bernard Haage there are some fragments of this work in German in a Nuremberg codex and in two Heidelberg codices, but I have not examined them. Peter gives his instructions for the constructin of an astrolabe in the *De motu* (MS Venice, Museo Correr, Provenienza Cicogna 2289, sec. XV, ff. 111v–112r). Regarding other mss of the *De motu*, see my monograph cited in Note 30 above, p. 288. On the subject of medieval astrolabes see Thomson, R. B., *The Astrolabium Planum of Jordanus of Nemore*, Toronto, 1978 – Thomson's approach is exclusively from a mathematical/astronomical point of view and he does not discuss their astrological uses and usefulness. See also the various relevant studies by J. D. North. See Haage, B. D., 'Ein Handschriftenfund zum *Astrolabium planum* des Petrus von Abano', *Litterae ignorae*, 50, 1977, Goppingen, pp. 95–108, and idem. 'Dekane und Paranatellonta des *Astrolabium planum* in einem Nürnberger Fragment', *Archiv für Kulturgeschichte*, 60, 1978, pp. 121–39. For Blasius of Parma's (Biagio Pelacani) commentary on the *Sphaera*, see my *Astrologia e scienza*, Firenze, 1979, p. 365.

42 Savonarola, M., *Libellus de magnificis ornamentis regiae civitatis Paduae*, edited by Segarizzi, A., in Muratori, L., *Rerum italicarum scriptores*, 25 (15), Citta di Castello, 19022; Scardeoni, B., *De antiquitate urbis Patavii et claris civibus Patavinis*, Basel, 1560; Gaffarel, J., *Curiositates inauditae*, Hamburg, 1676, p. 201; Ursati Sertori, *Monumenta patavina studio collecta*, Patavii, 1652, Paulus Frambottus, p. 26.

43 'Ut ea quae in tabulis equationum domorum caeli posita sunt elucidissime ad sensum appareant, subsunt in hac secunda parte figurae caeli ad medium sexti climatis, cuius latitudo est circa 45 gradus per duodecim domos verificatae; quibus imagines facierum praeponuntur

the functioning of the images which describe the characteristics of the native born under each particular ascendant image.

When interpreting this text astrologically, it is important to distinguish those born under a given ascendant with its various aspects to the birth planet, from those born under a given planet, the 'Planetenkinder' (studied by Warburg, Saxl, Klibansky and Panofsky),[44] who display the general characteristics of the human temperaments according to the influence of the sun in a given sign, or decan of that sign, at birth. The latter typology is generic and relates to the 'temperaments' of the planets; for example, in the natal horoscope of the *homo saturninus*, the sun is in Capricorn, the domicile of Saturn, and all men who have the sun in Capricorn will have saturnine characteristics. Similarly, the location of the sun defines the *homo solaris*, the *homo martialis* and so on. Peter's theory, by contrast, tries to provide an individual typology and takes full account of the dynamic relationship between the sun and the ascendant. He uses the astrological technique of natal horsocopes based on the *alcocodon* and the *hyleg*. He evolves a varied typology which, by astrologically collecting individual properties, attempts to describe individual rather than universal types.

To conclude, Peter of Abano's attitude towards astrology was neither superstitious nor irrational; he was very much the philosopher and scientist of his day. Armed with Aristotelian philosophy and other doctrinal, medical and astronomical traditions derived from Byzantine and Arab culture, he inserted astrology into a rational and natural scientific context. He separated it from demonic magic and reinstated it between astronomy and mathematics whence certain philosophers and theologicans had removed it. In other words, he purged medieval astrology of its superstitious magical and demonic trappings and stressed its scientific and mathemaical character. At the same time, however, he acknowledged the inadequacy of certain strictly astronomical hypotheses (such as ibn Qurra's fanciful trepidation of the eighth sphere), the erroneousness of the available astronomical tables (which needed constant revision) and the unreliability of observational data which needed constant correction. In these respects Peter pushed astronomy-astrology another step down the road of science.

As Bruno Nardi observed, Peter should be seen as a proponent of the independence of secular medical, philosophical and astronomical thinking

---

graduumque omnium signorum, imagines ab excellentissimo viro medicinae facultatis doctore, experto Petro de Abano elaboratas, in medio figurarum caeli proprietates earundem imaginumque figurationes appositis locantur' (cf. *Astrolabium planum*, edited by Johannes Engel, Augusta, 1488, f. 44v).

44 See Klibansky, R., Panofsky, E. and Saxl, F., *Saturn and Melancholy, Studies in the History of Natural Philosophy and Art*, New York, 1964; Wittkower, R.-M., *Nati sotto Saturno*, Torino, 1968 (Italian translation of ed. first pub. London, 1963); Aurigemma, L., *Il segno zodiacale dello Scorpione nelle tradizioni occidentali dall'Antichità greco-latina al Rinascimento*, Torino, 1976. For Peter's theory of the *alcocoden* see Nardi, B., *Studi su Pietro Pomponazzi*, Firenze, 1965, p. 282; and Kunitzsch, P., *Mittelalterliche astronomisch-astrologische Glossare mit arabischen Fachausdrücken*, München, 1977, p. 12.

from theological interference.[45] He was an innovator. Instead of responding to the metaphysical interrogation of nature by constructing a physics, he did his utmost to forge a physical science which would respond to the empirical interrogation of nature. His colossal erudition, the wealth of information he presents and the originality of his philosophical ideas make Peter the representative par excellence of all the multifarious tendencies, excepting the demonological, of the late thirteenth- and early fourteenth-century natural sciences. And at the conceptual core of his epistemology lies the astrological notion of the natural causality of the heavens.

45 See Nardi, op. cit. p. 78; regarding the contrary and contradictory interpretations of Peter's thought – magus and occultist on the one hand, physician and rational scientist on the other – in the historiography of the sixteenth and seventeenth centuries, see Piaia, G., 'L'immagine di Pietro d'Abano nella storiografia del sei-settecento', *Vestigia philosophorum*, Rimini, 1983, pp. 199–214. See also Marangoni, P., 'Per una revisione del pensiero di Pietro d'Abano' in *Il pensiero ereticale nella Marca trevigiana e a Venezia dal 1200 al 1350*, Abano Terms (Padova), 1984, pp. 66f, for a picture of Peter which differs from that presented by Bruno Nardi. Marangon re-analyses the reasons for his condemnation, without, however, referring to unpublished works like the *Lucidator* and the *de motu* which provide important new clarifications of the controversial aspects of Peter's philosophical and astronomical thinking, such as whether or not he believed in the eternity of the world. On this particular question and the astronomical and astrological grounds for believing in the world's cyclical renewal, see my own 'Pietro d'Abano e le fonti astronomiche greco-arabe latine (a proposito del *Lucidator dubitabilium astronomiae*)' in the proceedings of the conference *Primo umanesimo e filosofia a Padova* (Padua, 1985), where it is shown that Peter had astronomical reasons – namely, equinoctal precession – for believing that 'motum cum mundo incoepisse'.

ASTROLOGY AT THE ENGLISH COURT IN THE LATER MIDDLE AGES

Hilary M. Carey

Astrology and divination has enjoyed a traditional association with kings and courts. To name just a few, King Arthur had his Merlin, the Emperor Frederick II had Michael Scot, Catherine de Medici had Nostradamus and Elisabeth I of England patronised John Dee. In the era before cheap printed books, astrology can aptly be described as a 'Royal Art', affordable only to the aristocracy and with its own particular glamour. If this was the tradition, it remains to be seen how real and extensive was the authority of the court astrologer. In this paper I propose to examine the influence of astrologers at the English court from roughly the reign of Edward III to that of the first of the Tudors, Henry VII, with a view to establishing the extent of their power.

It is not difficult to understand why secular rulers have been so attracted to astrology, as to any system which claimed to predict the future, avert disaster and hasten the overthrow of the enemy. In fact we may wonder why the allures of astrology were overlooked for so long by the members of the English court. Edward III seems to have been the first English monarch, with the possible exception of Henry II,[1] to have taken any positive interest in astrology. Almost certainly, the main factor in this neglect was the church's repudiation of astrology which, it was argued, defied both the omnipotence of God and the freedom of the will. By the thirteenth century the official canonical position was firmly against astrologers, including even the priest who used his astrolabe to recover property stolen from the church.[2]

Edward III may have been better informed of the moral dangers of astrology than many of his successors. After Edward's triumph at the Battle of Crécy on 26 August 1346, Thomas Bradwardine delivered a victory sermon in the presence of the king in which astrologers were singled out for

A fuller treatment, and more extensive documentation, of the issues covered in this article are to be found in my thesis, *Astrology and Divination in later Medieval England*, University of Oxford, D.Phil., 1984, especially chapters 4–8.

[1] Note the horoscopes relating to Henry II in London, Brit. Lib., MS Royal Appendix VI. J. D. North has suggested that these may be the work of Adelard of Bath.

[2] The literature on the astrology debate is very extensive, but for a useful summary of the main issues see G. W. Coopland, *Nicole Oresme and the Astrologers*: A study of his 'Livre de Divinacion' (Liverpool, 1952). For the case of the priest and the astrolabe, which comes from the *Liber Extra* issued in 1234 by Gregory IX, see Edward Peters, *The Magician, the Witch and the Law* (Philadelphia, 1978).

attack.[3] It is God alone, Bradwardine urged, who created the sun, moon and stars and all that is on the earth, and His power over creation is demonstrated in scripture. Addressing the king directly Bradwardine demanded:

> What astrologer could predict this? What astrologer could judge that this would happen? What astrologer could foresee such a thing? Indeed, most beloved, here is one prediction which will never be proved false: whatever God wishes to happen or to be done, that is done; whoever God wishes to be victorious, he is victorious; and whoever God wishes to reign, he will reign. Although therefore the heavens and the earth, and all things under the heavens should be against you, if God is for you, what can harm you? And although the heavens and the earth and all things under the heavens should be for you, if God is against you, what can help you?[4]

With this eloquent plea Bradwardine could have been speaking on behalf of the medieval Church to a large proportion of the courts of France, Italy and Germany, which were even then in the grip of a fashionable enthusiasm for all forms of divination, including astrology.

Edward III himself may have been attracted by the claims of astrology. He is known to have possessed a copy of the occult manual known as the *Secreta secretorum*, written in the form of a letter from Aristotle to Alexander, which was attributed in the Middle Ages to Aristotle.[5] In his copy there are a number of illustrations which show a king surrounded by astrologers, some of whom hold instruments, while others point to the planets above.[6] Is this king merely a representation of Alexander, it might be asked, or could he be intended as Edward III in the guise of Alexander? The same thorny problems of identity must be faced when considering the text. The essential importance of the advice of an astrologer is one of the central themes of the *Secreta*. The matter is put most succinctly in chapter 22:

> O most clement king, if it is at all possible, you should neither rise up nor sit down nor eat nor drink nor do anything without the advice of men learned in the art of astrology.[7]

Yet if this was an ideal to be strived for, Edward III showed little inclination to invite astrologers to attend the court.

[3] H. A. Oberman and J. A. Weisheipl, 'The *Sermo Epinicius* ascribed to Thomas Bradwardine (1346)', *Archives d'histoire Doctrinale et Littéraire au Moyen Age* 25 (1958), 295–329. I have accepted the authors' convincing arguments as to Bradwardine's authorship of this sermon.

[4] *ibid*., 309.

[5] M. R. James ed., *Walter de Milemete's De nobilitatibus sapientiis et prudentiis regum* (Oxford, 1913).

[6] *ibid*., fols 31v, 51v, 53v.

[7] Robert Steele ed., *Opera hactenus inedita Rogeri Baconi* (Oxford, 1909–40), fasc.5, 60. 'O Rex clementissime, si fieri potest, non surgas nec sedeas nec comedas nec bibas et nichil penitus facias sine consilio viri periti in arte astrorom'.

There was at least one English astrologer who would have been competent to provide the king with the level of advice demanded by the *Secreta*. John Ashenden, a younger contemporary of Thomas Bradwardine at Merton College, Oxford, was the most prolific of all medieval English writers on astrology.[8] Ashenden claimed, on somewhat dubious grounds, to have predicted the great plague of 1349, which included Bradwardine among its victims. Although Ashenden was never offered any civil post, he was an ardent patriot, and several of his treatises included references to the king's prospects in his encounters with the French and the Scots. Ashenden's writings were widely distributed and survive in remarkable numbers, so there is a good chance that King Edward knew of his particular skills, but chose to ignore them. Ashenden was not the only scholar in Oxford with an interest in astrology, and the writings of Geoffrey Chaucer remind us that other members of the court of Edward III shared that interest. However, it is necessary to assume that the hardnosed business of employing astrologers to obtain political and tactical advantage over potential adversaries remained a covert affair. A great deal of attention has been given to assessing the level of Chaucer's faith in the efficacy of astrology.[9] The ambiguity of Chaucer's position is probably a good reflection of that of the court as a whole. The much quoted reference to astrology in the *Franklin's Tale* sums things up nicely by seeming to dismiss the art as,

> . . . swich folye
> As in oure dayes is nat worth a flye, –
> For hooly chirches feith in oure bileve
> Ne suffreth noon illusioun us to greve.[10]

Quite simply, the Church would not put up with it. Rational evaluation of the efficacy of astrological predictions did not come into it, or at least only seldom. It was assumed that astrologers, like the medieval practitioners who drew heavily on astrology to provide the theoretical foundation of their own practice, knew what they were doing.

Edward III probably had more protection against the 'folye' of astrology than many medieval English kings. It is well known that beliefs in supernatural powers of causation flourish in times of adversity.[11] As a relatively successful king, Edward III may have found it easier to retain his confidence in God alone, who always brings us the victory, as Bradwardine insisted. The same cannot be said of Edward's successor, Richard II. A man endowed with cultured and exotic tastes, erratic temperament and uneven fortune, it is not surprising to find that Richard II appears to have been the

[8] For further information on Ashenden see Carey, *Astrology and Divination*, especially chapters 3 and 4.

[9] At greatest length by Chauncey Wood, *Chaucer and the Country of the Stars* (Princeton, 1970).

[10] F. N. Robinson ed., *The Works of Geoffrey Chaucer*, 2nd ed. (Oxford and London, 1974), 139. *Franklin's Tale*, ll. 1131–1134.

[11] For the early modern period this thesis is defended by Keith Thomas in his *Religion and the Decline of Magic* (London, 1971).

first English king to have taken a positive interest in the occult sciences. In 1391, Richard commissioned one of his Irish clerks to compile for him a book of divination.[12] The presentation copy is now held in the Bodleian Library and it consists of four different texts:

1 *De quadripartita regis specie*.[13] This is a tract on the virtues of a good king taken from the *Secreta secretorum* with some additional material. (fols 1–3).

2 *Phisionomia Aristotelis*. The physiognomy, that is the art of discerning character by an analysis of the physical features, again taken from the *Secreta secretorum*. (fols 3–5v).

3 *Philosophia Visionum cum sompniis Danielis*, the commonly encountered dream-book attributed to Daniel, with an alphabetical index ascribing significance to things seen in dreams. (fols 6–8v).

4 *Liber Judiciorum*, a geomancy with accompanying tables. (fols 9–89v).

The geomancy is by far the longest work in the manuscript, which is very finely executed.[14] Besides miniatures of the sixteen geomantic figures, the geomancy is headed by a representation of the king on fol. 9 and includes full indexes, chapter headings and other aids for the reader. It was a book designed to be used, though unfortunately there is no sign that Richard II ever chose to do so.

The anonymous compiler revealed something of the circumstances of his commission in the bombastic preface to the geomancy of fol. 9:

> I have compiled this present book of geomancy, in as brief a form as I was able, at the special request of our most excellent Lord Richard, the most noble king of the realms of England and France, who governs in sublime fashion not so much by force of arms as by philosophy and the two laws; and indeed he has not declined to taste the sweetness of the fruit of the subtle sciences for the prudent government of himself and his people.[15]

This would seem to put the matter beyond doubt. Richard II ordered this book to be made in order that he might learn somthing about the occult sciences. It is also a restatement of the traditional theme of the essential service provided by an astrologer or wiseman to the prince in the good government of his realm. Richard is characterised as a wise ruler, not, like his

[12] Oxford, Bod. Lib., MS Bodley 581. There is a near contemporary copy of the geomancy in London, Brit. Lib., MS Royal 12 C.v.

[13] This tract has been edited by Jean-Philippe Genet in *Four English Political Tracts of the Later Middle Ages*, Camden Soc., 4th ser., 18 (London, 1977), 31–39. Genet also gives a full discussion of the manuscript.

[14] For a description of the manuscript see Fritz Saxl and Hans Meier, *Catalogue of Astrological and Mythological Illuminated Manuscripts of the Latin Middle Ages* III (London, 1953), I.311–312.

[15] Bodley 581, fol. 9a; Cited by Genet, *Four Tracts*, 23.

father, a warrior prince, but one learned in the noble arts. The extent of Richard's commitment to the occult sciences as demonstrated by this book, must be called into question when we consider why he should have chosen to possess a geomancy, rather than a book on astrology. One answer is provided by the compiler, namely, because geomancy is easier:

> Since the science of astrology is both of great difficulty and is time-consuming to learn, for which the present life is scarcely sufficient, I have compiled this present little book of geomancy, not from my own views, but from the rules and precepts of established authorities in this art, up to the year of Our Lord 1391, in the month of March.[16]

A book of astrology, such as Al-kindi's *Introductorium*, or the short treatise that Pèlerin of Prussia prepared for Charles V of France, or the translations of Ptolemy made for the same king, perhaps by Nicole Oresme, were evidently considered too hard for King Richard at this stage.[17]

Two other reasons may have figured in Richard's decision to call for a geomancy, rather than some other astrological work, for his library. There was still the difficulty of the Church's intolerance of the deterministic implications of any form of fortune telling. The compiler of Richard II's geomancy, referring to a tract known as the *Speculum astronomie* commonly attributed to Albertus Magnus,[18] alleges that geomancy escaped the general ban. It was not geomancy, he argues, but *astrologia* which lays down that all judgments are necessary, and as Ptolemy affirms in his *Quadripartitum*, 'The wise man rules the stars'.[19] Evidently Richard II showed some of the same caution displayed by his father when it came to patronising astrologers.

The other reason why a king might dissociate himself from anything which might bring the taint of heterodoxy arose from the cut-throat realities of later medieval court intrigue. The tyrant or the heretic could provide the justification for his own deposition. Richard II's contemporaries regarded it as a disgraceful thing for a king to spurn his older and wiser counsellors, and resort to pseudo-prophets, necromancers and younger men. Such a king was as foolish and dangerous as Saul, who sought the advice of the Witch of Endor, the *pythonissa*, when Yahweh had deserted him for David, and might reasonably be replaced. In Walsingham's unsympathetic accounts of the last years of Richard's reign there are many allegations that the king showed an unhealthy regard for prophecies, omens and irregular sources of advice. He is said to have taken to hoarding his treasure, and surrounding himself with

16 ibid.

17 These tracts are contained in Oxford, MS St John's College 164. For this and other astrological books associated with Charles V see E. Poulle, 'Horoscopes princiers des XIVe et XVe siècles', *Bulletin de la societé des antiquaires de France* (1969), 63–9.

18 Lynn Thorndike, *A History of Magic and Experimental Science* 8 vols (New York and London, 1923–1958), I.692–717.

19 Bodley 581, fol. 9v: Nam astrologia, nec ista sciencia, ponit omnia judicia fore necessaria, et hoc affirmat Ptholomeus in *Quadripartito* suo ubi dicit, Vir sapiens dominatur astris.

pseudo-prophetae who spurred him in his ambition to become the new Holy Roman Emperor, and one of the greatest princes of the world.[20] Whatever the truth of the matter, Walsingham does not seem to have had in mind the practice of the sophisticated and complex art of astrology, but rather the heraldic prophecies attributed to Merlin, Bede and other worthies, which were gaining a wide circulation at this time. According to Jean Creton, an attendant at the court of Charles VI of France, Richard's father-in-law, the English were notoriously gullible about prophecies of this kind:

> They were of such a nature in their country, that they would believe very perfectly in phantomes ('fanthonnes') and sorcery, and they would resort to them very willingly. But it seems to me that this was not well done, but rather it was a great fault in their faith.[21]

This comment, coming from an eye-witness account of the deposition of Richard II in 1399 should alert us to the danger of accepting Walsingham's unsubstantiated gossip at face value. Allegations of an ill-defined but slandarous intent, ranging from sexual irregularity to blasphemy and the practice of sorcery, were the stock-in-trade of political propagandists throughout the period.

Rather than linking Richard II with the foolishness of the Old Testament King Saul, Richard II's finely illuminated book of geomancy allies him with a number of his royal contemporaries. Charles V of France and his brother John, duc de Berri, as well as Wenceslaus II of Bohemia, Holy Roman Emperor and the father of Richard II's first wife, Anne, are all renowned for the brilliance and sophistication of their courts. What these kings also have in common is the possession of a number of beautifully illuminated books on the occult sciences. Charles V may well have served as a model for Richard II in these matters. When he died, in 1380, there were more than a thousand books in his library, and over ten per cent of these concern astrology, geomancy, chiromancy and necromancy, with a further seventy volumes devoted to astronomy, the essential companion study to judicial astrology.[22] Like Bodley 581, many of Charles V's extant books are adorned with his picture, and on occasion he is represented in his study, surrounded by books.[23] In an Oxford manuscript, St John's College 164, he is shown with what appears to be an armillary sphere. In the Prologue to the French translation of the *Quadripartitum*, it is explained that even before his accession the future Charles V had wished to possess:

[20] *Thomas Walsingham: Annales Ricardi Secundi, 1392–1399*, H. T. Riley ed., R.S. (London, 1866), 233.

[21] Translated and edited by John Webb, *Archaeologia* 20 (1824), 1–423. For Creton see J. J. Palmer, 'The authorship, date and historical value of the French chronicles of the Lancastrian Revolution', *Bulletin of the John Rylands Library* 61 (1978–9), 145–181; 398–421, at pp. 151–154.

[22] For the library of Charles V see Léopold Delisle, *Recherches sur la librarie le Charles V* (Paris, 1907) 3 vols; idem, *Le Cabinet des manuscrits de la Bibliothèque Imperiale* Vol. I (Paris, 1868).

[23] C. R. Sherman, *The Portraits of Charles V of France* (New York, 1969), includes plates of all the surviving portraits.

> des livres in françois de la plus noble science de cest siècle, c'est vraie astrologie sans superstecion, et par especial ce que en ont composé les philosophes excellens et approvés.[24]

Apparently Richard II wished to follow this example.

Like Edward III and Richard II, Charles V also received advice concerning the theological objections to the practice of judicial astrology, and the proper attitude of a prince toward the art. Nicole Oresme addressed this subject in his *Tractatus contra astrologos*, written before 1361, with a French translation appearing a few years later.[25] Although he wrote ostensibly to repudiate the study of astrology, in the way of most medieval critiques of the science, Oresme reserved his main attack for predictions concerning individuals or particular events, which flew in the face of the orthodox doctrine of free will. Where the *Secreta secretorum* had advised the prince to do nothing without consulting an astrologer, Oresme is less enthusiastic. It is probably a good thing, he argues, for a prince to acquaint himself with the science of astrology, but it would be imprudent for him to devote too much attention to the technical details. If necessary a prince should provide support for students of approved branches of astrology, but they must not be elevated to positions of public responsibility unless they demonstrate the necessary aptitude. Superstitious diviners ought to be shunned for the false, worthless and dangerous deceivers that they were.[26]

The ownership of finely illuminated books on the occult sciences, and more especially geomancy, quickly seems to have become something of a court fad. In the generation after Richard II geomancies were owned by Humphrey, duke of Gloucester[27] and his brother, John, duke of Bedford. Bedford also commissioned a long and elaborate *Summa* of physiognomy from Roland Scriptoris, the author of his geomancy.[28] Henry VII had a fine edition of a number of astronomical and astrological texts drawn up, including some tables taken from the geomancy of Humphrey, duke of Gloucester, as well as the geomancy of Alpharinus.[29] In Italy, books on the occult sciences seem to have had an even wider dissemination than in France or England. The Dukes of Urbino and Ferrara and the libraries of the Gonzaga, Visconti, Sforza and Medici families all reveal an interest in the study of astrology, alchemy and the related arts.[30]

24 Paris, Bib. nat., MS fr.1348. Quoted by Delisle, *La librarie de Charles V*, 114–5.

25 Coopland, *Nicole Oresme*. See also the article by Dr Stefano Caroti in this volume on Oresme's other astrological polemic, *Quodlibeta*.

26 Cited by Thorndike, *History of Magic*, II.416.

27 London, Brit. Lib., MS Ashmole 66.

28 The geomancy occurs in London, Brit. Lib., MS Royal 12 C.xvi, MS Sloane 3487 and Oxford, Bod. Lib., MS Ashmole 434. The physiognomy occurs in Brit. Lib., MS Royal 12 C.xv and 12 G.xii. For Scriptoris see Thérèse Charmasson, 'Roland l'Ecrivain, médecin des ducs de Bourgogne', in *Actes du 101e Congrès national des Societés savantes* (Lille, 1976). Sciences, fasc, III.21–32.

29 London, Brit. Lib., MS Arundel 66. Thorndike *History of Magic* II.121.

30 Pearl Kibre, 'The intellectual interests reflected in libraries of the fourteenth and fifteenth centuries', *Journal of the History of Ideas* 7 (1946), 257–297, especially 285–7.

For Richard II, geomancy and astrology were probably recognised as having a useful, but nevertheless peripheral place in the fashionable and innovative court life he tried to foster. There is no evidence that he created any such post as 'court astrologer', or that astrological considerations impinged in any way on his political judgment, such as it was. For this development in England it is really necessary to wait for the fifteenth century and the reign of Henry V. The dominating feature of the political history of fifteenth century England is the recurrent war with France. Blows and arrows were not the only commodities exchanged in these encounters, and it is probable that the French taste for astrological consultants was acquired at this time by the English nobility. There was no shortage of well qualified astrologers to provide this service. In the course of the fourteenth and fifteenth centuries the study of astrology had been institutionalised in many Italian and German Universities by its incorporation into the syllabus of study of the Faculty of Medicine.[31] In Paris, in 1371, Charles V had established a College of Astrology and Medicine within the University, endowing it with a fine collection of books and instruments, and two scholarships.

Symon de Phares, who is the most important source for the activities of astrologers during the period of the Hundred Years' War, probably drew on the resources of the College set up by Charles V to compile his *Receuil* of famous astrologers.[32] Symon wrote his treatise in the year of the death of Charles VIII (d. 7 April 1498), after he had been subjected to a determined but ultimately ineffectual attempt to forbid his practice of astrology by the Archbishop of Lyon, the Parlement de Paris and the Faculty of Theology of the University of Paris. In spite of the fact that Symon had been consulted in November 1490 by Charles VIII himself, of 26 March 1494 the Parlement directed that his books be confiscated and that he be handed over to the Bishop and Inquisitor of Paris for further examination. Symon's *Receuil* was written in the confident expectation that an account of the astrologers throughout history, starting with Adam and continuing to the present day, who had given faithful service to princes would inspire the new king to take up his case and ensure the return of his books. Obviously, given these circumstances, it is necessary to use Symon's evidence with great caution. Many of the astrologers he identifies cannot be attested from any other source. No doubt this is partly because he had access to a mass of ephemeral astrological literature which has not withstood the ravages of time, horoscopes cast or the outcome of some particular siege, for example, or for some noble captain's bithday, eagerly read at the time of composition, but soon discarded, left unbound and sent to oblivion. But Symon is also guilty at times of wishful thinking and exaggeration bordering on deceit. A small army of ineffectual Cassandras would seem to have predicted such major

[31] Richard Lemay, 'The teaching of Astronomy in Medieval Universities principally at Paris in the Fourteenth Century', *Manuscripta* 20 (1976), 197–217.

[32] *Receuil des plus célèbres astrologues et quelques hommes doctes faict par Symon de Phares du temps de Charles VIII*, Ernst Wickersheimer ed. (Paris, 1929). For Symon see also Thorndike, *History of Magic*, IV.544–561.

events as the Battle of Crécy in 1346, the capture of John II of France at the Battle of Poitiers on 18 September 1356, the merciful deliverance of the same king from the hands of the English four years later, the death of Edward, the Black Prince, in 1376, and the deposition of Richard II in 1399.

A typical example, drawn from Symon de Phares, of the political and military uses to the prince of an accomplished astrologer is the case of Master Yves de Branchier, who was in the service of the Constable of France, Bertrand du Guesclin.[33] Master Yves was an expert in the branch of astrology known as 'elections', by which it was possible to choose an auspicious moment in which to encounter the enemy. On one notable occasion, he caused the overthrow of 30,000 English troops led by John de Montfort and Henry de Grosmont, duke of Lancaster, because he had taken the precaution of calculating their horoscopes. A secret weapon indeed. Needless to say, I have not been able to verify this particular military catastrophe.

A much more secure foundation for the analysis of the practice of astrology at the English court in the early part of the fifteenth century is provided by the few extant remains of the astrologers' work in the form of horoscopes written on the occasion of some historically verifiable event. These horoscopes provide good corroboration of the general impression, if not the precise details, of the realms of France and England in the period of the Hundred Years' War, as described by Symon de Phares. Horoscopes of the nativities of Henry IV, Henry V and Henry VI of England and their associates all figure in three French collections of horoscopes discussed by Emmanuel Poulle.[34] At one point, the births, birthdays and coronations of Philip, duke of Burgundy (d.1467), Henry VI of England (d.1422), and Charles VII of France (d.1461) are considered together with the conjunction of 1467, an event which was known to presage the change of dynasties, the rise of religions, the coming of prophets and other great events. The horoscopes of three other key English participants in the French wars, John, duke of Bedford, Thomas Montague, earl of Salisbury, and Sir John Fastolf, also appear to have been examined in a hostile light for all signs of weakness.

As the fields of battle lay in France, so it seems did the greater number of practising astrologers, and in England there is no equivalent to these French collections until much later. Nevertheless, there was nothing to discourage French astrologers from travelling to England to try their luck at court. I strongly suspect that one such adventurer, the instrument-maker Jean Fusoris, is responsible for the treatise on the nactivity of Henry V which survives in just one contemporary copy in the Ashmole collection in Oxford.[35] Fusoris may have composed his treatise on his visit to the English court with the French embassy of May 1415. In the previous year Fusoris had sold a number of astronomical instruments to Richard Courtenay, bishop of Norwich, one time Chancellor of the University of Oxford, and a close personal friend and advisor to Henry V. He was hoping to recover payment

33 Symon de Phares, *Receuil*, 229.
34 Poulle, 'Horoscopes'.
35 Oxford, Bod. Lib., MS Ashmole 393, fols 109–111. There is a 17th century copy in MS Ashmole 192, fols 26–36.

of the debt and receive a promised introduction to the king himself. We have a detailed record of all these events because Fusoris had the misfortune to be tried for treasonable correspondence with the enemy after a clerk was intercepted carrying a letter from Courtenay to Fusoris.[36] From the records of the trial we learn that Richard Courtenay had a passionate interest in astrology, and sought out Fusoris's expert opinion on the scheme of the king's nativity, as to when and by what means the king might die.[37] While in Winchester Fusoris also encountered other members of the English court with an interest in astrology. Although Fusoris professes a very low estimate of their skill, there is no reason to suppose that England had no native masters of the art. The more likely explanation for his comments is that Fusoris felt the need to protect his own patrimony. Hence, when asked by a certain 'English doctor of theology' whether there were many astrologers in Paris, Fusoris replied that there were only a few students of that science because it was not a 'sciencia lucrativa', that is, there was not much money in it.[38]

The poor returns was not the only thing to discourage the scholar with some skill in the science of the stars from seeking court patronage. There was the ever present danger of becoming implicated in some court scandal and suffering the drastic consequences of a patron's fall from favour. Sorcery was regarded as a standard way to attempt to dispatch the sovereign, a treasonable offence. Men of some learning, including the king's physicians and confessors, were especially liable to face charges of conspiring to assist some court figure to bring about the death of the king by sorcery.[39] On three separate occasions during the fifteenth century we learn of astrologers in the enjoyment of noble patronage solely because of their subsequent trial and disgrace. This would suggest an improbable connection between treason and judicial astrology, and it would seem more likely that these cases are merely the symptoms of the more intimate involvement of astrologers in court affairs generally.

The trials of astrologers mark most decisively the transformation in the perception of astrology in the fifteenth century at the English court from that of mere entertainment, to a potent source of trouble. As the enquiries made by Richard Courtenay to Jean Fusoris demonstrate, one of the chief uses of astrology was to determine when, and in what fashion, someone was going to die. In the fifteenth century such predictions had a way of solidifying into fact, and it is not surprising that legislation was enacted making it an offence to attempt to predict the death of the king by any form of divination.

The first case is that of a certain Friar Ralph or Randolf, or, according to the Parliament rolls, Friar John Randolf, a Franciscan who served both Humphrey, duke of Gloucester, and the dowager queen, Joan of Navarre. In one manuscript Friar Ralph is attributed with the authorship of a set of tables

36 Léon Mirot, *Le Procès de Maître Jean Fusoris*, in *Mémoires de la Societé de l'Histoire de Paris* 27 (1900), 137–287.

37 Mirot, *Procès*, 173–4; 231–2.

38 *ibid.*, 243.

39 W. R. Jones, 'Political Uses of Sorcery in Medieval England', *The Historian* 34 (1971–2), 670–687.

with aspects of the planets, written at the request and with the full support of Humphrey, duke of Gloucester.[40] In the canon to these tables he refers to Oxford as his 'alma mater',[41] though he seems to have escaped the notice of Emden. Elsewhere he is described as, 'Sacre Theologie Professor', and 'Maistre of Dyvynyte', and Randolf's tables leave no doubt that he was also a competent astrologer.[42] By his own account he sympathised with those who laboured under the burden of calculations necessary for those who wished to analyse the present, past and future effects of the planets.[43] In May 1419, Randolf was implicated in the charges laid against the widowed Queen Joan, who was accused of having attempted to use sorcery to destroy the king, 'en le plus haute et horrible manere qui l'en purroit deviser'.[44] Astrology is not mentioned, and in the account of the affair in the Rolls of Parliament, Friar Randolf is only named as Queen Joan's accuser, not her collaborator. Fearing for his life nevertheless, Randolf fled to Guernsey but was captured, brought before the king at Nantes and locked up. That he was spared an immediate execution for treason suggests at least that there was some doubt as to his guilt. In 1425, Randolf was again embroiled in scandal when he was implicated in a serious dispute between the Chancellor, Henry Beaufort, and Randolf's patron, Humphrey, duke of Gloucester. Part of the complaint against Gloucester was that he had removed Friar Randolf from the Tower of London, perhaps feeling an urgent need of his professional advice on some matter. Ultimately Randolf was returned to the Tower where he is said to have been murdered by a sun-struck priest in 1429.[45]

The second case concerned two eminent Oxford scholars, Roger Bolingbroke, at one time Principal of St Andrew's Hall, and the physician Thomas Southwell, a leading member of the medical profession.[46] In May 1423, Southwell was one of a group of physicians and surgeons who petitioned the Mayor and Aldermen of London for the establishment of a College, 'for the better education and control of physicians and surgeons practising in London'.[47] It is particularly ironic that Southwell, who was to die in such a dishonourable fashion, should have put his name to a petition intended in all probability to restrict the illicit practice of medicine by unqualified quacks. In 1440, both Southwell and Bolingbroke were named as accessories in the trial for sorcery of Eleanor Cobham, duchess of Gloucester, with the 'Witch of Eye', Margery Jurdemayne.[48] Eleanor did not deny that she had sought some

40 London, Brit. Lib., MS Sloane 407, fols 223–226. See also Cambridge, Univ. Lib., MS Ee.III.61, fols 108v–116 (Canon), 108v–120 (Tables).
41 Cambridge, Univ. Lib., MS Ee.III.61, fol. 108.
42 For contemporary accounts of Randolf, see *Chronicle of London*, C. L. Kingsford ed. (Oxford, 1905), 73, 80, 273 and 298, n. 5; *Rotuli Parliamentorum* IV.118.
43 Cambridge, Univ. Lib., MS Ee.III.61, fol. 108.
44 *Rot. Parl.* IV.118.
45 *Chronicon Rerum Gestarum in Monasterio S Albani*, H. T. Riley ed. (London, 1870), I.38.
46 For Southwell and Bolingbroke see A. B. Emden, *A Biographical Register of the University of Oxford to A.D. 1500*, 3 vols (Oxford, 1957–1959), 214–5, 1734–5.
47 *City of London Letter Book K*, R. R. Sharpe ed., 11; Cited by Emden, *Biog. Reg. Oxford*, 1735.
48 For the numerous contemporary accounts of this trial see G. L. Kittredge, *Witchcraft in Old and New England* (Cambridge, Mass., 1929), 81–4; Jones, 'Sorcery', 683–4. I have relied on the version in the *English Chronicle*, J. S. Davies ed., Camden Soc., old series, 64 (1856), 57–60.

form of astrological advice from the two clerks, but claimed that she had merely sought some assistance in conceiving a child for Duke Humphrey, and had no interest in harming the king. After a spectacular trial, Eleanor was exiled for life, Margery Jurdemayne was burnt as a relapsed heretic on 27 October 1441, and Southwell died, 'per dolorem', according to one source, in the Tower of London.[49] As for Bolingbroke, there is a very sympathetic account of his end in which he appears as the tragic, even messianic, hero of this nasty business, in a chronicle found among the manuscripts of the antiquarian, William Worcestre:[50]

> And a certain clerk, one of the most renowned in all the world in astronomy and the magical arts, master Roger Bolingbroke, was arrested and publicly in the cemetery of Saint Paul's, with the vestments and waxen images of his magic, and with many other magical instruments, he sat in a certain high throne, so that people from everywhere might see his work; afterwards he was hanged, drawn and quartered, and his head was placed on London Bridge. This master Roger was one of the most notable clerks in the whole world, and he was accused on account of the aforesaid Lady Eleanor, to whom he was advisor in the magic art, and after his death many lamented exceedingly greatly.[51]

Despite the regrets of the chronicler and Eleanor Cobham's protestations, there can be little doubt that Bolingbroke and Southwell were guilty as charged. Bolingbroke is the author of at least one treatise dedicated to the Duchess of Gloucester which gives a brief introduction to judicial astrology, including questions concerning how a man might die.[52] If this treatise is incriminating evidence, utterly damning is the surprising survival of the actual horoscope cast by Bolingbroke and Southwell to determine the time and manner of the king's death.[53] The effect on Henry VI of discovering that a horoscope cast by two expert astrologers predicted his imminent demise was understandably great, and brought a swift reaction. Two of the king's most trusted servants, John Langton and John Somerset, were directed to see to the composition of an alternative reading of the king's horoscope which would assure the badly shaken monarch of his continued assurance of good

[49] *Six Town Chronicles*, Ralph Flenley ed. (Oxford, 1911), 102.

[50] For Worcestre see *William Worcestre: Itineraries*, J. H. Harvey ed. (Oxford, 1969); Emden *Biog. Reg. Oxford*, 2086–7. K. B. McFarlane, 'William Worcester; A Preliminary Survey', *Studies Presented to Sir Hilary Jenkinson* (Oxford, 1957), 196–221, at 206–7, demonstrates that Worcestre is not the author or the scribe of the bulk of this manuscript, as Hearne rashly assumed. The anonymous chronicle is printed as *Annales Rerum Anglicarum* from Hearne's edition of MS College of Arms, Arundel 48 by Joseph Stevenson in *Letters and Papers Illustrative of the Wars of the English in France* II, pt. ii (London, 1864), R.S., 762–3.

[51] *Letters and Papers*, 763.

[52] MS Gloucester Cathedral 21, fols 100–104v. For a discussion of the manuscript, which once belonged to the physician John Argentine, see Carey *Astrology and Divination*, 271–6.

[53] Cambridge, Univ. Lib., MS Ee.III.61 (1017), fols 160v–161. For the complex relationships between the six existing horoscopes of the nativity of Henry VI see Carey *Astrology and Divination*, 264 ff and Appendix III, nos 12–15.

health.[54] The author completed his work on 18 July 1441, and delivered it in person about a month later on 14 August to the king's household at Sheen.[55] Three months after this delivery, both Bolingbroke and Southwell were dead, and Eleanor Cobham had been committed to life imprisonment.

The Eleanor Cobham scandal marks something of a watershed in the history of astrology at the English court, when it ceased to be regarded as a genteel diversion and a respectable occupation for courtiers and scholars alike, and became instead a deadly game. The continued threat of sharing the fate of Bolingbroke and Southwell as the scapegoat in some grisly political scandal effectively deterred respectable scholars from the practice of judicial astrology. That this threat remained very much alive is attested by the trial of two Oxford clerks, John Stacy and Thomas Blake. On 19 May 1477, Thomas Burdett of Arowe, a close friend of George, duke of Clarence, the brother of Edward IV, was accused of treason.[56] It was alleged that Stacy and Blake, on 12 November 1474, had been employed by Burdett, 'to calculate the nativities of the King, and of Edward, prince of Wales, his eldest son, and also to know when the king would die'. Clarence attempted to defend his friend, but was himself arrested soon after. Both Stacy and Blake were members of Merton College, Oxford, the cradle of an earlier generation of mathematicians.[57] Undoubtedly they were incidental victims in the downfall of the Duke of Clarence. Their involvement here suggests that astrologers had become a regular, if necessarily discrete, feature of English court life by the early part of the fifteenth century.

Although it is possible to demonstrate that astrologers were in the habit of attending the English court, it is much more difficult to show that they exercised any political influence. All the evidence suggests that the main function of the astrologer, as distinct from a prince's more regular sources of advice, was to tell his clients, in an acceptable form, what they already believed, or wanted to hear. Even Ptolemy had admitted that astrology was not an exact science, and the range of possible interpretations of any astrological scheme was almost infinitely flexible. Most horoscopes of individuals are not accompanied in the manuscripts by any clues as to their interpretation, which was probably provided by the astrologer on the spot. All this is similar to what we know of the practice of Renaissance astrologers who have been more generous with the details of their methods of practice.[57] Where more details are provided, the information concerning individuals tends to the conventional or the flattering. Thus, in the treatise of the nativity of Henry V,[58] we learn that Mars, the primary significator in the nativity,

[54] Cambridge, Univ. Lib., MS Ee.III.61, fols 160–171. This is the autograph manuscript of Lewis of Caerleon, for whom see Pearl Kibre, 'Lewis of Caerleon; Doctor of Medicine, Astronomer and Mathematician', *Isis* 43 (1952), 100–108.

[55] Indictments of Thomas Burdett, esq., John Stacy and Thomas Blake of constructive treason in *Third Report of the Deputy Keeper of the Public Records* (London, 1842), Appendix II.213–4; *Rotuli parliamentorum* VI.193–5.

[56] Emden, *Biog. Reg. Oxford*, 197, 1949.

[57] Of these, the most generous was probably Simon Forman, see A. L. Rowse, *The Case Books of Simon Forman: Sex and Society in Shakespeare's Day* (London, 1974).

[58] Oxford, Bod. Lib., MS Ashmole 393, fols 109–111.

makes the native fortunate in war and supports him in the overthrow of his enemies.[59] In a later chapter it is added that the native will acquire riches through war, and through those matters pertaining to Mars. This would be a particularly apposite prediction for the subduer of the Welsh, the victor, in 1415, of Agincourt and the conqueror of Normandy. Even in his own lifetime, Henry V had a reputation for piety, and the incident of Friar Ralph and Joan of Navarre indicates he had little sympathy for the practitioners of the occult sciences. The author of this treatise seems to have perceived this, and warns the king against the children of Mercury, including clerics, messengers, and other artificers such as alchemists, magicians and those of similar nature. The author neglects, however, to mention that an astrologer was as true a child of Mercury as any alchemist. The understandable wish to please the king also seems to have dictated the author's calculations of the king's time of death. We know that Henry died of dysentery at Bois de Vincennes on 31 August 1422 in his 36th year, if like the calculator of this nativity we accept 1386 as the year of Henry's birth. The astrologer however, who after all had little to gain from forewarning the king of his untimely and unpleasant end, predicts that he will die after a full life at the age of 53. Later owners of this treatise were forced to remedy the resulting discrepancy in the manuscript.[60]

Overall, it must be admitted that English kings and courtiers seem to have paid remarkably little attention to the advice so freely proferred by astrologers. Partly, this may have been due to the quality of advice that was available. In the reigns of Henry V and Henry VI, no doubt largely because of the notorious fates of Friar Randolf, Bolingbroke, Southwell and other clerks branded with the name of sorcerer, English practitioners of the occult sciences acquired very cautious habits. It is curious that all of the men mentioned above, and Stacy and Blake after them, were Oxford graduates. Perhaps it is no accident that at the same time in Cambridge there developed what might be called a 'Cambridge School' of scientific astrology. In this group it is possible to include John Holbroke, the Master of Peterhouse, John Argentine, Provost of King's College, and his student Thomas Scalon, and the physician, Lewis of Caerleon.[61] The distinguishing feature of this small group, who shared a mutual admiration and collected each other's works, was a special delight in computation and the mechanical compilation of astrological tables, and the belief, as it seems to me, that the accuracy of astrological predictions depended on the accuracy of astronomical data: the planetary positions, the ascendant, and the cusps of the various mundane houses, which form the basis of any horoscope.

This rather bizarre and pseudo-scientific attitude is most evident in the treatise on the nativity of Henry VI, which survives in one manuscript in the

[59] Ashmole 393, fol. 110.

[60] Carey, *Astrology and Divination*, 244–6.

[61] For Holbroke see Emden *A Biographical Register of the University* of Cambridge to A.D. 1500 (Cambridge, 1963), 309. D. E. Rhodes, 'Provost Argentine of King's and his books', *Trans. Camb. Bib. Soc.* 2 (1956), 205–12; idem, 'The princes in the Tower and their doctor', *Eng. Hist. Rev.* 77 (1962), 304–6; Kibre, 'Lewis of Caerleon'.

hand of Lewis of Caerleon.[62] The author seeks to confound the horoscope of Bolingbroke and Southwell with another version attributed to the mathematician, John Holbroke. It is asserted, although there is no evidence to support this statement, that Holbroke was chaplain to Henry V at the time of the birth of his son, the future Henry VI. Because he was obliged to remain in France, the king sent Holbroke, accompanied by certain famous relics of the saints, to witness the birth. In fact Holbroke had the honour to be the first man to see the infant prince.[63] If this treatise had not been written for the critical eyes of Henry VI himself, it might be suspected that the whole incident had been contrived to add a touch of authenticity to Holbroke's computation of the ascendant at the time of the king's birth, a notoriously tricky business. It is then argued at length that Bolingbroke and Southwell had based their horoscope on the Oxford tables of William Rede for the latitude of 51°50′, whereas Holbroke had used the Paris tables of Jean de Linières for the latitude of 51°. Since the birth actually took place at Windsor, latitude 51°20′, Holbroke's figures are the more accurate, and, the author claims, *this makes all the difference* to the predictions drawn from the horoscope. What the practical value of quibbles of this kind might have been at the king's conference table, I leave you to decide.

By the reign of Henry VII, there seems to have been something of a stand-off between the court and any potential astrologers. Those with some acquaintance with the subject continued their work as physicians, astronomers and mathematicians, and kept their advice to themselves. Astrology as a distinct and viable profession does not appear to have emerged at this time. The stalemate, if we can call it that, was only broken effectively by the arrival in England of a number of astrologers who had received their training in Italy. John Argentine of King's College, Cambridge, is something of a link between the two traditions, for although he collected the works of English astronomers and astrologers, including Roger Bolingbroke and John Holbroke, he also took out a degree in Medicine from some Italian University, perhaps Padua.[64] Argentine's books and manuscripts indicate that he returned to England imbued with enthusiasm for the new humanist learning and other fashions of the Italian Renaissance, not least its infatuation with the occult sciences. When a son, Arthur, was born to Henry VI on 19 September 1486, Argentine was appointed as the young prince's physician, and he earlier attended Edward, the tragic eldest son of Edward IV. Given this background, it is not surprising to discover that Argentine drew up the horoscopes of the nativities of Edward IV and Edward V.[65] Sadly, neither prayers, nor the ministrations of Doctor Argentine, nor the foresight promised by the astrologers, was enough to save the young princes. Last seen alive by Argentine himself, they were probably murdered by September 1483.[66]

The point has been reached chronologically when it becomes difficult to

62 Cambridge, Univ. Lib., MS Ee.III.61.
63 *ibid.*, fol. 161v.
64 Rhodes, 'Provost Argentine'.
65 Gloucester Cathedral 21, fol. 9v.
66 Rhodes, 'The princes in the Tower'.

describe the period as 'medieval'. The attendance at the court of Henry VII of no less than two Italian astrologer-physicians, John Baptista Boerio,[67] and William Parron, whose story has been told by John Armstrong,[68] marks the physical arrival of the Italian Renaissance in England with some force. It is particularly significant that it was left to the Italian, Parron, to exploit the printing presses to market the cheap annual prognostications that are such a feature of the following centuries. No doubt John Ashenden and John Holbroke would have turned in their graves to see the royal art of astrology so vulgarised. But this was the new age, and the ancient, ever adaptable and pragmatic art of astrology, having dragged her feet through the mire of fifteenth century English court politics, seems to have felt no sense of impropriety in selling her charms to a wider market.

[67] C. H. Talbot and E. A. Hammond, *The Medical Practitioners in Medieval England: A Biographical Register* (London, 1965), 117–9; G. Portigliotti, 'G. B. Boerio alla Corte d'Inghilterra', *L'Illustrazione Medica Italiana* 1 (1923), 8–10.

[68] C. A. J. Armstrong, 'An Italian Astrologer at the Court of Henry VII', in *Italian Renaissance Studies*, E. F. Jacob ed. (London, 1960), 433–54.

THE TRUE PLACE OF ASTROLOGY IN MEDIEVAL SCIENCE AND PHILOSOPHY: TOWARDS A DEFINITION

Richard Lemay

1 *The Problem*

Proper understanding and consequently the right methodology to account for the role played by astrology in medieval science and philosophy remains a sorely absent feature of contemporary as well as of earlier historiography. The reasons for this blight are numerous and of varying inspiration. Historical, conceptual, societal and cultural factors in turn or in conjunction all work against a proper view of the medieval situation.

1 Historically the scientific revolution of the sixteenth and seventeenth centuries created a radical breakaway from earlier millenial conceptions of the scope of science. The new ordering of the cosmos after Copernicus and Galileo, the new method of reaching scientific data through mathematical evidence, obtained not through mere sense perception but via observation and more reliable instrumentation, and submitted to verification through relentless experimentation, these are historical factors which combined to produce a new conception of scientific advance and which rendered nearly wholly obsolete the older scientific apparatus inherited principally from Aristotle and his followers.

Perhaps still more significantly, the focus of historians who set out to understand and explain the place of astrology in ancient and medieval science was directed the wrong way. Stimulated by the Renaissance favoritism toward Greek classics, historians have stressed the capital role of Ptolemy's *Tetrabiblos*, which was advanced by nearly all of them as a substitute, on the one hand for the more basic and dominant works of Aristotle in the Greek tradition, and on the other for the role of Abu Ma'shar in the Arab tradition, in providing the 'scientific' background of ancient astrology throughout the Middle Ages and the Renaissance.

We have had Aristotle's works in reliable transmission for a fairly long time. But, it may be asked, have these works been read in the proper light, in particular with respect to their furthering the principles of astrological science? Who from among the modern commentators of Aristotle, for instance, has pointed to the *De Caelo*, or to the *De Generatione et Corruptione*, or the *Meteorologica*, as constituting the charter of 'scientific astrology'? To a careful observer of the evolution of science after Aristotle, there can be little doubt that these works supplied the scientific background of astrology during 2000 years after him. And therefore astrology, during all that time, formed an indispensable and intimate part of physical science and cosmology. Until the sixteenth century, no scientist could ever afford, or even be tempted, to reject Aristotle's ground principles of physics: Aristotle, for all of them, was science incarnate, and to expect any scientist in these times to dismiss Aristotle's cosmological principles is tantamount to assume that a devotee of science could reject the very basis of science. Hence the effort to understand medieval attitudes toward astrology by applying to this science our contemporary paradigm of science (to use Kuhn's convenient term) – the usual approach to the problem – seems to foreclose in advance all avenues leading to the medieval mind, to its structural framework, and to the contents of its own different paradigm. Failure to connect every and all expression of medieval 'lore' to its foundations in Aristotelian cosmologay and physics is like locking the entrance to the garden and throwing away the key. We find here an in-built obstacle, entirely the work of modern scientific orientation, to the proper understanding and interpretation of the place of astrology in medieval outlook.

As for the other pole of the medieval paradigm supporting its extraordinary indulgence in astrology, namely the role of the Arabian astrologer Abu Ma'shar, who had the historical merit of fusing with Aristotelian science the enlarged apparatus of astrological lore inherited from the Alexandrians, the Persians, the Babylonians and the Indians, it cannot be sufficiently emphasized first, that the Arabian's contribution in this momentous process stands at the apex of the evolution of all Greek and ancient science in the Near East, for which his works surpassed and truly outshone his immediate predecessors, contemporaries and successors.[1] His achievement resulted in a

[1] The eminent role of Abu Ma'shar as leading authority in astrology, although clearly recognized by medieval scholars, does not transpire clearly in modern historiography. An excursus, as short as possible, will be necessary here, in view of what we have to say about him later. We must deal with Abu Ma'shar's predecessors, his contemporaries and his early successors. As for his predecessors, we must list them in two categories: the ancestors, and the immediate predecessors: the ancestors (Hermes, the Babylonians and Persians, the Indians, and especially Kankah who came to Baghdad ca.770 and immediately found two arab imitators or translators, Ya'qūb at Tariq and al Fazārī) were all well known to Abu Ma'shar who incorporated their best views in his own works.
The immediate predecessors: the Naubaht brothers, Ma'shallah and his disciples Sahl ben Bishr and at Tabari, Abū 'Alī al-Hayāṭ; Umar ibn Farruhan, Tūfil ben Tūma, created the astrological vogue in Baghdad but with fragmentary works which Abu Ma'shar used in part to build his overarching synthesis.
Among his contemporaries: al-Kindī, Sanad b.'Alī, Yaḥyā ibn Abī Manṣūr, ibn Bazyar, al-Abaḥḥ, Muḥ. ibn al-Jahm, al-Khwārizmī, the Banū Mūsā etc. Abu Ma'shar seems to have known them all personally and was well aware of their production; in fact he was unjustly accused of plagiarizing some of them (we have tried to refute this charge in a communication to the *Second International Symposium of the History of Arabic Science*, Aleppo 1979, and in the *Albert-le Grand Lecture* for 1981 in Montreal). The fact seems to be that Abu Ma'shar synthesized all the known astrological literature of his time.
Of his followers and successors we may mention among the most preeminent ones who merely elaborate on the data supplied by Abu Ma'shar, either developing or criticizing them: al-Nairīzī, al-Saimarī, ibn Hibintā, al-Hamdānī, al-'Imrānī, al-Hāshimī, al-Qabīṣī, al-Sijzī, Kūshyār ibn

universally shared vision of the life of the cosmos dominated by astrology. And secondly, it must be recognized that Abu Ma'shar's works are pretty nearly inaccessible to the modern student of those ages, either in their original arabic form, all of which still lie in unexplored manuscripts,[1a] or in their Latin tradition (or Byzantine, for that matter). The traces of the medieval Latin tradition which were made available through incunabula editions prove to be vastly inadequate for a minimum comprehension of Abu Ma'shar's role and influence on astrological science, both East and West. Add to this bibliographical handicap the gruesome fact that his personality and intellectual achievements have become the butt of so many intemperate, uninformed, but nonetheless injurious estimations[2] that it appears unlikely

---

Labbān, al-Tanūhī. One may consult D. Pingree, *The Thousands of Abu Ma'shar* (London, The Warburg Institute 1968) to realize the extent of the influence of Abu Ma'shar's doctrines on many of them. Some were known in medieval Latin translations (Hali al-Embrani, Alchabitius, Anaritius). One particular astrologer from Fustat (Cairo), Aḥmad ibn Yūsuf (Abu Ja'far), whom we have described as the 'forger' of the famous *Kitab at-Tamara/Liber Fructus (Centiloquium* of Ps.-Ptolemy) [Aleppo, *First Intern. Symp. . . . Arabic Science, 1976*, Aleppo 1978, Proceedings, Vol. II, pp. 9–107. F. Sezgin's contention (*Geschichte des arabischen Schrifttums* VII, p. 9 and passim) that the *K. at-Tamara* existed before and was translated at least four times from the Greek is a monstrous historical error, resulting from his misreading of al-Birumi's manuscript text *fi sair sahmay*. The four translations mentioned by al-Biruni concern Ptolemy's genuine *Kitab al-arba'a/Quadripartitum*] shows himself vastly familiar with Abu Ma'shar's works which he pillages uninhibitedly. Two later Arabian astrologers well known in the West: Ibn Abi al-Rigial (Abenragel) and Ali ibn Ridwan (Abenrodoan) also knew, and even criticized some of Abu Ma'shar's points. In the West, the reputation of Abu Ma'shar was unrivalled since the time of the early translations (ca.1130) until the *Speculum Astronomie* of Albertus Magnus (ca.1250). After this work, some scruples were felt about depending so intensively on an 'infidel' author like Abu Ma'shar, and the condemnations of 1277 in Paris, where astrology was dominant (v.g. the prop. : 'if the sun would stop its course, earthly fire would not burn any more') reinforced this feeling. After which western devotees of astrology had the bright idea of composing their own *Summae* of astrology: Leopold's (son of the Duke of Austria) *Compilation* before 1300, and above all John of Eschenden's *Summa anglicana* (ca.1348) which could be quoted within a seemingly Latin tradition; but a study of the contents of these compilations shows that they merely repeat Abu Ma'shar, with his name only discretely mentioned here and there. The process had already started earlier with Guido Bonatti's *Decem tractatus astrologie* (before 1250) and with Michael Scot's *Liber introductorius* (Scot died ca.1235). In Italy during the fourteenth and fifteenth century, Abu Ma'shar's credit remained unimpaired (v.g. Biagio Pelacane di Parma and studies by Graziella Federici-Vescovini; Paola Zambelli and her studies on Abu Ma'shar and Richard of Fournival). It merged with the so-called averroism of the late medieval italian Universities. Pomponazzi in the fifteenth-early sixteenth century is still very well informed about Abu Ma'shar's theories: cf. my *Pomponazzi, Libri quinque de Fato* . . . Lugano/Zurich 1957, passim.

1a Now see the facsimile edition of *Ms. Istambul, Suleymaniye, Garullah 1508* by F. Sezgin, at the Frankfurt am Maine 'Institüt für die Geschichte der arabisch-islamischen Wissenschaften'.

2 Otto Loth, in 1875, 'al-Kindi als Astrolog' tried to prove that Abu Ma'shar had plagiarized al-Kindi's *Risala* on the Duration of the Arab Empire according to astrological conjunctions, whereas it seems to us that the situation was exactly the reverse. George Sarton in his famous *Introduction to the History of Science*, followed unintelligently (because hastily read and misunderstood) a statement of Aldo Mieli to the effect that Abu Ma'shar *Great Conjunctions* included a portion (Otto Loth's contention) which was plagiarized from al-Kindi; Sarton unhesitatingly stated that Abu Ma'shar *Great Conjunctions* in its entirety was a plagiarism of al-Kindi and this work should be restored to al-Kindi's credit. But the most disparaging efforts came from

that without a drastic, urgent revision of these unenlightened views, his actual role throughout the medieval period in the defence and diffusion of the science of astrology based on peripatetic philosophy is likely to escape for a long time still the attention it deserves from the historian.

2 Conceptually, the effect of the scientific revolution was to render nearly impossible for historians henceforward to grasp the different scope of ancient science, of its epistemology and orientation. Yet a close study of earlier paradigms remains theoretically possible, through patient and open-minded examination of texts and their context. This approach alone can reveal the key concepts and tools that buttressed ancient science, as well as the structure of thought elicited through them. Inevitably these greatly differ from our modern apparatus in lexical, semantical and structural fashion all at once. In fact, the disjunction operated by the scientific revolution currently justifies holding the view that astrology was indeed a hopelessly misguided, groundless, false direction taken by the scientific curiosity of man. To reassure oneself, one only has to stress the irrelevancy of the premisses of astrology in the perspective of modern science. If any humanistic contents remains attached to historical astrology, it is not usually seen as pertaining to astronomical science, but rather to psychology and social science.

3 Now as to the sociological and cultural dimensions of our problem, the historian's effort to understand the role of astrology in medieval and Renaissance times is further hampered by the existence, at all times, of a nearly unbridgeable, quasi-biological gap between two classes of adepts of astrology. At the theoretical level, there were those who sought knowledge for its own sake and as the highest good in their culture; but, on another level, the practical one, there were those to make use of the received knowledge for the benefit of their contemporaries, and for their own personal benefit in the bargain. This latter group included not merely the 'quacks', as easily indentifiable in their time as in our own, but it also reckons among its members a sizeable proportion of social activists, including princes, ecclesiastics, leaders in all fields of social endeavour who, confident in the value of the best scientific expression of their time, ended up by maintaining toward the astrologers and their science a seemingly inexhaustible credit of efficacy and practicality in the social texture. Even a superficial observer of the part played by astrological predictions through the complex web of relations in medieval society: politics, medical profession, commercial and family relations, sexuality, war games, etc., cannot fail to notice the eminent, indeed at times the leading role played by astrologers in

D. Pingree, both in his *The Thousands of Abu Ma'shar* where our man is called an impostor, and in his article in the *Dictionary of Scientific Biography*, New York 1971, where the combination of misinformation and prejudice against a medieval astrologer reaches incredible proportions.

determining policies[3] and more,[4] social attitudes in private and in public throughout the medieval period, in Christian as well as in Islamic society. All active layers of medieval society were, in fact, involved in an unquestioned allegiance to astrology for its usefulness, since it made available for the benefit of social needs the discoveries of the scientists concerning the cosmological and ecological setting of their lives. True enough, as a matter of sheer social determinism, there were plenty of abuses of the credit and favour, as there has been at all times in all societies, whatever the structure of the social network that prevails. However, this peculiar dimension would provide no more decisive ground to judge of the place of astrology in medieval society, than would pointing to the abuse of computers in our own times to evaluate the role and importance of computer technology in our society.

But, at the other end of the social structure, there was the growing class of savants – genuinely and wholly dedicated to the search of truth ascertainable from past experience and knowledge – who did indeed provide the theoretical support for the practice of astrologers. Among the confused mass of surviving texts and documents that can be identified and categorized from these earlier ages, the scrupulous historian of astrology as a science must

[3] As far as medieval 'politics' are concerned, it must be realized that particularly from the thirteenth century on, along with Michael Scot's intimacy with the Emperor Frederick II, and with Bonatti's testimony that most Italian courts, including the ecclesiastical ones, included astrologers whose services were expected for political and military decisions, astrologers had become the political advisors (special assistants or counselors) of kings, nobles, bishops and even popes. We notice from approximately the year 1359/60 at the University of Paris the rise of interest in Aristotle's *Politics* as an academic subject (first taught by Albert of Saxony). The reign of Charles VI in France, which started ca.1364 after the death of his father John I who died as an English prisoner in London as a casualty of the Hundred Years War, brought the climax of the astrologers' influence at court, since Charles V's interest in astrology was so paramount. But after his death in 1380, under his feeble son and successor Charles VI who soon became mad and 'absent' (understand in the terms of the *Centiloquium* Prop. 59: 'de absente' or the mad one), the astrologers' influence at court came into sharp conflict with the tenants of 'rational' politics based on Aristotle. The marvelous picture of this tension given in the *Songe du Vieil Pelerin* by Philip of Mézières (ed. G. W. Coopland, London 1969) in the middle eighties points to the intensity of the conflict between astrologers' 'politics' and that of the Sorbonnards.

[4] As far as *mores* under astrological guidance are concerned, we should like briefly to allude to a lasting dichotomy about the basis of human love which seems to have existed from the treatise of Andreas Capellanus *De Amore* (ca.1180) [specifically condemned as heretical at Paris in 1277; cf. Mg. M. Grabmann, 'Das werk *de Amore* des Andreas Capellanus und das Verurteilungsdekret des Bischofs Stephan Tempier von Paris vom 7. Marg 1277', Speculum VII (1932), pp. 75–79] until the time of Boccacio at least (cf. R. Hollander, *Boccacio's Two Venuses*, N.Y.: Columbia University Press, 1977, pp. 35–40). The intrusion of a planetary, irresistible influence of Venus upon love in human relations, came to contradict the older classical tradition of Venus as the capricious goddess of love since Ovid. There was nothing capricious in Venus the planet's influence upon human love, and modern historiography, having massively disregarded the role of astrology in medieval outlook, got lost in imaginary explanations about these unsettled conditions. Capellanus' work, in our opinion, injects the newly acquired consciousness of the influence of the planet Venus upon the manifestations of love, which decidedly differs from Ovid's more urbanlike, fickle notion of seduction in the marketplace or the circus. For a medieval man, you can tamper with Venus planetary influence upon human love only within the dictates of astrological rule.

inquire about the higher levels of intellectuality and genuine concern with scientific knowledge that may transpire and matter more. He must endeavour to sort out and scrutinize with an impartial eye the productions that evidence a serious mental effort to understand, preserve and transmit, often with increased precision or clarification, the scientific lore of preceding ages.

There is naturally a great difference in intellectual vigor between these two sectors of devotees: the scientists on the one hand, and practitioners of the art on the other. The data demands from the conscientious historian that he define his categories clearly, and that he apply to each the appropriate criterion. When considering, therefore, medieval and Renaissance astrology, it is now natural that a fully repugnant, discountenancing, but nevertheless well-documented picture could be drawn of it as of an inexcusable aberration affecting nearly every sector of society, as if sharing a general madness. Yet these generations without exception received, on perfectly scientific grounds within their own paradigm, astrology as a science, and practiced it in good faith as service to the enlightened community. Hence, despite the wide gap between these sociological groups, the ones attached to intellectual values on the one hand, and those operators on the other, who honoured science by meticulously but blindly following its instructions, one may easily be led to general but questionable conclusions for failing to observe the necessary distinctions. In observing popular practices in particular, and dismissing their assumedly learned foundations to concentrate on their near charlatanism, it has proved all too tempting to lump up learned astrology together with obvious charlatanism and quackery, and to dismiss the whole endeavour as unredeemable aberration, plain superstition, and a social plague from which the scientific revolution has saved us. Such a historiographic premiss can only blur the actual realizations of astrology and astrologers in the past. Furthermore, since within the emerging paradigm of modern astronomical science astrology found little support, if any, it has soon been discarded as a valid concern of the historian of science. Yet, however radically cut off from the new scientific roots, astrology can still be perceived as a valuable slice of immemorial lore, and can only be dropped to his disadvantage by the historian of science. In favour of its continuing vitality as an object of study stands the impressive apparatus of hoary authority and of majesty conferred upon it by the Babylonians and Egyptians of old, reinforced by the respect of the Greeks and the Arabs, the medieval scholars and yes, even the Humanists with their veneration for 'pristina theologia'. In addition, its assumed effectiveness through several millenia of supposedly 'verified' experience (not experimentation), made astrology too valuable a tool of bettering social intercourse to be threatened with sudden disappearance in the same degree of fatality that beset ancient science as a whole. As concerns astrology in particular, the disjunction between its social role and its theoretical, scientific foundation since the seventeenth century adds a further handicap for the historian who wishes to account for the role of astrology in ancient times. Without a deliberate and thorough divestiture from the mentality issued from this revolution, it remains extremely perilous to venture into an investigation of the true role of astrology in the history of science and of ancient societies. Too much of the superstition and quackery of later astrologers will always lurk along the way, tending to bend one's judgment about the stature of the immense, frequently remarkable efforts displayed in the major classics of ancient astrology.

As hinted to above, the cultural preparation of scientists and historians of our times inclines them to perceive such obsolete fields as astrology and most occult sciences of the past as not properly or directly belonging to science, and therefore neither to the history of science. If paid attention to at all – an imperative nevertheless imposed by the sheer bulk of surviving materials of this nature, its 'weird' connection with the greatest names and highest activities of past ages in the intellectual, political and social spheres – astrology perforce has to be acknowledged, but also indulgently forgiven as aberration due merely to the infancy of the human mind which has since learned to grow up to maturity. Such a linear conception may have its particular merits, but it is a handicap for the historian, since astrology was part and parcel of ancient scientific outlook. And it is among the most treacherous factors working against a serious, unprejudiced effort to understand the history of astrology in medieval and Renaissance periods.[5]

2 *Proposals for a solution*

Having dealt with these preliminary cautions, we should like to outline now what seems to us to be the proper methodology toward a truer grasp of the place of astrology in medieval science and society. We must first sincerely try and ascertain what medieval authors and practitioners conceived to be the 'science' of astrology. As it happens, this query leads directly to identifying the standard authoritative works which carried for them the doctrine and helped its diffusion. Finally, as a conclusion of this examination we shall have to stress the degree of accessibility and of reliability of these medieval classics at present.

(a) Looking for the medieval conception of astrology, we may first unregretfully dismiss the well known late Roman compilations, together with their even poorer epigones, which served to transmit the remnants of

[5] As an illustration of this point, we may refer to the article by E. S. Kennedy, 'Ramifications of the World-Year Concept in Islamic Astrology', *Ithaca-26 VIII-2 IX 1962.* Hermann, Paris, pp. 23–45. In his introductory remarks Kennedy stated: '. . . for the matter is so diffused, the elements so contradictorily compounded of religion, magic, and myth, and the sources so numerous that one often feels, as Bīrūni would say, that the only reason for studying the subject is to be able to warn the reasonable man away from it. But there is no easy way out; anyone who undertakes the serious study of ancient astronomy must, whether he like it or not, study ancient astrology with equal seriousness. And once the bars are down for pseudoscience, everything else creeps in'. To which the commentator A. P. Youschkevitch wittily retorted (*ibid.* p. 45:) 'Au commencement du texte écrit de son rapport on lit que la matière est si contradictoire, si diffuse qu'il a souvent pensé au cours de son travail que la seule raison de l'étudier était de détourner chaque personne raisonnable de s'en occuper. Evidemment c'est une plaisanterie de la part de M. Kennedy et par contre, on doit espérer que des savants raisonnables – et M. Kennedy en premier lieu – continueront les recherches dans cet intéressant domaine de l'histoire des découvertes et des échecs de la pensée humaine'.

ancient science through the early middle ages until the twelfth century. There is ample justification for this cavalier treatment. The total absence in them of any creative strand is one. In their approach to the ancient materials they are aiming at registering for posterity, they display a degree of uncritical attachment to truncated, sometimes distorted and contradictory theories that disqualifies them as living channels of the science they purport to deal with. None of these works, in addition, reflects any active participation, or growth, or refinement in the original old notions, preserved without discrimation by them and retransmitted in a confused state without a blink in the eyes. One may recall, for instance, the astronomical lore contained in either Macrobius or Capella, about which modern scholars are still undecided whether one or the other presents a platonic or an eudoxian world system, and how much of a taint of neo-platonism does colour their presentation. The issue in itself may be merely academic, since these authors themselves were probably unaware of such differences between genuine science of old and its present rhetorical wrappings. Their ambiguities, vagaries concerning the very definition of astronomy or astrology, can be taken as a token of their uselessness for a delineation of the scientific tradition they are supposed to belong to, either for their own times or with respect to their followers for centuries. In other words the compilations of Capella, of Macrobius, and even of Chalcidius and Manilius are detached from any practical world of science and float freely in the high atmosphere of rhetoric. Not even the *Tetrabiblos* or the *Almagest* of Ptolemy, these definitive classics of ancient astrology and astronomy survived in textual existence during this long period. The later inert, irrelevant and fanciful notions of Isidore of Seville about astronomy-astrology merely aggravated the impoverishment. Isidore may innocently use the label astronomy to designate astrology, or vice versa, and this state of confusion is still noticeable in the works of an important twelfth century naturalist like William of Conches.

However, beginning in the twelfth century, and stabilizing in the thirteenth, we find newly invented labels to designate *various* sciences to study the heavens. There is a general 'science of the stars' (scientia stellarum) as the discipline dealing with the knowledge of the whole heavens, and then the 'science of the movements' (scientia motuum) for astronomy, together with a 'science of the judgments' (scientia iudiciorum/judicial astrology) for astrology. The origin of these new appellations can be definitely traced to the twelfth century translations from the Arabic, at the same time that they accompany a revival of a genuinely scientific curiosity about the heavenly bodies, their movements and influences.

In Albertus Magnus' *Speculum Astronomiae*[6] a general statement is presented at the beginning to define the 'science of the stars'. It involves a neat distinction between astronomy and astrology, the one 'mathematical', the other 'judicial', but both declared interdependent and inseparable branches of

[6] All efforts recently displayed to deny Albertus' authorship of this work are spurious – lacking all serious foundations save for a desire to wash Albert's reputation from any taint of astrological inclination; it stands in the face of an overwhelmingly solid, unanimous medieval conviction starting from nearly the beginning.

the one science of the stars. Although ultimately traceable to Ptolemy's *Tetrabiblos*, which had been rediscovered in translation from the Arabic, the definitions given by Albertus are in reality culled directly from the opening chapter of Abu Ma'shar's *Introductorium Maius in Astronomiam*, twice translated into Latin during the first half of the twelfth century. The interest evoked by this work antedates, however slightly, interest in the *Tetrabiblos*. The *Introductorium Maius* was fully encyclopedic in nature, fusing astrology and peripatetic doctrines into one majestic edifice. Abu Ma'shar's strong reliance, openly asserted by him, upon both Ptolemy and Greek science in general (Aristotle in physics, Galen and Hippocrates in medicine, Aratus and the Almagest for the nomenclature of stars and constellations, Hermes and Strato for the *termini* of the planets), no less than his own erroneous assertion that the *Tetrabiblos* was not the work of the author of the Almagest, but of 'another Ptolemy king of Egypt', seem to have created among his early Latin readers a strong incentive to possess both the *Tetrabiblos* and the *Almagest*. And surely enough both were soon made available in Latin translations.[7] These early stirrings for new science from the Arabs resulted in great part from the strong, immediate impact caused by John of Seville's translations,[8] and in particular of Abu Ma'shar's *Introductorium Maius* of 1133. As early as

[7] The earliest translation of the Tetrabiblos appears to be that of Plato of Tivoli in 1136. As for the *Almagest*, two translations were made in the twelfth century, one from the Greek and the other from the Arabic. The one from the Greek was effectively done by Hermann of Carinthia in Palermo ca.1150, and the other from the Arabic by Gerard of Cremona, also around 1150. The traditional views about authorship and date of the translation from the Greek (the 'Sicilian translation' of Haskins) were misinterpreted by Haskins and the date must be placed nearer 1150 as the medieval tradition had it. As for Gerard's translation, the date of 1175 is unacceptable since its translation was the principal aim of Gerard in coming to Spain, the home of translations in his time; and Gerard must have arrived in Spain shortly before 1145: one can hardly understand why he should have waited thirty years to fulfill his aim, producing some seventy other translations in the meantime! The date 1175 is based on the subscription of one single manuscript, and it clearly refers to the transcription of this copy by one of Gerard's helpers, Thaddeus of Hungary, and not to the date of translation.

We are preparing a separate article establishing, against Haskin's patent errors, both the authorship of Hermann of Carinthia for the Sicilian translation from the Greek, and for the earlier date of ca.1150, since it was done for King Roger of Sicily who died in 1154. Now see: 'Itinéraire d'un médiéviste entre Europe et Islam', in *La diffuzione delle scienze islamiche nel Medioevo*, Roma, Accademia Nazionale dei Lincei, 1986.

[8] We may disregard any serious impact of Adelard of Bath's early translations in astrology since first, his early attempts were dismal failures: the *Centiloquium* with Incipit 'Doctrina stellarum', and to some extent his translation of Euclid, since it underwent two revisions after him, (Clagett classified Adelard's translation as Adelard I, II and III without envisaging the case of revisions by others than Adelard; Busard ascribes at least one of the revisions to Hermann of Carinthia); and second, Adelard's venture in translating other arabic works of astrology produced only the miserable book on amulets entitled *Liber prestigiorum elbidis*. Anyone who has practiced this work would only be astonished at the thought that it created a significant scientific stir, even in the twelfth century.

As for John of Seville's personality and works, the amount of unfounded, arbitrary, and even mischievous statements which have been issued on this subject is confounding. He was a well known savant of early twelfth century Spain, friend of known ecclesiastics and grandees [queen Theresa of Portugal], admired by Plato of Tivoli, Rudolph of Bruges, hired by archbishop Raymond of Toledo to do some philosophical translations with the help of Gundisalvi to put his mediocre Latin in good shape, etc. ... Alonso has so distorted the significance of manuscript suscriptions to John's works, and added his nonsensical theory of two John of Sevilles: one a

1139/40 Raymond of Marseilles expresses his admiration for this work and his determination to follow in its steps. By 1140, a second translation was produced by Hermann of Carinthia. This second translation, it is true, was generally set aside[9] in favour of John's own despite its crude Latin style.[10] Hermann's translation nevertheless survived in a number of medieval manuscripts, and unfortunately for the later history of medieval tradition, it was the only one printed in early modern times, in an incunabulum given by Erhard Ratdolt at Augsburg in 1489. Yet before its medieval demise, Hermann's translation had left plenty of traces in medieval lore, especially the prediction of the 'virginal birth' of Christ contained in Book VI, chap. 1, which can be seen quoted abundantly in all sort of media throughout the medieval period. In addition, Hermann's translation of the *Introductorium Maius* inspired his remarkable own *De Essentiis*[11] and led him ultimately to translate the *Almagest*, albeit from the Greek.[12] Both in its title [13] and in its

---

philosopher who did not know Latin, the other an astrologer who translated on his own. In fact the manuscript suscriptions so distorted and misunderstood by Alonso state unequivocally that John called himself John of Seville and of Limia, for he is residing in the region of Limia in the North of the future Portugal where a relative of his, Sisnando Davidiz (Ibn Daud), was a vizir for Ferdinand of Leon since 1065 till his death in 1095. John specifically states that he resides in Limia and translates in Limia. His career there must have been well known to every one since he was in the intimacy of Queen Theresa (who succeeded Sisnando Davidiz), and of bishop Maurice of Braga and Coimbre, future antipope Gregory VIII. (1118 A.D.)

Later called to Toledo (after the false pope's imprisonment and elimination?), John seems to have enjoyed the esteem and protection of archbishop Raymond and the friendship of two English ecclesiastics living in Spain: William and Gauco, to whom he dedicates one of his most interesting original work, the *Liber de diversis cursibus* (in fact an elaboration on Thabit ibn Qurra's *De motu octave spere* for its implications on astrological judgments). Given these realistic bits of information easily ascertainable by any serious scholar, one wonders why so much precious ink has been spent (M. Th. d'Alverny, 'Avendeut?' in *Homenaje Millas Vallicrosa*, Alonso, several articles in *Al-Andalus* etc. . . .) to present us with a different picture than what medieval contemporaries knew all too well!

9 Hermann's version of 1140 was superior in its Latin garb, but overloaded with rhetorical flourishes, and dismally inadequate, both for some omissions and for a large number of unwarranted interpolations where Hermann changes the letter and spirit of his model. This was noticed early by latin readers, and Hermann's translations seem to have been generally held in suspicion for these obvious motives.

10 The Library at Chartres before 1135 possessed some of the translations by John of Séville, in particular that of Zael (Sahl ben Bishr) and perhaps the Alchabitius. These manuscripts have been destroyed during World War II and we have only Charles Catalogue of the 1840s to attest to this fact.

11 Recently edited anew – if rather badly – by C. S. F. Burnett, Leiden 1982 after Alonso's first edition of 1946. Comillas de Santander, which still has a valuable preface, preferable in many instances to the elucubrations of Burnett's own preface.

12 See note 7 above.

13 The oldest and best manuscripts of John's translation carry the long-winded heading: 'Liber in quo est maior introductorius Abumasar astrologi ad scientiam iudiciorum astrorum, et tractatus eius super eadem iudicia, cum disputacione rationali et auctentica, et figure signorum atque nature'. One old Arabic copy, Istambul, Sulaimaniye Library, Garullah 1508, has a heading which is close but not quite the same. In the passage of Book I, chap. 1, (the accessus), the title of the book is described by Abu Ma'shar himself: Liber introductorius ad scientiam iudiciorum astrorum, which correspond exactly in John's version and in the Arabic. Hermann has here: 'Introductorium in astrologiam Abuma'xar Albalachi' which Ratdolt's printed edition transcribed *Introductorium in Astronomiam Albumasaris Abalachi*.

general vocabulary, Abu Ma'shar's *Introductorium Maius* in John's translation provided for the Latin scholars a new terminology and a clearer distinction for the different sciences of the stars: astronomy was the 'science of the [celestial] motions', while astrology was the 'science of the judgments of the stars', hence 'judicial astrology'. The introduction of the weird expression 'judicial' astrology into this semantic field was a mere accident of translation from the Arabic. The classical Greek term for astrological judgments was 'prognostication', a term which the Humanists will eagerly try to substitute for 'judgment', in line with their general rejection of Arabic influence upon Latin science in the preceding epoch. So, prognostication in Alexandrian scholarly works became 'taqdimatu'l-ma'rifa' – an exact rendering: knowledge in advance – in the Arabic translations. Since many practical sciences: medicine, weather prediction, military strategy, etc. needed to use the concept, the term prognostication/taqdimatu'l-ma'rifa spread into all these disciplines. As for the Arabic root 'ḥkm', it belongs semantically to the act of a specialist (an official, a learned man) rendering a definitive pronouncement, something like the Latin *declaratio*. From the time the judicial apparatus was consolidated in early Islamic society, and the more qur'ānic term of qāḍi was appropriated into it, the expression ḥakīm more and more acquired the meaning of competence in a field of learning, or a learned authority, and ḥukm/aḥkām became an authoritative pronouncement. Abu Ma'shar makes frequent use of the term: Ptolemy is sometimes called al-ḥakīm Baṭalmiūs, Galen or Hippocrates ḥakīm al-aṭibbā. Furthermore, when the context is clear: medicine, philosophy, astronomy, the term ḥakīm is frequently employed absolutely: in an astronomical context al-Ḥakīm is Ptolemy; in a medical context al-Ḥakīm designates Galen; in a philosophical one al-Ḥakīm indicates Aristotle. The use of the expression aḥkām an-nujūm in the title of Abu Ma'shar's *Introductorium Maius* shows the expression to have already become normal for 'authoritative pronouncements' or the equivalent of the Latin *auctoritates*. The Latin translators, especially John of Seville, merely translated aḥkām by judgments, which was a lexical convenience, but with potentially disturbing effects. This was a time in the Latin West when Adelard of Bath taunted his nephew and the Latins in general for their blind reliance on *auctoritates*, and when Adelard wrote his *Sic et Non*, again disparaging the *auctoritates*. We may even observe that as early as the 1140s, Hermann of Carinthia, one of the important translators from Arabic, prefers the expression *judges* to that of *auctoritates* in the field of astrological science, when he asked from his inseparable companion Robert of Chester to produce for him a translation of al-Kindi's *Judicia*, because both have already recognized in al-Kindi the most reliable *Judge* among Arab astrologers. Robert complied and the collaboration led to creating the first nucleus of the later famous astrological classic of the Latin middle ages: the *Book of Nine Judges*.[14] Hence also the generalization before

[14] The nucleus is contained in MS. British Library, Arundel 268, fols 75–76. C. S. F. Burnett, 'Arabic-Latin Translators working in the mid-12th century' in *Journal of the Royal Asiatic Society*, 1977, pp. 63–108 knows this manuscript, in which the prologue to the *Nine Judges* (he calls it the

long of the term judicial astrology, a term which however has no direct equivalent in Arabic works: only the slip from hakim/learned among the Arabs to hakim/judge among some of the Latin translators gives the historical clue.

Sharing by the side of al-Kindi in the intense intellectual activity at Baghdad in the early and middle ninth century, Abu Ma'shar had probably observed how aptly Greek physical science – the falsafa in fact – so much in favour under the patronage of the caliphs, offered reputable grounds for the linking of Alexandrian-eastern astrology with peripatetic doctrines. A keen mind could also perceive how deeply Ptolemy's *Tetrabiblos* was permeated with peripatetic physics and cosmology. The actual transfusion of Aristotelian physics (from the *De Caelo*, the *De Generatione et Corruptione*, the *Meteorologica*, and part of the *Physics* and *Metaphysics*) into Abu Ma'shar's *Introductorium Maius* cannot escape the careful reader. However, with a seemingly scant knowledge of the *Introductorium Maius*, and that mainly through the utterly unsatisfactory incunabula edition of Ratdolt in 1489,[15] D. Pingree claimed in the *Dictionary of Scientific Biography*[16] that Abu Ma'shar's peripateticism was a mere rehash of the Hermetic lore of the Harranians. Such a statement implies a gross ignorance of Abu Ma'shar's *Introductorium Maius*. Truly in this work, Hermes' authority is called upon, but it is always in separate, isolated chapters, and only after full treatment of the physical basis of astrology in Aristotle and Ptolemy has occupied the space of several anterior chapters. When adduced at last, in what is visibly the 'edulcorated' form meant by Pingree, the authority of Hermes merely serves

Three Judges) clearly sounds like Hermann addressing his friend Robert, including the standard formula of friendship between the two men 'mi karissime R[oberte]'. But finding another copy in Oxford, Bodl. 430 where the name of Michael, bishop of Tarazona appears, rather than mi karissime R[oberte], he prefers to ascribe the work to Hugh of Sanctalla. *Pace* Burnett, the whole gist of this prologue is about Hermann and Robert's friendship and their collaboration: Hermann had further requested from Robert a translation of the *Judicia* of al-Kindi, the most authoritative 'Judge' in Arabic astrology (cf. Robert's prologue to his Judicia, ed. in L. C. Karpinski, *Robert of Chester*), and the extracts from al-Kindi, in Robert's translation, precisely open the intended collection. If it stopped at only three authorities made use of (Burnett's Three Judges), it must have been because of some uncontrollable interruption, for the prologue announced a much longer list. The list finally became the famous Nine Judges in the course of time, resulting from additions along the proposed plan, and innumerable interpolations by later probably self-appointed contributors. Burnett seems unable to grasp the core of meaning in this prologue, and prefers to have recourse to an impressive number of textual parallels here and there, to justify his untenable hypothesis. No one who knows Hermann's and Robert's careers can fail to grasp Hermann's plea in this prologue. Hugh, like his other contemporary Gundisalvi who pillaged the *De Essentiis* of Hermann in his *De Processione Mundi*, has taken advantage of some trouble in the relations between Hermann and Robert, and Hermann's exile to Salerno to unscrupulously appropriate to himself a text of the absent Hermann.

15 Pingree remarks that the allusion by Abu Ma'shar to his *Astronomical Tables*, which occurs in the Latin text (he cites printed ed. 1489) does not appear in the Arabic manuscripts of the *Kitab almudhal al-kabir* he has consulted, and he cites Leiden, Or. 1051, and Istambul, Nuruosmaniye 2806. I don't know from this what is Pingree's notion of reading a manuscript, but his Leiden manuscript contains formally the allusion he claims is not there.

16 New York, Charles Scribner's Sons, 1971, ed. C. S. Gillespie, Vol. I, sub verbo 'Abu Ma'shar'.

to confirm with Eastern authority what has already been amply and forcefully established on the principles of Aristotle and Ptolemy. In fact, the hermetic chapters in the *Introductorium Maius* – three or four at most (II.4; V.4, 7, and perhaps passing references in VI.1 and in the 'partes' of VIII.5) and clearly labeled so, out of a total of 103 chapters for the whole work, can be no more than a mere concession by the author to the hoary Babylonian tradition surviving among the Harranians in Northern Mesopotamia. At any rate, just as Hermann of Carinthia felf justified in asserting to Thierry of Chartres that Abu Ma'shar had 'amplified' Ptolemy's *Tetrabiblos* (thus supplanting it to a certain degree), so did an annotator of a manuscript copy of Roger Bacon's *Perspectiva* (who may be Roger himself) blandly declared that the 'authority in the science of the heavens' is to be found not so much in the *De Caelo* of Aristotle as many think, but in Abu Ma'shar. This statement, which must be of the thirteenth century at the latest, reveals the paramount importance that the work of Abu Ma'shar had acquired in the minds of medieval schoolmen. Albertus Magnus even calls Abu Ma'shar one of the three important commentators of the *Physics* of Aristotle. More than in Aristotle's works touching upon this field but merely providing the logical integration of astrology into physics, and more than Ptolemy's snappy treatment in his *Tetrabiblos*, which consisted in long nomenclatures without elaborate justification, Abu Ma'shar's *Introductorium Maius* offered to the natural philosopher of the Latin Middle Ages an integrated view of the cosmos, describing in its theoretical portions and in strict syllogistic and experiential terms borrowed from Aristotle, the laws of motion and causality in the physical universe understood as a single vast machinery in which man is carried along. This machinery is described in its broader implications for human history in Abu Ma'shar's *Great Conjunctions*, also translated by John of Seville. It is further detailed at the level of individual men and social groups in Abu Ma'shar's numerous other works, all translated by John of Seville.

There ought therefore to remain no doubt that the new interest in astronomy-astrology experienced in the Latin world at the beginning of the twelfth century was awakened principally by John of Seville's translation of the *Introductorium Maius* in 1133, and of several other works of Abu Ma'shar. For decades afterwards, the field of new astrology is dominated by Abu Ma'shar's influence. The departure from stale Latin compendia of the late Empire was a radical one, for the Latin scholars who showed themselves actively interested in the new science massively followed in the footsteps of the Arabian astrologer. This can already be observed among Parisian scholars around 1190 when Alanus de Insulis in his *Anticlaudianus* listed Abu Ma'shar as the authority (Hakim) in astrology, right after Ptolemy in astronomy.

(b) Ptolemy's authority was also revived in parallel fashion and through the same channels as Abu Ma'shar's works. His *Tetrabiblos*, as well as his *Almagest* were soon made available in Latin translations. But in addition, many spurious works of astrology from the Arabs were attributed to him and in fact gained wider circulation than his genuine works. Thus we had a book on amulets, or astrological images carved under exceptionally favourable celestial influences. Adelard of Bath translated a *Liber Prestigiorum Elbidis* supposedly by Thabit ben Qurra, the ninth century Harranian astrologer and astronomer at the court of the abbasid Caliphs; but it is clear from the opening statement of this work that the principal authorities invoked are Ptolemy and Hermes. Similarly John of Seville translated a work supposedly of Ptolemy on *Images* (celestial influences captured in a carved stone) and in his prologue to this translation John alludes to the earlier partial translation done by a 'fatuus antiochenus' who seems to be Adelard of Bath. A spurious work on the science of the sphere is also attributed to Ptolemy which is mainly astrological, dealing with the properties of the signs and planets. But the most important spurious work of astrology ascribed to Ptolemy which had an influence far outstripping all others was the *Kitāb at-Tamara* or *Liber Fructus/Centiloquium* put up by the Egyptian astrologer and mathematician Aḥmad ibn Yūsuf (ca.920 A.D.). We have demonstrated in an article of 1976 the circumstances and success of this pseudo-ptolemaean work[17] and F. Sezgin's opinion to the contrary must be wholly rejected because founded on a sheer misreading of his manuscript source. The *Liber Fructus/ Centiloquium*, eventually being prescribed reading for the students in medicine in European Universities from at least the middle of the fourteenth century, together with the *Alchabitius* and Ptolemy's genuine *Quadripartitum*, became the indispensable vade-mecum of physicians, in addition to many other groups of social activists throughout the middle ages. No less than six translations or adaptations during the medieval period, and several others during the Rennaissance were produced, a tradition to be continued by vernacular translations in the seventeenth and eighteenth centuries. The *Centiloquium* was accompanied by a commentary (the real purpose of the forgery) by Aḥmad which had the favour of the medieval astrologers and physicians and all other practitioners, unfortunately under the false ascription of Haly the commentator. We have explained in our 1976 article the origin of this confusion. But the Renaissance translations made from the Greek (itself a translation from the Arabic) made by George of Trebizond and Gioviano Pontano in particular substituted their own commentaries of the 100 propositions to that of Ahmad. Thus the later printed editions which to the exception of the two incunabula of 1484 and 1493 at Venice all contain these Renaissance translations deprive the modern researcher of an access to the medieval tradition of this incredibly successful text of medieval astrology.

Other works of medieval astrology, all derived from the Arabic, which obtained wide currency, still could not rival in doctrinal influence and mentality the works of Ptolemy or Abu Ma'shar, of whom Abu Ma'shar ranks the first by far. Other popular works merely condensing the same doctrines were those of Zael (Sahl ben Bishr), Umar Tiberiadis ('Umar ibn Farruham at-Tabari), Ma'shallah, al-Khayat (Abohali). All these merely

17 R. Lemay, 'Origin and Success of the Kitab at-Tamara, Liber Fructus . . .'. *First International Symposium of the History of Arab Science, Aleppo 1976, Proceedings*, Vol. II. *Contributions in European languages*, pp. 91–107.

brought more water to the mill which was principally running on Abu Ma'shar and Ptolemy. We must make a slight exception for Alchabitius (al-Qabisi; Aleppo tenth century, translated by John of Seville in the 1120s) who was enhanced first by a commentary appended to it in 1331 at Paris by John of Saxony, and later prescribed for students of medicine, along with Ptolemy's *Quadripartitum* and the pseudo-*Centiloquim*. The question therefore of what did medieval men think of astrology can be reasonably and accurately answered only by examining these basic 'classics' of medieval astrology. More yet, since these works were extensively studied and applied in practice by medieval professionals of all sort of trades and crafts, the understanding of a medieval text, and in particular the requirements for its critical edition would be uniquely enhanced by a familiarity with the contents and orientation of these classics. Too often one finds stated in modern studies about medieval physicians, for instance, that this one or that one is singularly free of astrological orientations, when the presence of extensive quotations or paraphrases or medieval astrological classics explodes in virtually every page. This error of perception stems from the ignorance of the contents of medieval astrological texts.

(c) But here is where the shoe hurts. What is the present accessibility to scholars of these medieval classics of astrology? Two major factors of historical importance must be recalled, that have operated as a barrier to such access for the modern researcher. One is the crucial fact of the 1277 condemnations of several 'dangerous' propositions at the University of Paris. Many modern scholars (Duhem, E. Grant) have pretended to see in these condemnations the birth certificate of modern science starting with the mathematical revolution of the fourteenth century. As far as astrology is concerned, several propositions included in 1277 concern directly astrological doctrines (such as the one stating that 'if the sun stopped, earthly fire would not burn anymore': a tenet of astrology that earthly fire draws all its reality from the movement of the sun; but more to the concern of medieval ecclesiastics, the danger of fatalism involved in astrological doctrines was clearly aimed at in these condemnations). That the gist of this astrological trend was of Arabian inspiration escaped no one; witness the *De Erroribus Philosophorum* of Giles of Rome in the 1280s, where most 'dangerous' philosophers listed are Arabian. We have tried in two past articles[18] to show that this reaction was fully shared by Dante who saw an unacceptable danger to Christian faith in the wholesale acceptance of the influence of Arab science. This is not merely a matter of Averroism as it has been deemed generally, but a genuine revulsion against the Western dependence on the Arabs in the field of science and natural philosophy since the beginning of the twelfth century. The generation immediately after Dante (Petrarch, Boccacio) joined,

[18] R. Lemay, 'Le Nemrod de l'Enfer de Dante et le Liber Nemrod', *Studi Danteschi*, Firenze, 1963, pp. 57–132. R. Lemay, 'Mythologie païenne et chrétienne chez Dante. Le Cas des Géants', *Revue des Etudes italiennes*, Janu.-Sept. 1965. *Dante et les Mythes*, pp. 236–279. Now see: 'Itinéraire . . .', Roma 1986, quoted in note 7 above.

perhaps unwittingly, in this movement by taking advantage of the flow of Byzantine scholars who were beginning to pour into Italy under the advance of the Turks. Except in rare, daring cases like those of Cecco d'Ascoli, Biagio Pelacane of Parma, or John of Legnano in Bologna, astrologers tended less and less to quote directly former Arab sources as they did in fact dispose of Christian Latin substitutes, such as Guido Bonatti, Leopold of Austria and John of Eschenden. The century-long eclipse of Arabian astrological texts ensued, ever more darkening, and the availability of manuscript texts of their works, but above all the prominence given to their authority faded into a distant corner. By the time of Symon de Pharès, for instance, near the end of the fifteenth century, it even becomes dangerous to possess in your private library too many of these astrological works from the Arabs.

The second factor to account for the eclipse of medieval astrological classics was the invention of printing and the early diffusion of cheap editions of medieval astrological authorities. Astrology as an accepted science was still in its heyday in the Renaissance.[19] With the introduction of the printing press it received what may be now considered as an inordinate share of attention from the publishers, and of course from the learned public for whom they were working. Some early printers even specialized in this field, like Erhard Ratdolt at Venice till 1488, and afterwards at Augsburg under the aegis of the Fuggers. From Ratdolt's presses at Venice poured out edition after edition of astrological classics, and he continued his activity at Augsburg after 1488. After he left Venice, his plant and apparently some galleys passed into the hands of Boneto Locatelli, and later apparently of the Ottaviani Scoti who continued the tradition until the 1520s. In Basel, Heinrich Petri, others in Nuremberg and other locations joined the movement resulting in a vast number of publications of astrological works from the past. Since, however, this corresponded to a public demand, a kind of consumer market, the publishers who were in this for the material benefits it brought would usually select the cheapest manuscript of the most popular authors and reproduced them without any serious critical approach. Since the most venerable of them had gone through a long process of manuscript transmission, a cheap manuscript of the fifteenth or sixteenth century would be full of errors of transcription and of omissions and interpolations and the resulting edition would represent even a lower degree of scholarly value. When there existed several translations, the one selected would usually be the one most easily manageable and at the lowest cost.

In the cases studied above, the intervention of the printing press in the two major texts, Abu Ma'shar's *Introductorium Maius* and ps-Ptolemy's *Centiloquium* produced catastrophic results for the modern researcher of medieval astrology. In the case of Abu Ma'shar, the text printed by Ratdolt at Augsburg in 1489 was the version of Hermann of Carinthia, generally discarded by medieval scholars in favour of that by John of Séville. My own

[19] Yet ca.1535 in Paris, Michael Servetus failed to convince the Faculty of Medicine of the necessity of astrology for a physician. Cf. R. Lemay, 'The Teaching of astronomy . . .'. *Manuscripta* XX (1978), Festschrift P. Kibre.

research and critical study of the fate of Abu Ma'shar's text through the middle ages has shown that John of Seville's version outfavoured that of Hermann in the proportion of four or five to one (over 40 manuscripts of John's as against some ten of Hermann's); furthermore, the printed text of Ratdolt contains more errors and omissions than any previous manuscript. And it has been virtually the sole access since to the fate of Abu Ma'shar's text through the middle ages! Concerning the *Centiloquium* a worse fate still attended its transmission through the printing era. Of the at least six medieval versions from the Arabic, one alone was printed by Ratdolt in Venice in 1484 and reproduced there in 1493 by Boneto Locatelli. Controlled on the best medieval manuscript tradition, this edition is larded through and through with interpolations by medieval scribes or students, many of these interpolations can be traced in the margins of medieval manuscripts. Since the text was prescribed reading for students in medicine, one can imagine the number and variety of comments elicited by this study, comments which charge the manuscripts with an indescribable bush of extra materials. And there were at least five other versions of this text which never appeared in print, and thus are removed from the scholar who would work only from this printed text. In addition, the Renaissance translations of Trebizond and of Pontano soon obtained the favours of later scholars. Some fifteen editions were produced during the sixteenth century, all carrying these humanistic translations, thus further hampering the search for its medieval tradition.

We have mentioned only two major texts of medieval astrology on which some valid approach to the medieval conception of astrology must be based in the first instance, and it is clear that the materials are not easily accessible to modern scholars. Text-critical and philological spade work on a vast scale must be performed before we may hope to understand medieval astrology on its own, historically ascertainable terms.

NICOLE ORESME'S POLEMIC AGAINST ASTROLOGY IN HIS 'QUODLIBETA'

Stefano Caroti

The importance of Nicole Oresme's polemic against astrology lies in its systematic and philosophical approach, whose depth attracted the attention of Pico della Mirandola at a time when he too was trying to refute astral divination. This hypothesis advanced by the modern editor of Pico's *Disputationes adversus astrologiam divinatricem*, Eugenio Garin,[1] is confirmed by the fact that Pico quotes – in a more correct and comprehensible Latin – from Oresme's *Questio contra divinatores horoscopios*: 'scribit enim Nicolaus Orem, gravis philosophus et mathematicus, plerumque se vidisse factam salubriter in pede sanguinis missionem, cum Luna foret in Piscibus, et in brachiis item cum Geminos illa peragraret'.[2]

Oresme's *Questio* is a far cry from Pico's elegant style and maintains the distinct features of scholastic dispute: 'quare non est ista consequentia ita bona: plures phlebotomantur Luna existente in Geminis et bene succedit, ergo bonum est tunc etc.; sicut ista: alias Luna in Geminis phlebotomato male successit, ergo malum est tunc etc. Unde a pluri debet fieri denominatio . . . *unde multos vidi et consideravi lesos in pede Luna infortunata in signo Piscium et tamen cito fuerunt etc., et alios Luna existente alibi qui tarde, igitur sibi non est imputanda causa*'.[3] The observance of astral melothesia in medical prescriptions, and in particular the attention paid to the Moon's position in the signs of the zodiac (to which a special power on human limbs was attributed) is a topos in astrological hand-books. Ascribing to Oresme this very common *experientia*, Pico confirms unequivocally his borrowing from Oresme's *Questio*, the only writing (to my knowledge) where such a record is mentioned.

The survival of Oresme's arguments against astrology in Pico's more successful *Disputationes adversus astrologiam divinatricem* widely compensates for the small circulation of his ideas amongst his contemporaries.

[1] G. Pico della Mirandola, *Disputationes adversus astrologiam divinatricem*, ed. E. Garin, Firenze, Vallecchi 1946–1952 (Edizione Nazionale dei Classici del Pensiero Italiano, II e III), vol. I, p. 632; E. Garin, *Lo Zodiaco della vita. La polemica sull'astrologia dal Trecento al Cinquecento*, Bari, Laterza 1976, p. 98.

[2] G. Pico della Mirandola, *Disputationes*, II, p. 420.

[3] Nicole Oresme, *Questio contra divinatores horoscopios*, ed. S. Caroti, 'Archives d'Histoire Doctrinale et Littéraire du Moyen Age', XLIII (1976), p. 231.

Oresme's attack on astrology, culminating in his *Questio contra divinatores horoscopios*,[4] is insistent and thorough-going. He looks beyond the daily practices of popular astrology which, in any case, should have been more worrying for religious and civil authorities than for scientists. He realises that astrology constituted a genuine philosophical system with a speculative tradition of its own – and a tradition sufficiently vigorous to accomodate itself comfortably to a wide range of philosophies and cosmologies.

In his *Quodlibeta*[5] too Oresme repeatedly warns his audience of students and doctors against the dangerous philosophical consequences of accepting astrology's claims. The theme in question – the *mirabilia nature* – is particularly germane, since one of the bulwarks of astral determinism was the belief that natural philosophy could not account on its own for certain phenomena deemed to be marvelous: 'ut autem aliqualiter pacificentur animi hominum . . . aliquorum que mirabilia videntur causas proposui hic declarare. Et quod naturaliter fiant sicut ceteri effectus de quibus communiter non miramur, nec propter hoc oportet ad celum tamquam ad ultimum et miserorum refugium currere nec ad demones nec ad Deum gloriosum, quod scilicet illos effectus faciat immediate plus quam alios, quorum causas credimus nobis satis notas'.[6]

From its beginning Oresme's attack on astrology is clearly part of a broader strategy, which was to investigate nature *iuxta propria principia*, without dangerous excursions in search of universal agents and causes beyond the limits of natural sphere. Among doctrines which contend with natural philosophy for the right to explain physical events, astrology is undoubtedly the most insidious. Arguments which have recourse to demons or *potentia Dei absoluta* are in fact easily refuted *naturaliter loquendo*; astrology, on the contrary, is not so easily dismissed because its field of action is not limited to the marvels of nature. Indeed, within its purview are all physical phenomena and, ultimately, religious and moral deeds as well.

Throughout astrological thought we find this noteworthy attempt to reduce moral, religious, political and physical events to the same causal system. It was to appeal strongly to the 'esprits forts' of seventeenth century free-thinking. But behind this enterprise Oresme perceived a theoretical

[4] S. Caroti, *La critica contro l'astrologia di Nicole Oresme e la sua influenza nel Medioevo e nel Rinascimento*, Atti dell'Accademia Nazionale dei Lincei, CCCLXVI. Memorie della Classe di scienze morali, storiche e filologiche, s. VIII, vol. XXIII, fasc. 6, Roma, Accademia Nazionale dei Lincei 1979.

[5] B. Hansen, *Nicole Oresme and the Marvels of Nature. A Critical Edition of his 'Quodlibeta', with English Translation and Commentary*, Ph.D. Diss., Princeton University 1974 (for the four chapters and the 'Probleumata per modum tabule'); see now B. Hansen, *Nicole Oresme and the Marvels of Nature. A Study of his 'De causis mirabilium' with Critical Edition, Translation and Commentary*, Pontifical Institute of Mediaeval Studies, Toronto 1985 (Studies and Texts, 68). E. Paschetto, *Demoni e prodigi*, Torino, Giappichelli 1978 and 'Linguaggio e magia nel De configurationibus di N. Oresme', in *Sprache und Erkentniss im Mittelalter. Akten des VI. Internationalen Kongresses für Mittelalterliche Philosophie der Société Internationale pour l'Etude de la Philosophie Médiévale 29 August–3 September 1977*, II Band, Miscellanea Medievalia, XIII/2, Berlin/New York, Walter De Gruyter 1981, pp. 648–656.

[6] Firenze, Biblioteca Medicea Laurenziana, MS. Ashburnham 210, f. 21rb. Quotations are from this ms.

ambiguity about the kind of causality which astrology was proposing, not to mention the irreconciliability of any such assumptions with religious faith of whatever kind.

This unificatory scheme was, moreover, a serious challenge to the accepted definitions of what were the proper fields of the various disciplines. It also threatened to nullify painfully acquired beliefs in the regularity of nature, in natural laws, which themselves rested on a clear-cut distinction between physical agents and voluntary causes. The latter, of course, could hardly be contained within any rigidly predictive system.

St Thomas' opinions on these accounts are for Oresme very dangerous; his insistence on the power of human inclination in determining men's behaviour could in fact foster an untenable superiority of the natural complexion on the spiritual part of human being. In the *Quodlibeta* St Thomas is not mentioned, but Oresme probably refers to him in arguing against those who stressed the importance of human inclinations on men's will. In the *Questio contra divinatores horoscopios* St Thomas is introduced by Oresme as one of the main supporters of this opinion: 'et ideo male dicit beatus Thomas, scilicet quod homines sequuntur inclinationem suam, ymmo dico recte oppositum'.[7]

In the *Questio* and more insistently in the *Quodlibeta*, Oresme stresses that astrology's most radical claims cannot be sustained from a religious viewpoint. It is not possible to reduce all events, including miracles, to the undifferentiated level of a physical change, guided by the movements of the planets: 'Dominus autem Yehsus Christus facebat mirabilia ut cognosceretur et ut converteret ad etc. sed non propter pronosticationem futurorum'.[8] Such a reduction abolishes the miracle of Christ's incarnation and the exemplariness of His life and actions, by subjecting even these to cast-iron astral laws, along with everything else in the sublunary sphere. Though as a rule reluctant to elude philosophical discussion, in this case Oresme was ready to resort to ecclesiastical censures – that is the famous Paris condemnation -: 'tertio quia viderunt aliquos aliqua facere et multa fieri, quarum causas voluerunt habere naturales, que tamen erant supernaturales, sicut de sanctis prophetis (ms.: philosophis) et factis eorum a Deo glorioso missis. Unde causas talium celo et eius constellationi attribuerunt, ut patet in eorum libris. Unde propter hoc Avicenna in sua *Metaphysica* et precipue in suo VI *Naturalium* particula quarta posuit intellectum habere mirabilem potentiam . . . quod tamen est articulus condempnatus et posuit quod pure naturaliter possunt esse tales prophete sicut Christus et Moyses et in hoc fuerunt decepti. Et consimiliter voluerunt salvare de magicis qui tempore Moysi coram pharaone etc. unde dixit quod tales virtutes haberent a celo, sicut etiam Haly

[7] N. Oresme, *Questio*, p. 237. In the XXIXth *questio* of the *Quodlibeta*, where elections are discussed, we find echoes of this passage: 'et si dicetur non omnes utuntur libertate, sed utuntur plus inclinatione naturali, istud non valet, quia non est aliquis quin ex libertate multa fecerit' (f. 62va–b), as well in the XXXVth *questio* ('quare homines communiter sunt mali'): 'ad tertium dico quod anima rationalis quamdiu est in corpore non per se solum exercet opera, sed dependet ex etc., et illa dependentia, quamvis naturaliter non dicitur mala, tamen moraliter dicitur etc. Per idem ad quartum: celum bene calefacit vel iuvat ad talem complexionem, sed habentem liberum arbitrium non necessitat', f. 64va.

[8] F. 46va.

super *Centiloquium* narrat de quodam qui habebat virtutem etc. et est purum mendacium . . . Et ita dixerunt quod a celo, scilicet qui tali hora, tali signo celi ascendente etc. talem virtutem haberet quod est pure falsum, quia in virtute divina supernaturali. Et tunc fuerunt moti ad talia faciendum, ymmo etiam et sibi imponunt quod per artem notoriam sibi naturaliter acquisitam sciebat que sciebat et non per gratiam infusam etc. ymmo de ipso et Christo turpia dicunt, ut patet de Averroy et Abraham in suo tractatu *De coniunctionibus*'.[9]

The anti-astrological arguments in the *Quodlibeta* follow a precise plan. First the position of the major philosophical authorities vis-à-vis astrology is defined; then a precise definition of the influence of the heavens over the sublunary world follows. With differing degrees of emphasis, this influence had come to be unanimously accepted by the Aristotelian scholastic tradition – to such an extent that it had become a topos in certain areas of commentaries on Aristotle's works. This more general analysis of celestial influence is then reinforced by a scrutiny of some specific applications of astrological theory (elections, interrogations, nativities and images), where the theoretical side is clearly ancillary to the eminently pragmatic exigencies of astral divination.

In the first *questio* expressedly concerned with astrology, Oresme turns his attention to this practical, operative aspect of astrology. He proclaims its foreigness to the philosophical tradition of celestial influence – at least in the minute and precisely defined way which the astrologers propose: 'utrum Aristoteles et alii philosophi notabiles consentierant iudiciis astrologorum quantum ad iudicia particularia, ut de nativitatibus et electionibus et de interrogationibus, et quod futura contingentia et particularia seu singularia per astrologos possunt presciri per regulas datas super hiis ab astrologis'.[10]

In his reply to this *questio* Oresme just mentions philosophers' silence on the topic, Holy Bible's prohibitions and the dangerous consequences on faith, and, finally, the unfortunate fate of kings and princes who put themselves in the hands of astrologers (such an argument is diffusely developed in his *De divinatione*). The first argument, though only sketched out, establishes the lack of 'authority' – in the medieval sense of the word – for practices such as nativities, elections and interrogations.

The importance of such a clear preliminary statement is obvious. First and foremost Oresme wants to refute the claim that astrology is aligned with

[9] F. 47ra–b. 'Fuerunt ergo et adhuc sunt multi qui omnia Deo immediate attribuunt aut demonibus et fuerunt alii, ut Avicenna et Averroys et Agazel qui omnia causis naturalibus imposuerunt et nulla mirabilia etc.', f. 47rb. Oresme is convinced that these two apparently divergent attitudes of mind can be both rooted in a single misunderstanding, which is the constant target of the *Quodlibeta*: the neglect of proximate causes. In the XXXth *questio* Oresme refers to the subordination of religions to the stars: 'quod est contra Aristotelem et etiam contra fidem omnem et legem, patet quia de Moyse et lege sua seu quam dedit et de Christo et lege quam dedit turpia dicunt et dicunt quod naturaliter per coniunctiones que precesserunt eos talia debebant venire et sic mori et sic finiri, ita quod non libere. Et dicunt se scire quando leges eorum scilicet cessabunt et deserentur (ms.: deseruentur), de quibus iam sunt mentiti, vide super hoc libros eorum', f. 63ra.

[10] F. 39rb.

Aristotle's natural philosophy – a claim that in any case relies on a less than rigorous and coherent interpretation of certain passages of Aristotle – and in so doing, Oresme hopes to deny astrology's feigned antiquity, which did so much to guarantee its authoritativeness. The novelty and effectiveness of Oresme's attack can be seen in his use of 'ex auctoritate' argument. He does not bother to juxtapose extracts of philosophical or theological works; he merely points out that philosophy is completely silent on nativities, elections and interrogations, in order to move swiftly to a rational confutation of astrology's basic principles.

Since an astrological judgement presupposes a knowledge of the stars and of our natural constitution, Oresme has to define the limits of what can be known in both fields astronomy/astrology and physics. He stresses first how impossible it is to know with sufficient precision the *proportiones* and *complexiones* of the various elemental compounds regulated by generation and corruption. What we know about heaven is even more limited: it is restricted for Oresme to some knowledge of the magnitudes and movements of the heavenly bodies, and those too are not known with any certainty: 'non certitudinaliter et punctualiter sicut patet scientibus aliquid de illis'.[11]

When the uncertainties regarding terrestrial agents are added to the conjectures about presumed celestial causes, the sum cannot be less than fatal for the specific judgements claimed by astrology: 'non igitur volo negare quin celum et partes eius agant per lucem etc. hic inferius, sed volo negare quod hoc scire dearticulate sit possibile, scilicet usque Sortem vel pedem Sortis vel digitum'.[12]

The only part of astral forecasts which Oresme finds noteworthy is slight and scarcely scientific in any serious sense. The examples he gives of what is acceptable are nothing more than the fruits of ordinary daily experience, such as the observable relationship between changes in the weather and changes in certain mental and physical conditions.[13] And even these are conjectural and far from rigorously predictable. So much so that when Oresme discusses weather forecasting in relation to comets, he uses *persuaderi* instead of *sciri*: 'dico quod per cometas aliqua generalia possunt persuaderi tam ratione cause quam ratione concurrentie. Est enim cometa ex materia calida et sicca et igitur aerem et alia illa materia habet calefacere et ventos causare et inde que sequuntur ad ventum'.[14]

The high level of vagueness and the uncertainty of such forecasts are due to

[11] F. 46va.

[12] F. 46vb. Not even Moon's action on tides is granted: 'unde vides quod fluxus maris augetur et diminuitur secundum alias dispositiones inferiores quam secundum lunam solum; non enim quodlibet mare fluit aut refluit unum sicut aliud . . . Ex quo demonstrative patet quod qui nescit dispositiones materiales et inferiores passorum nescit etiam qualem effectum planete inducent', f. 46vb.

[13] 'Dico igitur quod ex agregatione planetarum et lucis earum quandoque mutantur qualitates in elementis et etiam in mixtis . . . et ideo quidam quandoque dolent humores aut nervos ante coniunctionem vel post vel prope, et precipue quando aer mutatur in qualitatibus oppositis illis morbis vel etc.; quia non solum in coniunctione solis et lune vel alterius planete, sed multi dolent quando tempus mutatur', f. 46vb.

[14] F. 46va–b.

the complete lack of precise rules – or rather, the lack of a specific theoretical model which bestow on each terrestrial phenomemon its own astral cause. So, even if the intimate connection of earth and heavens is granted, the impossibility of identifying the essential elements of that connection means that any claims for divination are merely question-begging and, at the practical level, inconsequential.[15]

In order to set astrology apart from acknowledged sciences, and consequently to refuse its claims to be one of them, Oresme is even ready to grant some privilege to divination. In replying to the fifth *questio*,[16] his conviction in astrology's complete lack of authority seems to waver: he looks likely in fact to admit astrology's antiquity and its at least original reliability: 'dico etiam quod homines boni ingenii et totaliter dediti speculationi et philosophie, precipue antiquitus, ultimo studebant in astrologia, scilicet in parte illa de qua est dictum etc. Et erant tales homines summe morigerati et fundati in philosophia tam naturali quam morali quam etiam generali et speciali; et igitur uti consilio talium prudentium erat bene recommendandum qui per bonas coniecturas tam in celo quam impressionibus metheorum quam elementis quam moribus hominum quam etiam dispositione inferiorum multorum sciebant previdere et provide agere. Non igitur erat intentio Aristotelis intelligere quod homines uterentur consilio talium quales sunt qui modo aliquid scire de astrologia mentiuntur imprudenter nec assentire talibus adventiciis penitus rationi contradicentibus et veritati'.[17] Oresme however resorts to this rhetorical expedient of confronting ancient wisdom with modern divination only to strengthen the unreliability and the poorness of contemporary astrologers.

The definition of heavenly influences on the sublunary world in the eighth *questio* ('Quomodo celum dicitur causa istorum inferiorum et in quo genere cause'[18]) preserves a degree of vagueness about the third factor – the form of the heavens[19] – which is generally classed as *influentia* in the *Questio contra divinatores horoscopios*. In both the *Quodlibeta* and the *Questio*, however, Oresme quickly drops this third factor and refers only to motion and light. This lets him confute any claims for divination, while at the same time

15 'Sciuntur etiam coniunctiones planetarum et solis et lune ex quibus secundum dispositiones recipientium et passorum, et precipue in elementis, fiunt quedam transmutationes de calido ad frigidum et de sicco ad humidum vel e converso, nec super hoc habetur certa regula, et hoc patet ad sensum', f. 46vb. The charge against astrology for not having a rigorous and specific theory of astral influence is rather dangerous, on account of Oresme's sceptic attitude. His reply to such a criticism from diviners is in fact somewhat weak and conventional: 'respondeo quod nullus homo sufficit ponere nec conclusionem nec regulam in astris experiri propter (ms. om.) brevitatem vite (ms. om.) et diversitatem requisitorum . . . Et si dicas quod ita diceretur de theologia, philosophia morali et precipue medicina, respondeo quod non, quia in illis sunt regule universales vere et bene probate, quas homo diligens potest contrahere ad particulares', f. 47va.

16 F. 47vb.

17 F. 47ra. 'Unde dico si Aristoteles videtur laudare et recommendare considerationem de celo et astris, hoc autem est quantum ad motum et magnitudinem et paucas virtutes ut lucem etc . . ., ut expresse ponit XI *Animalium*, ut patet etiam ex suis dictis in *De celo*', ff. 46vb–47ra.

18 Ff. 47vb–48ra.

19 'Respondeo igitur quod per motum et lucem seu lumen celum agit hic inferius et etiam per suam formam, si formam debeat habere, seu per se (ms. om.) si ipsummet sit forma', f. 47vb.

retaining the notion of astral action, though in such a way that it can not be used as a foundation for any system as broad and yet specific as astrology.

Oresme's choice is crucial for the history of astrological thought. Apart from establishing a prestigious 'auctoritas', it offered a new approach to one of the main issues in medieval physics. We all know how the renewed natural philosophy of twelfth century was accompanied by a new attention to astral causality.[20] This attention was certainly not discouraged when newly available Aristotelian works on physics seemed to corroborate the hypothesis. Let one example suffice. In Albertus Magnus' *Explanatio* to the second book of *De generatione et corruptione* Aristotle is read as clearly favouring a non-generic relationship between the superlunary and sublunary spheres: 'est autem periodus mensura quae ex circulo coelesti imprimitur vel influitur rei causatae a circulo in inferioribus . . . Notandum quod nullatenus solus accessus solis et recessus facit periodum . . . sed periodum facit relatio ascendentis signi super horizontem ad omnia alia ista signa circuli cum suis stellis et planetis in hora conceptus vel nativitatis rei inferioris, quae creatur vel concausatur a circulo celesti'.[21]

The astrologising reading of this passage – and of others mainly from *Metereologica* and *Physica* – became a topos of scholastic commentary and teaching, and lent authoritative support to astrology's scientific claims. Perhaps this is why Oresme had to admit some degree of heavenly influence – albeit so indeterminate as to exclude divination – and did so on the basis of the most rigorous possible analysis of the causal process. Indeed, this analysis is the main strength of Oresme's polemic.[22]

[20] Bibliography on this topic is wide. I just recall some of the most important items: M. T. d'Alverny, 'Astrologues et théologiens au XIIe siècle' in *Mélanges offerts a M. D. Chenu Maitre en Théologie*, Paris, Vrin 1967 (Bibliothèque Thomiste, XXXVII), pp. 31–50 and 'Abélard et l'astrologie' in *Pierre Abélard, Pierre le Vénerable. Les courants philosophiques, littéraires en occident au milieu du XIIe siècle*, Paris, CNRS 1975, pp. 611–630; T. Gregory, *Anima Mundi. La filosofia di Guglielmo di Conches e la Scuola di Chartres*, Firenze, Sansoni 1955 (Pubblicazioni dell'Istituto di Filosofia dell'Università di Roma, III), 'L'idea di natura nella filosofia medievale prima dell'ingresso della Fisica di Aristotle. Secolo XII' in *La Filosofia della Natura nel Medioevo. Atti del III Congresso Internazionale di Filosofia Medievale. Passo della Mendola-Trento, 31 Agosto–5 Settembre 1964*, Milano, Vita e pensiero 1966, pp. 27–65 and 'La nouvelle idée de nature et de savoir scientifique au XIIe siècle' in *The Cultural Context of Medieval Learning*, ed. by J. E. Murdoch and E. D. Sylla, Dordrecht, Reidel 1975, pp. 193–218.

[21] Alberti Magni, *De generatione et corruptione*, ed. P. Hossfeld, *Opera omnia*, V., 2, Muenster, Aeschendorff 1980, p. 206.

[22] In his *Questiones super de generatione et corruptione* when discussing the problem 'utrum quodlibet corruptibile habeat aliquam determinatam peryodum sue durationis', though perfectly aware of the astrological interpretation of the topic ('oppositum patet per Aristotelem capitulo penultimo, ubi dicit quod omnia terminantur quodam peryodo in duratione, et dicitur "peryodus" certum tempus et determinatum per quod corruptibile durat. Et dicitur a "peri" quod est certum, et "odos", quod est circularis, quia illud tempus est determinatum per circulationes certitudinales ipsius celi'), Oresme proposes a different solution. While dedicating only a fleeting remark to the supporters of the astrological reading of Aristotle's text, he insists on the physical distinctive features of natural bodies and on proximate agents of corruption. S. Caroti, *'Peryodus' e limiti di durata nelle 'Questiones super de generatione et corruptione' di Nicole Oresme*, Cahiers du Séminaire d'épistemologie et d'histoire des sciences, Université de Nice, XIX (1985).

For Oresme the kind of causality most appropriate to so generic and undifferentiated an action is efficient causality: 'dico igitur quod in genere cause efficientis celum dicitur causa'.[23] Other classes of cause, in fact, require a more scrupulous investigation of the relation between agent and patient, especially the 'formal' relationship.

Not bothering with a detailed confutation Oresme discusses it thus: 'fuerunt tamen multi qui posuerunt Deum, scilicet ultimum motorem et principaliorem formam omnium, esse formam totius mundi, sicut et anima intellectiva ponitur anima totius corporis, et sic non solum esset in genere cause efficientis sed etiam et formalis'.[24] The soul/body relationship paradigm is improper because of the evident ontological difference between soul and heaven: 'verum est quod exemplum de corde etc. non est totaliter simile etc., quia celum est corpus simplex non habens qualitates peregrinas. Et Deus forma simplicissima et etiam ab eo plus et per alium modum dependet celum ab eo quam cor ab anima etc.'.[25]

Even if, in the former quotation, Oresme seems to refer mainly to divine causality, the nod towards heavenly movers and the soul/body relationship hints at a persistance from earlier centuries of notions dear to astrology: the twelfth century *anima mundi* and the thirteenth century's idea of the spheres with their movers. It is reasonable to assume that these are not casual hints, but intentional – though in some way subtle – references to the most philosophically evolved forms which nourished medieval astrology.

Oresme evidently wanted to warn philosophers to be careful when analysing the connection between primary and secondary causes. In the perennial vagueness which envelops this philosophical conundrum Oresme detects in fact the germs of a doctrinal confusion which could only favour astrology. It was normal practice in commenting on Aristotle, even for writer remote from astrology, to approach this problem in an entirely unsystematic fashion, without making any of the necessary distinctions about the roles of divine intervention and celestial influences. The fact that Aristotle's natural philosophy recognised astral action – even if only in a generic and illustrative way, and not as a genuinely speculative position – nourished the possibility of an alliance between astrology and physics, an alliance whose consequences Oresme finds particularly dangerous.

The use of two typical astrological arguments in commenting on Aristotle's natural works bears witness to the connection between astrology and natural philosophy: the consequences of possible arrest of the motions of the heavenly bodies on generation and corruption, and the relationship between

[23] F. 47vb.

[24] F. 47vb. Oresme refers to a similar position in his *Questiones super de celo*: 'probatur primo ex bonitate ipsius Dei quia secundum philosophos ipse Deus est anima et vita totius mundi non per informationem sed per continuam influentiam et conservationem et efficientiam, sicud sol conservat lumen. Et adhuc magis propter hoc dicitur primo huius quod ab ipso dependet celum et tota natura et XII *Metaphysice* quod ipse est agens totum et secundum hoc est ubique, modo ipse est optimus et mundus est optimus effectus corporalis ipsius Dei', C. Kren, *The 'Questiones super de celo' of Nicole Oresme*, Ph.D. Diss., University of Wisconsin 1965, vol. I, p. 59.

[25] F. 48ra.

astral and terrestrial qualities and powers. Oresme debates both topics in the *questiones* of the *Quodlibeta* where he attacks divination. In the first, well known for having being condemned in Paris by Etienne Tempier,[26] the direct subordination of physical events to astral motions is unequivocally stated.

Oresme's open approval is for the persistence of generation and corruption; his choice rests mainly on four arguments. In the first he clearly distinguishes corruption from annihilation (*annichilatio*), the latter being only God's prerogative. He can so easily state the inconsistency of astrologers' position: even imagining a sudden arrest of heavenly motions they must admit the persistence at least of corruption (and of generation as well, for the necessary connection between both), the hypothesis of annihilation being untenable.[27] In the second and fourth arguments – resting upon his definition of astral action – Oresme notices that the absence of one of the basic agents (motion and light) can be at least partially counterbalanced by the other.[28]

The second topic too ('utrum in celo sunt tot virtutes seu influentie specie distincte quot (ms.: quod) sunt hic inferius effectus specifice distinctos'[29]) is clearly rooted in astrological doctrines. Oresme denies this close relationship between elementary and heavenly qualities, mostly referring in his eleven arguments to acknowledged heaven's non-elemental nature, sensitively attested by circular motion. The first and the second arguments are, however, less conventional. In the first, where the non-existence of such a relationship is inferred from the necessity of avoiding an actual infinite, Oresme appeals to the hypotheses of world's eternity and of incommensurability of heavenly motions.[30] Even granting the birth of a new species on earth at every new astral conjunction, the probable incommensurability of heavenly bodies' motion leads to the untenable admission of an infinity of new species.[31]

[26] 'Est etiam articulus Parisius condempnatus scilicet dicere quod celo quiescente ignis non ageret in stupa', f. 48ra–b. R. Hissette, *Enquête sur les 219 articles condamnés à Paris le 7 mars 1277*, Publ. Univ.-Vander Oyer, Louvain-Paris 1977 (Philosophes Médiévaux, 22), pp. 142–143.

[27] 'Et si dicatur corruptio fieret, sed non generatio, contra si homo corrumperetur generaretur cadaver; nec valet dicere quod annichilaretur simpliciter, quia annichilatio convenit ipsi Deo', f. 48ra. Oresme criticizes as well the avicennian *dator formarum* for its evident astrological consequences, or, at least for the common astrological interpretation: 'ex quo concludo quod non oportet recurrere ad datorem formarum per modum quem ponunt multi, bene autem concedo quod ad Deum', f. 48ra.

[28] The possibility of the state of rest of heavenly bodies, untenable by nature, depends on God's absolute power: 'aliqui autem dicent quod implicat contradictionem quod celum quiescat, ideo hoc posito sequuntur et contradictoria. Sed clarum est quod Deus qui ipsum fecit potest ipsum facere quiescere, nec motus localis sic est sibi qualitas essentialis quin etc.; nec valet quod naturaliter implicat, sed non supernaturaliter', f. 48ra.

[29] F. 39rb.

[30] 'Et si dicitur ita possunt esse in celo sicut individua in terra aut elementis, respondeo quod falsum est quia sic de facto essent ibi infinite, cum de facto fiunt finita individua, ponendo mundum eternum', f. 48rb.

[31] 'Et ita diceretur de speciebus, quia possibile est, sicut alibi scilicet super *De celo* est probatum, quod si sint in celo aliqui motus incommensurabiles quod infinite fuerunt coniunctiones et erunt, quarum una non erit similis alteri, et sic etiam nove species', f. 48rb. E. Grant, *Nicole Oresme and the Kinematics of Circular Motion. 'Tractatus de commensurabilitate vel incommensu rabilitate motuum celi'*, The University of Wisconsin Press, Madison, Milwaukee, London 1971.

In the second argument Oresme reminds that differences in the effects do not necessarily involve a plurality of causes.[32] This is the strongest and most pertinent argument in Oresme's debate; the others, in fact, rely upon hypotheses (such as the incommensurability of heavenly motions and the eternity of world) or the authoritativeness of Aristotle's statements. Oresme is perfectly conscious of the inadequacy of his arguing, so much so that he refers to it as to persuasions ('persuasiones'), just to emphasize its conjectural nature: 'et quamvis predicte persuasiones non demonstrant quin in celo sint plures qualitates et virtutes, tamen sufficienter persuadent quod in celo non sunt tot quot hic inferius species vel individua vel saltem quod non est necesse quod in celo sit una virtus que causat muscas et alia que causat ranas et alia que etc., sed celum corpus simplex cum luce etc. mediantibus diversis coagentibus diversa potest etc. Ex quo sequitur quod qui nescit dispositiones intermedias et coagentia, ipse etiam nescit quis vel qualis effectus fiet'.[33]

As we have seen, much of the inspiration – if not the specific details – of Oresme's anti-astrological polemic came from his deep and genuine scepticism about humankind's ability to know enough about astronomical and physical factors. Despite possessing a philosophically well-developed theory of causality, research in physics, especially into the links between terrestrial change and celestial dynamics, was doomed to impotence, because there was no reliable data to which the theory could be applied. The proponents of astral divination, however, erred not only in claiming to be certain about data which, according to Oresme, are humanly unknowable: they erred also in exempting divination from the prescriptions of a specific theory of causality.

In the eleventh *questio* Oresme surveys the various kind of causality and he deals with the relation between primary and secondary causes – which is crucial, because astrology is fundamentally ambiguous on the subject. The diviners, in fact, claim precedence for celestial over elementary agents, but they do not investigate the relationship between them, nor do they consider other levels of causality. They banally affirm the superiority of celestial agency simply because they want claim precedence for their science over the others, like physics and methereology.

It is against this ambiguity that Oresme directs his discussion of the celebrated proposition: 'causa prima plus influit in effectum quam secunda' – a proposition which echoes the opening of the *De causis* and which was often invoked as a mainstay of astral influence. First Oresme shows that the kind of causality must be determined, so that final causes are not confused with material, formal with efficient. The primary/secundary cause nexus presupposes, moreover, a series of subordinate causes whose hierarchy is to be established according to rigidly fixed criteria: 'si igitur prima causa sit prima

[32] 'Secundo quia ab eodem agente simplici potest provenire diversa operatio (ms.: opera) mediantibus diversis, ideo non propter pluralitatem istorum inferiorum oportet in celo ponere etc.', f. 48rb.
[33] F. 48va.

quod per se, quod principalis, quod necessario requisita, quod alias causas dirigat et a nulla dirigatur vel quod alias producat et a nulla earum producatur (ms.: producitur), quod effectum conservet vel quod maneat in effectu, quod quasi idem est, quod prius incohet seu presupponatur et sit prior aliis in aliquo modo prioritatis et etiam secundum illam intendatur effectus, bonitas aut etc. tunc concedo quod illa plus influit in effectum quam secunda, et hoc adhuc in eodem genere quo dicitur sic causa prima, scilicet in genere materie aut efficientis vel etc. Aliter autem nego quod causa prima plus influit in effectum quam secunda'.[34] If the relationship between primary and secondary causes cannot be defined in these terms, the relation is impermissible and it is a nonsense to assert the priority of the primary cause. In the contrary case – that is, when we can identify a true series of subordinate causes – precisely the same criteria make it unlikely that the primary cause will have the simple precedence over the secondary cause which astrology wishes it to have.

In the twenty fourth *questio* Oresme looks at the same problem from another angle. He defines the modalities of action of the primary cause proper, that is of the universal cause: 'respondeo quod quantum est in prima causa universali equaliter influitur naturaliter loquendo, quia a causa perpetua et immutabili equaliter et uniformiter semper influitur etc.'[35] The modalities of the universal cause exclude the possibility of particular causes standing in subordinate relation to it, at least in the form already described. Oresme is explicit about this: 'et cum hoc stat quod multe possunt esse eque perfecte . . . et quamvis una sit perfectior alia, non propter hoc sequitur quod sibi subordinetur'.[36]

What we have here is an important stage in Oresme's argument: the possibility, or otherwise, of justly calling itself a science depends on the attribution of celestial influence to one or other of these two kinds of causality. If the priority of celestial action over elementary action can be taken back to that of the action of the primary cause in a series of subordinate causes, then astrology's attempt to investigate the innermost workings of terrestrial phenomena is highly worthwhile. If, by contrast, the priority of celestial action can be shown to be of the same kind of that of universal on particular causes then there is no foundation for the claim that all change in the elemental sphere be attributed to the heavens. Obviously Oresme chooses for the latter option, and not only for polemical motives. This option, in fact, is amply justified in the tradition which – as we shall see – identified astral influence with the *causa* or *natura universalis*.

Oresme's polemic goes on taking advantage from the obtained results: the general influence of the universal cause is received in different ways according to patients' inclinations; it cannot anyhow account for the changes of physical events, which depends ultimately on the various combinations of proximate agents.

[34] F. 49va.
[35] F. 59rb.
[36] F. 59rb.

One of the main features of Oresme's *Quodlibeta* is exactly his constant call for a closer inspection into causality's laws. Some *questiones* following the twelfth are consecrated to this topic; in one of them – the fifteenth – Oresme expresses his perplexities about conservative causes: 'adhuc aliter dico quod (ms. om.) per causas essentialiter ordinatas vel subordinatas non aliud intelligo nisi causas quarum una dependet ab alia et hoc in esse et productione aut saltem in productione. In esse et productione sicut lumen dependet a luce; in productione sicut calor manus dependet a calore ignis, vel sicut Sortes dependet a patre suo etc. Et de hoc dixi prius quod michi difficile est intelligere quomodo una res absoluta dependeat continue et conservetur postquam est producta ab alia, nisi sicut apodiamentum sustinet domum declivem'.[37]

It is just the relevance given to the uniqueness of every physical event that makes the regress to universal causes destitute of any philosophical sense. It requires, on the contrary, an examination as deep as possible of proximate agents.

It is easy to understand why Oresme insists in denouncing the absurdity of astrology's explanatory system. If the starting point of both Oresme and astrology was a specific event and there was even some consent in considering heaven as an universal cause, their actual explanations of terrestrial events were nevertheless diametrically opposed. As a matter of fact their consent on universal causality was purely nominal; it is, on the contrary, this very point that makes Oresme's and astrology's positions incompatible. Astrologers in fact stress the universality of heavenly bodies' action in order to maintain their superiority on proximate agents, while Oresme considers such an universality as an impediment to a serious research in the physical events.[38]

To be more incisive, Oresme's anti-astrological work is not restricted to an investigation of the philosophical bases of divination. It also analyses the operative aspects of astrology – elections, interrogations and so on. The passage from theory to practice reveals more closely what is astrology's preferred field of action, though its claims to explain the marvels of nature – and consequently to be a necessary complement for natural philosophy – were manifest since the beginning of *Quodlibeta*.

The ambition to create a necessary complement for physics conceals intentions of a very different kind. Just how different emerges when we

[37] F. 51rb–va.

[38] The complete negligence in considering the proximate causes is not exclusively astrology's distinctive feature; Oresme, as we have already mentioned, sharply criticizes the fideistic attitude of ascribing everything to God's power. One of the rare references to Antichrist's reign in the *Quodlibeta* is significatively connected with such an attitude: 'istam autem consequentiam ita prolixe deduco, quia videtur michi quod faciliter credere est et fuit causa destructionis philosophie naturalis et etiam in fide facit et faciet magna pericula et erit causa recipiendi Antichristum . . . si autem vellem omnia concedere scire vel esse que Deus potest et que facit, mira haberem concedere etc.; ymmo ex tali argumentatione periret tota philosophia, ymmo et theologia, quia sic dubitarem utrum ignis calefacit vel Deus et de articulis fidei, quia sic haberem dubitare utrum Deus faciet novos articulos', ff. 45va and 46ra.

scrutinise the kind of answers which are expected from diviners. For instance, elections and interrogations – certainly the most widespread practices – are exclusively or almost exclusively concerned with events which involve the will of either the interrogator or of those about whom the interrogation is made. Therefore even though lip-service is always paid to it, free will is violated in practice.

Oresme's twenty-ninth *questio* ('utrum sit aliqua hora melior alia ad incipiendum agere vel facere aliquid, ut equitare, domificare, novos pannos induere vel etc.'[39]) shows how this voluntary component of electional actions compels the diviners to make a dramatic choice. Either they radically mitigate the certainty of any judgement they make or they must admit that the event in question is not absolutely subject to freedom of will.[40]

Finally, there is no shortage of internal contradictions. Oresme recalls, for instance, the traditional charge of the useless of elections in presence of the horoscope,[41] and, regarding the latter, the perennial failure to take into consideration all its really significant elements.[42]

The thirtieth *questio* is concerned with astrological images. First we have a series of arguments which stress the straightforwardness of the heavens and the complete discontinuity between sculpted images and the effects claimed for them.[43] Oresme then gives his opponents the right to reply. The first counter argument betrays in a fascinating way Oresme's defensiveness and often confessed perplexity about existence of demons.[44] Here he invokes Augustine's authority in declaring astrological images to be of demonic provenance, but he is immediately taxed with incoherence about powers which should be ascribed to demons. This rebuke causes clearly some disappointement in Oresme, who justifies his reservations in such a way as to exclude the faintest shadow of heterodoxy.

39 F. 59rb.

40 'Et sic oportet quod omnia ponas necessario evenire aut quod iudicium tuum nihil valeat', f. 62vb.

41 'Secundo quia secundum divinatores iam in nativitate tua fuit ordinatum quid per totam vitam accidet tibi, igitur sive vadas sive non, sive induas tunicam sive non etc. ita accidet etc., nec valet dicere quod ex electione in illa hora potes impedire aut confirmare, ita quod si nativitas significaverit tibi malum et cum hoc in mala hora incipias, tunc peius accidet etc., hoc non valet, quia ex nativitate debes scire utrum tali hora eliget seu faciet illud quod non faciet', f. 62va.

42 We may resort to a very common example in the astrological litterature: when Socrates, walking to the market meets with an enemy who kills him; or, alternatively, meets with Plato and, taking advantage that the latter is drunk, gets a real bargain from him, the bad or the good fortune does not depend on Socrates' nativity but rather on that of the people's who begged him to go to the market, or even on that of Plato: 'igitur nullus potest precognosci effectus etc. nisi etiam omnium concurrentium et coagentium precognoscatur nativitas', f. 64rb.

43 'Secundo concesso quod in celo fiat talis virtus quod propter hoc, scilicet quod tunc in auro aut cera aut etc. si fiat talis figura influatur et inducatur talis virtus, est pure ficticium et sine aliqua ratione, quia si ratione figure, ita bene si fuisset facta per annum ante deberet tunc induci illa virtus in illa que tunc fit ymo adhuc melius in vere a natura figuratis sicut in vero leone vivo, si vis facere illas figuras. Et si dicitur quod ratione talis motus qui tunc fit, quero a te modum per quem ille motus iuvat quod illa virtus descendat de celo sicut quasi in instanti', f. 65vb.

44 S. Caroti, *'Mirabilia' scetticismo e filosofia della natura nei 'Quodlibeta' di Nicole Oresme*, Annali dell'Istituto e Museo di Storia della Scienza di Firenze, IX (1984), pp. 3–19.

Astrology's reply relies upon experience (confirming, supposedly, talismans' power), Ptolemy's and Haly's authority and finally on astrological explanation of Mose's turning Arons' rod into a serpent. Oresme emphasizes the danger of such an explanation, that takes a divine miracle back to the level of purely physical events. As far as authority is concerned, Oresme denies Ptolemy's paternity (as the author of the *Almagest*) for the astrological works circulating under his name: 'nec mireris me negasse Tholomeum, primo quia ille qui fuit philosophus et qui fecit *Almagesti* non fuit ille qui scripsit tales abusiones'.[45] It is much easier for him to refute the argument 'ex experientia', because it confirms the complete ineffectiveness of astrological images in healing, notwithstanding some avowal of the power of imagination.[46]

The conceptual weakness of astrology's rationale, and its enormous success at all levels, calls for an analysis which is not based only on philosophical considerations. A number of *questiones* – for instance the sixth and the thirtieth – move in this direction to approach the problem from a psychological point of view. In the sixth ('propter quid homines moti fuerunt ponere iudicia talia in astrologia si non sint vera et quid movit eos recurrere ad celum?'[47]) Oresme notes that the principal motive for turning to astrology is the individual's total ignorance about the more proximate causes of events. He elucidates his belief by resorting to the very dangerous comparison with people attributing erroneously everything to God's power.[48]

The response to the thirtieth *questio* is more precise. As well as insisting on the incompatibility between astrologers' answers and all kinds of canon and civil law, Oresme considers why the diviner should be embarassed when faced with narrowly specific questions. Clearly the first explanation is to be sought in the unfounded nature of the art to which the question is addressed. And any answer to it must be ambiguous enough to withstand multiple interpretations, if its futility is not to be revealed immediately.

Nor is financial gain the main motive for astrologers' determination to preserve their credibility at any cost. Indeed, their main motive is ambition and a thirst for vain glory, often nourished by easy successes at the expense of a gullible clientele. They in fact maintain their reputation by deception only, by ambiguous answers ('ad duos vultos') which play upon the

[45] F. 66rb.

[46] This *questio* ends with an interesting list of the most widespread supersticious practices: 'dico quarto quod (ms. om.) multos clericos et multos alios valentes invenires qui tibi iurarent se fuisse expertos aliqua que tamen sunt pura mendacia, sicut de geomantia, sicut de diebus positis in kalendario periculosis, sicut de dispositione anni quando kalende ianuarii incipit tali die, sicut quando legitur Evangelium Marci quod pluit, sicut in die veneris communiter fit mutatio plus quam etc., sicut de cruce que fit dum legitur passio Domini die Pasche, sicut etiam de annulo qui fit de denario primitus oblato die veneris et in die nativitatis, in prima missa et sicut quasi infinita que et catholice et naturaliter et astrologice sunt pura mendacia', f. 66rb.

[47] F. 39rb.

[48] 'Dico igitur quod ad hoc fuerunt moti primo et principaliter propter ignorantiam quia nesciebant causas propinquas. Et adhuc aliqui moderni precipue laici et multi clerici non fundati in philosophia omnen effectum non consuetum si videatur eis bonum immediate actribuunt Deo glorioso et si malum (ms.: malus) actribuunt immediate demonibus . . .', f. 47ra.

credibility of their customers, by pretending to divine secrets which they had already found out by other devious means, and finally by surreptitiously bringing about the very actions which they had pretended to predict.

Oresme hints also to a perverse constituent element in the questioner/diviner relationship. Vain curiosity about future events is actually rooted – besides in eagerness for money – in an anomalous physical constitution: 'quia communiter qui querunt et confidunt in eis sunt melancolici aut in aliquo miseri et timidi . . . quilibet enim ita ardenter cupit futura scire et precipue unde lucrum emergit, quod quamvis homo sciat oppositum illius quod querit . . . nihilominus querit nec potest vulgus propter hoc cessare talia libenter audire'.[49] Once astrology's scientific basis had been thoroughly undermined, arguments such as these are the most effective way of refuting an art which is a parasite on human frailty.

Oresme is fully competent to give us a faithful account of astrologers' backstage machinations, because, as he tells us on several occasions, he had studied their methods assiduously,[50] and in his youth had even attempted – unsuccessfully – to become one himself: 'nec in eis inveni veritatem et tamen ita diligenter studui et famosos in illis frequentavi, sicut forte ipsi qui me reprehendunt et si nihil scivi vel potui iudicia eorum intelligere sicut et alii layci et forte aliquantulum magis advertentes in eis cognovi eorum sophysticam responsionem . . . unde millesies consideravi horas electionis a multum reputatis et michi amicis et tam pro me quam pro aliis volui experiri, et in iuventute credebam et assentiebam sicut et multi, sed numquam potui veritatem invenire'.[51]

Even though one must always be very cautious about autobiographical reports, especially in medieval scholastic literature, there is no evident reason to distrust Oresme's youthful assent to astrology, all the more so since this avowal certainly cannot be used as a *captatio benevolentiae* in a work bitterly rebuking human frailty towards divination.

We find in the *Quodlibeta* echoes of less personal experiences connected with alleged magical or diabolical events. In addition to their vainglory and stupidity, Oresme has the courage to point out that confessions of demonic possessions are to a great extent extorted with torture.[52]

[49] F. 63ra–b.

[50] In the *explicit* of his *Questio contra divinatores horoscopios* Oresme avows to have studied astrology and tested its powers to devine future events: 'et sic finitur questio contra divinatores . . . quam non feci causa alicuius invidie nec causa apparentie, sed ut se corrigant et advertant, quos detinuit error devius quia sepe in astrologia studui et codices eorum revolvi et cum actoribus contuli et ad experiendum musavi', N. Oresme, *Questio*, p. 310.

[51] Ff. 47rb and 62vb. Oresme reports the unsuccessful attempt to give an astrological explanation to his inability: 'et cum de hoc loquerer, respondit quod a nativitate mea hoc contingebat quod in astrologia veritatem non invenirem; tunc dixi sibi: hoc non prius dixisti, cum tamen figuram nativitatis vobis (ms.: nobis) monstrassem', f. 62vb.

[52] 'Ad quintum respondeo quod multi per violentiam tormentorum confitentur que numquam fecerunt . . . et ego hoc vidi, quia de una dicebatur quod faciebat et quod ipsa fuerat confessa, et ego rogavi propositum quatenus promitteret me alloqui illam, quod michi concessit. Sed cum in presentia propositi et aliorum sim (ms.: sibi) loqutus etc. et petivissem etc. sicut timida effunditur: 'vere nescio quid dico nec quid dixi' et multa alia et quilibet percepit quod non esset (ms.: esse) nisi truffa etc. Dico etiam quod per tormenta fatemur etc.', ff. 45va and 70vb.

Throughout the *Quodlibeta* we find allusions to other astrological topics or to problems which astral divination regarded as its special province, such as monstrous births[53] and spontaneous generation.[54] There is, however, only a single mention of the topic which was to have a special importance in the history of astrology in thirteenth and fourteenth Centuries, and of which echoes are found later in the *Commentarii Conimbricenses*.[55] In the twenty-second *questio* – one of the lengthiest and most interesting – Oresme considers, among other things, the problem of the void and the counter-natural, though not violent, motion of certain bodies in order that a void should not be formed. During the discussion he observes that 'ista debent tractari quarto *Celi* et octavo *Physicorum*. Et sic patet quod frivolum est dicere quod hoc fiat a natura universali'.[56]

This reference to those who use the *natura universalis* to account for the counter-natural motion of bodies, in cases where natural motion would lead to the formation of a void, is by no means as extraneous to the topic of this paper as might be thought. And this for two reasons: first and foremost, it reveals that Oresme is scupulously careful to avoid any recourse to extra-natural causes; and second and more generally, it reveals a shift in philosophical lexicon which probably arose from Oresme anti-astrological polemics.

As befits his extreme caution when dealing with problems that do not have easily verifiable solutions, Oresme relies only upon conjectures, but in such a way that they can be traced directly back to Aristotelian natural philosophy. Since the void is a strongly disruptive factor in the universe, a general principle of harmony, sometimes called *ordinatio universi*, is employed to make sure that a void does not arise, *e.g.* through the upward motion of heavy bodies. Such motion, though counter-natural with respect to the place, is not violent, because the priority is to maintain the continuity of the universe's components. Oresme's brief reference to the 'frivolousness' of those who support the *natura universalis* is presumably a reference to a point of view he did not share. The *Quodlibeta* unfortunately are too generic for us

53 'Propter quid et a quo generantur monstra unius speciei in alia ut in porca quandoque canis et quandoque e converso vel utrum a constellatione, ut dicunt divinatores. Sed quod non patet ad sensum et hoc dicit Aristoteles XVIII *Animalium* quod in speciebus parientibus plures fetus sepius sunt monstra . . . secundo quia canis ut in pluribus generat canem et porcus porcum etc. igitur non a constellatione; consequentia est optima bene intelligenti et alteri non loquor', f. 41rb. See also S. Caroti, *'Mirabilia' e 'Monstra' nei 'Quodlibeta' di Nicole Oresme*, History and Philosophy of the Life Sciences, VI (1984), pp. 133–150.

54 Oresme bitterly censures those who support the possibility of spontaneous generation for perfect species such as human: 'tunc dico quod Avicenna in tractatu *De diluvio* contra Averroym solvit istam questionem dicendo quod homo et animalia perfecta possunt generari ab alio quam ab homine etc. et in alio quam in muliere etc. nullam tamen rationem efficacem ad hoc imponit, nisi quod ipse dicit quod celum et planete sic possunt disponi quod disponerent naturam elementorum vel mixtorum sic quod virtus maris et femine etc. Nescio autem si hoc sompniavit, quia debiliter hoc probat', f. 59rb.

55 *Commentariorum Collegii Conimbricensis S.I. in VIII libros 'Physicorum' Aristotelis*, IIa pars, Venetiis 1606, L.IV, cap. IX, art. 3, p. 88.

56 F. 56va–b. S. Caroti, *'Ordo universalis' e 'impetus' nei 'Quodlibeta' di Nicole Oresme*, Archives Internationales d'Histoire des Sciences, XXXIII (1983), pp. 219ff.

to be able to precisely identify Oresme's polemical target or targets, so we are restricted to straightforward suppositions.

Oresme limits the explanatory hypotheses of the *natura universalis* to two: identification with God – and this is Oresme's preferred solution, under the influence perhaps of Buridan – or with celestial influence, which is less defensible even than the first option.[57] Once again we encounter a retrogression in favour of remote causes due to an insufficient awareness of more proximate agents – the attitude which Oresme regards as pre-scientific and pre-philosophical.

Of the two possible solutions, we are more interested in the second, whose origins can be found since the first emergence into medieval culture of Aristotle's writings on natural philosophy. In his *Explanatio* of Aristotle's *Physics*, Albertus devoted a suitable *digressio* to the *natura universalis*, which he identified with the astrological *virtus peryodi* of the second book of *De generatione et corruptione*: 'et haec natura est proportio virtutis motuum coelestium, secundum quod sunt periodus una omnium naturarum particularium . . . Hoc etiam modo in via corruptionis praeter cursum naturae particularis est mors: eo quod particularis natura non movet ad illam, sed universalis, secundum rationem illam quae est mensura vitae per periodum, sicut melius explanabimus in secundo libro *De generatione et corruptione*: sic enim secundum cursum naturae universalis est successio in generabilibus et corruptibilibus secundum quod a superioribus diversimode movetur materia, cum tamen natura particularis non proprie intendat successionem, nisi in quantum tangitur a virtute naturae universalis'.[58]

Albertus' speculations certainly did not fall on deaf ears and it is quite likely that Oresme is addressing thinkers under this influence. Evidence for this supposition is provided by Pierre Duhem[59] and, more recently, by Edward Grant,[60] both of whom have studied the question of the void from the angle of the history of science. Grant in particular has drawn our attention to the solution proposed by Walter Burley – certainly not shared by Oresme. Burley defines in fact the *natura universalis* as superintending the harmony of the universe and preventing the formation of voids through astral influence:[61] 'ymmo videtur quod oportet dicere quod virtus regitiva

57 'Aliud a Deo non intelligo per naturam universalem et si dicas quod a celo adhuc esset magis frivolum', f. 56vb.

58 Alberti Magni, *Explanatio super Physicam Aristotelis, Opera omnia*, ed. A. Borgnet, vol. III, Paris, Vives 1890, pp. 101b–102a.

59 P. Duhem, *Le système du monde. Histoire des doctrines cosmologiques de Platon à Copernic*, VIII, Paris 1958, Hermann, pp. 121–168.

60 E. Grant, *Medieval Explanations and Interpretations of the Dictum that 'Nature Abhors a Vacuum'*, Traditio XXIX (1973), pp. 327–355 now in *Studies in Medieval Science and Natural Phulosophy*, Variorum Reprints, London 1981, VIII. See also E. Grant, *Much Ado about Nothing. Theories of Space and Vacuum from the Middle Ages to the Scientific Revolution*, Cambridge University Press, Cambridge (Mass.) 1981, pp. 67–70. For Bacon see M. Schramm, 'Roger Bacons Begriff von Naturgesetz' in *Die Renaissance der Wissenschaft im 12. Jahrhundert*, Artemis Verlag, Zürich 1981, pp. 197–209.

61 For Burley, who likes *agens superceleste, agens celeste, agens universale, virtus regitiva universi* better than *natura universalis* see E. Grant, *Medieval Explanations*, p. 334, n. 13.

universi vel aliquod agens superceleste regens ordinem universi faciat aquam moveri sursum ne fiat vacuum; et similiter impediat aliquando aquam moveri sursum ne fiat vacuum'.[62] This hypothesis relies on Burley's more general conviction of the superiority of celestial agents: 'quod autem agens celeste habet regere hec inferiora patet <per> Commentatorem VII *Methaphysice* commento 32, ubi dicit quod impossibile est ut agens corporeum transmutet materiam nisi mediantibus corporibus supercelestibus. Similiter primo *De celo* commento 24 dicit quod corpora celestia conservant elementa'.[63]

On the evidence of this quotation alone we can not tell whether Oresme is referring specifically to Burley. The restricted circulation of Burley's *Questiones super Physicam* could tend to rule him out. But a point of view similar to Burley's must have been fairly common – we even find echoes of it in John Buridan's *Questiones super de celo*, which Oresme certainly knew. Even though Buridan entrusts God with the task of maintaining the *ordinatio universi*[64] in the seventh *questio* of the fourth book, in the second *questio* of the same book he appears to credit celestial influence with a decisive role in maintaining the natural arrangement of the elements: 'debemus ymaginari a toto caelo unam inflentiam continuam usque ad centrum; tamen illa influentia prope caelum et remote habet aliam proprietatem et virtutem, et propter illam influentiam sic virtualiter diversificatam superius et inferius ordinant se gravia et levia in hoc mundo inferiori. Et non debet negari ex eo quod illam influentiam non percipimus sensibiliter, quia etiam non percipimus illam quae de magnete multiplicatur per medium usque ad ferrum, quae tamen est magnae virtutis'.[65]

To be sure, we cannot interpret this kind of recognition of astral influence as an explicit endorsement of astrology proper. But Oresme realised that divination derived much support from this kind of generic affirmation that terrestrial changes depended on celestial dynamics. The exhaustive analysis of causality in the *Questio contra divinatores horoscopios* and the precise definition of the primary agent in a series of subordinate causes in the *Quodlibeta*, are unequivocal proof of Oresme's conviction that the question of causality offered the best arena in which to defeat astral divination. This conviction was shared by Henry of Langenstein, whose *De habitudine causarum* and *De reductione effectuum* appear to take Oresme's programme to its logical terminus.

In the first *conclusio* of *De habitudine causarum* terrestrial effects not explicable in terms of natural elemental processes – including therefore the upwards movement of heavy bodies in order to prevent voids – are

[62] E. Grant, *Medieval Explanations*, p. 335, n. 16.

[63] E. Grant, *Medieval Explanations*, p. 336, n. 17.

[64] 'Et credo quod motus ad replendum locum ne sit vacuum, non est naturalis secundum appropriationem ad aliquam naturam corpoream specialem, sed est naturalis communiter omni corpori sine appropriatione; et hoc provenit a primo ordinatore totius universi et omnis naturae; ab illo enim omne corpus habet per suam naturam quod moveatur sive supra sive infra, ad replendum locum si corpus sibi contiguum auferatur', Johannis Buridani *Quaestiones super libris quatuor De celo et Mundo*, ed. E. A. Moody, The Mediaeval Academy of America, Cambridge (Mass.) 1942, p. 269 (L.IV, q.7).

[65] J. Buridani, *Quaestiones*, p. 250.

cathegorically excluded from the influence of heavenly bodies: 'ad hoc quod in aliquo passo aliquis effectus fiat vel dispositio preter naturam particularem non oportet aliquam influentialem qualitatem ab aliqua intelligentia, stella vel constellatione in ipsa hora ibi specialiter diffundi'.[66]

Even Henry does not completely rule out some degree of celestial influence over the sublunary world. Like Oresme, he tends to grant the heavens a general and constantly uniform influence, which can not be used to account for sublunary mutability.[67]

One of the most important achievements of this philosophical attempt to define the causal relationship between celestial and terrestrial agents is to be detected in an important change in the philosphical terminology. The *natura universalis* of Albertus and of Roger Bacon – rendered suspect by its more or less implicitly astrological implications, or at least by interpretations favourable to the notion of the dependence of the sublunary on the superlunary spheres – this *natura universalis* is replaced by Henry of Langenstein's more neutral *natura communis.*

He excluded from it any link whatsoever with the heavens, along lines similar to Oresme's refusal to consider celestial action as the equivalent to that of the primary cause in a series of subordinate causes: 'opus nature communis in istis inferioribus non est opus alicuius intelligentie nec alicuius concathenationis essentialis earum'.[68]

The distinction between *natura communis* and *natura particularis* is to be found entirely within the elements and therefore it provides a reliable guarantee against any attempt to smuggle in remote causes such as celestial influences. An echo of this conceptual shift is to be found in a contemporary of Henry's, Albert of Saxony. In his *Questiones super de celo* Albert discusses the counter-natural motion of bodies to prevent voids and he uses *natura communis* and *natura particularis* in such a way that he can be credited with a probable approval of Henry's proposal.[69]

[66] Henry of Langenstein, *De habitudine causarum*, concl. I, Firenze, Biblioteca Medicea Laurenziana, Ashb. 210, f. 145vb.

[67] 'Eadem stella et pars orbis semper eandem influencialem qualitatem retinet quam modo habet', Henry of Langenstein, *De habitudine causarum*, supp. IV, f. 145rb.

[68] Henry of Langenstein, *De habitudine causarum*, concl. 17, f. 153rb.

[69] Albert of Saxony, *Questiones super de celo*, L.I, q.1 (Antonius de Carchano, Pavia 1481, f.aIIIva): 'utrum cuilibet corpori simplici insit naturaliter tantum unus motus simplex . . . istam solutionem alii dicunt sub aliis verbis quod unius corporis simplicis solum est unus motus naturalis secundum speciem conveniens ei secundum naturam propriam; nihilominus nihil prohibet alicuius corporis simplicis secundum speciem esse plures motus distinctos secundum speciem, quorum unum esset ei naturalis secundum naturam propriam, alius autem esset ei naturalis secundum naturam seu inclinationem communem sibi et aliis'.

TOWARDS A POLITICAL ICONOLOGY OF THE COPERNICAN REVOLUTION

Keith Hutchison

1 Introduction

In this paper, I do not directly attack a question from the history of astrology proper. I seek rather to apply our tentative understanding of the currents of renaissance astrology and related schema of thought, to what is perhaps the central problem of sixteenth and seventeenth century astronomy. Why was the geostatic cosmology of Aristotle and Ptolemy abandoned in favour of the heliostatic cosmology of Copernicus and his successors? This, of course, is a much hackneyed topic: indeed, it has become a standard testing-ground for philosophical accounts of the rational criteria for evaluating competing scientific theories. It thus stands as an important target for those who wish to enquire, alternatively, how 'extra-rational' social processes affect the character of the cosmologies accepted in different societies, and it is this sociological target that I shall be aiming at here. I seek to show that attention to the political messages which renaissance observers drew from the heavens provides us with good grounds for thinking that changing conceptions of the way society should be ordered were instrumental in effecting the re-ordering of cosmology we know as the Copernican Revolution. It was – I suggest – the growing notion that society ought to be tightly organised around a central authority, instead of being loosely dominated by a dispersed nobility, that made heliostatic astronomy succeed.

To appreciate my argument, it is desirable to analyse the Copernican Revolution into what are widely seen as its four critical components: firstly, the abandonment of the Aristotelian distinction between the Heavens and the Earth; secondly, the recognition that the Tychonic system was the best version of the Ptolemaic arrangement of the planets; thirdly, the discovery that planets do not move on circular orbits but ellipses; and finally, the belief that it is the Sun rather than the Earth which sits immobile in space. Traditional 'rationalistic' accounts have no difficulty in dealing with the first three of these components,[1] but they make no headway with the fourth. For

[1] The fact that they have no difficulty does not mean however that a social cause is not in operation. I shall suggest below, for example, that the attractiveness of the Tychonic over all other forms of the Ptolemaic system is in fact a social phenomenon in rationalistic disguise, and I shall also observe that the demise of the Aristotelian doctrine that the Heavens are radically

they are faced with the peculiar fact that, prior to the end of the seventeenth century, there was a severe lack of evidence to support the idea of a moving Earth. Indeed, it is fairly widely accepted that important supporters of this idea, such as Kepler, Galileo, Descartes and Newton, had a strong commitment to the heliostatic vision of the Solar System which preceded the production of evidence which would support the belief. That is, they believed in, and pursued, the Copernican system on grounds which would not normally be viewed as 'scientific'. Hence extra-rational motivations were highly instrumental in generating the abandonment of geostatic cosmology. So our understanding of the Copernican Revolution – and of major scientific change in general – should be greatly enhanced if we can discover some of the potent 'non-scientific' reasons for insisting that the Earth was in motion.[2]

So much for where we are going. Now we must consider how we shall get there. To render my claims plausible, I shall attempt to establish the following four sub-theses:

a That political symbolism was attached to astronomy.

b That such symbolism was taken seriously, and was more than a mere mode of speech.

c That through this symbolism the Ptolemaic and Copernican systems represent different models of political organisation, the former medieval and decentralized, the latter baroque and highly centralized.

d That this symbolism was causally implicated in effecting the Copernican Revolution.

The first of these four theses is extremely easy to support. The second is less straight-forward, but appears beyond doubt. The last two are far more difficult to substantiate, and on these I can only make a tentative and indirect beginning: the reader who is awaiting a definitive argument is doomed to disappointment.

---

distinct from the Earth is likely to reflect the blurring of the distinction between aristocrat and commoner. It is quite plausible that rational arguments were devised both consciously and unconsciously to support cosmological attitudes which were ideologically useful or which, for any other reasons, were felt aesthetically appropriate.

[2] In presenting an argument that these reasons are political in character, I might seem to be coming very close to the radical claim that scientific theories are nothing but ideological artifacts. I do not accept this radical thesis, so it is worth noting in advance that my argument will only support it partially: I accept indeed that Kepler's elliptical orbits were adopted by the scientific community on the rational (as opposed to ideological) grounds that they agreed with observation; and I accept that good evidence severely discredited the Aristotelian notion that the Heavens were immutable. So my case only supports the weaker thesis that ideology becomes important in selecting for attention a subset of the totality of theories which meet the criteria of rationality. A stronger argument would be needed to insist that these criteria do not exist.

2 The Sun as a Political Symbol

By the time of the renaissance (as we shall eventually see), the idea that the Heavens as a whole symbolise political authority was a well-established chiché of European thought. For the purposes of the present paper, one portion of this symbolism is of particular importance, and that is the idea that the Sun is an especially appropriate metaphor for a King,[3] and vice versa, that a King is an especially appropriate metaphor for the Sun.[4] Another common, and closely related, image of royal power was the wheel from the chariot which the Sun-god Helios or Apollo traditionally drove across the sky. This 'cosmic wheel', sometimes identified with the wheel of fortune, was a pair of circles modelling the universe, connected by spoke-like radii representing the divisions of the zodiac.[5]

Various messages were conveyed by this symbolism. No doubt different observers interpreted it differently, but one central idea of great importance was the notion that a King is a force for order: he harmonises competing elements of the body corporate which, without his governance, would generate social chaos like the disorder produced by the four horses of a chariot which lacks a driver to hold th reins of power. Thus the Apollo-king was also represented as a harpist, harmonising the disparate strings of his instrument, as conductor of an orchestra, or identified with some solar hero such as Hercules or Michael who defeats the multi-headed dragon of social chaos. Shakespeare's Ulysses, for example, portrays the Sun as a source of temperance in the universe. Its astrological function, as moderator of the harshness of the numerous planetary influences, reflects the role a King is supposed to play in his Kingdom, restraining discordant excess on the part of the aristocracy.[6]

[3] Strictly speaking, it is not the Kingship which is at issue here, but superior political power of the type held by monarchs. Some of my evidence will refer to the asserting of such power on the part of non-kings.

[4] For some examples of special clarity, or special relevance to this paper, see: literature on Louis XIV, the 'Sun-King' *par excellence*, e.g. Mousnier, 'Le Roi-Soleil'; Hawkes, *Man, passim.*; Visconti, *Hours, passim.*; the 'Regimentsspiegel' triptych in the Swiss *Landesmuseum*, Zurich, discussed Gyr, *Zunfthistorien*, pp. 79–82; miniature zodiac wheel from the *Vatican Ptolemy*, as reprod on p. 89 (fig. 16), Rice, *Dark Ages*; Copernicus, *De Revolutionibus*, p. 50; Regiomontanus, as cited in Zinner, *Entstehung*, p. 160; Kepler, as quoted in trans. in Burtt, *Foundations*, p. 48, *Harmonies*, p. 1081, *Epitome*, p. 873; Brahe, *Recent phaenomena*, p. 259; Desaguliers, *System, passim.*; Harvey, *Movement*, pp. 3–4.

[5] See L'Orange, *Iconography, passim.*; Smith, *Dome, passim.* and *Architectural symbolism, passim.*, but esp. p. 89 n. 51; zodiac wheels, from *Vatican Ptolemy* (as in note 4), and from Neptune's chariot in the Piazza della Signoria, Florence; early 14c. wall-painting of King and wheel in Langthorpe tower, Northamptonshire, England, detail reprod. as fig. 61, Fowler, *Age*, p. 122; Prince Philip's Entry into Antwerp, ceremonial gateway of 1549 pictured in mid-16c. Woodcut reprod. as fig. 84 in Strong, *Splendour*, p. 105; 'Ommeganck Procession, 1615' by Denis van Alsloot, painting in Victoria and Albert Museum, London, detail reprod. as fig. 33, Trevor-Roper, *Age of Expansion*, p. 90; 'Alexander in Darius' tent', late 17c. Gobelins tapestry designed C. Le Brun, depicted in fig. 50, p. 112 of Viale, *Tapestries*.

[6] Quotation from Shakespeare, *Troilus and Cressida*, Act 1, Sc.III, lines 78–136. See also: Frankfort, *Before Philosophy*, p. 34; L'Orange, *Iconography, passim.*; Graves, *Greek Myths*, I, pp. 156–7, 232, 357, II, pp. 9–10, 13; Tacitus, *Annals*, [xii], p. 320; Dio Cassio, *History*, pp. 144–5; zodiac wheel from *Vatican Ptolemy* (as in note 4); Heninger, *Cosmographical Glass*,

> The heavens themselves, the planets, and this centre
> Observe degree, priority, and place,
> Insisture, course, proportion, season, form,
> Office, and custom, in all line of order:
> And therefore is the glorious planet Sol
> In noble eminence enthroned and sphered
> Amidst the other; whose medicinable eye
> Corrects the ill aspects of planets evil,
> And posts, like the commandment of a King,
> Sans check, to good and bad. But when the planets
> In evil mixture to disorder wander,
> What plagues and what portents! what mutiny!
> What raging of the sea! shaking of earth!!
> Commotion in the winds! frights, changes, horrors,
> Divert and crack, rend and deracinate
> The unity and married calm of states
> Quite from their fixture!

This symbolism was overt and public. A solar image was adopted as a personal motif by such rulers as Louix XIV, Richard II, and the Visconti family. The Sun, and other celestial images, occur frequently in renaissance pageantry. They are often chosen for the *impressae* decorating jousting shields used in the courtly tournaments which were revived in the sixteenth century, and were elaborately prominent in public festivities, such as that welcoming Katherine of Aragon to London at the beginning of the sixteenth century or the celebrations organised in Paris in 1581 by Henri III to mark the wedding of the Duc de Joyeuse. They were thus an integral part of the vast propaganda campaign mounted in the baroque age to magnify the respect due to kings.[7]

As has already been mentioned, and as closer scrutiny of the above examples would confirm, it was not simply the sun which had political and

---

p. 179; De Tolnay, *Michelangelo*, vol. 3, pp. 98–102; Veronese, ceiling fresco in Stanza dell' Olimpo, Villa Barbaro, Maser, Italy, reprod. on p. 15 of Pignatti, *Veronese*; Brown, *Palace*, pp. 156–61, 272–3; Utz, '*Hercules*', pp. 356–61; Viviani, 'Gallic Hercules'; 'Ommeganck Procession, 1615' as in note 5; Apollo in Royal Palace, Amsterdam, depicted Fremantle, *Town Hall*, fig. 46; Mossakowski, 'Copernicus's seal', (where however no political interpretation is placed upon the symbolism: that, however, a reference to kingship is implicit in Copernicus' choice of seal is clear from other cases of similar iconography, e.g., the Apollo representations in figures 14 and 15, Panofsky, *Hercules*). In making these remarks about the 'meaning' of the symbols I should not be taken as implying that these were the only meanings, or the original meanings. It is clear they were not. Throughout this paper, I shall make no attempt to embrace the range of alternative meanings or symbolisms associated with my subject-matter.

[7] See, for example: Visconti, *Hours, passim.*; Kantorowicz, *Two-Bodies*, p. 32 n. 18, fig. 4; Mousnier, 'Roi-Soleil'; Anglo, *Spectacle*, pp. 68–85; Yates, *Astraea*, pp. 65nl, 107–110, 154–66; 'Ommeganck Procession 1615', as in note 5; 'George Clifford' ca.1590, miniature by N. Hillyarde in Maritime Museum, Greenwich, England, reprod. as pl. v, Winter, *Miniatures*; Strong, 'Homage'. For an especially important example, see the 'heliocentric' cosmos depicted in Prince Philip's Entry, as in note 5.

social connotations attached to it: such meanings were attached to all celestial objects. On different occasions different inhabitants of the heavens were singled out for special emphasis. Thus kingship was sometimes represented by Jupiter or by Saturn, and it was also very commonly attached to the Moon. A good deal of dispute indeed had once been generated by the question whether it was the Moon or the Sun which correctly represented secular kingship. Proponents of Papal supremancy argued that it was the Pope who was symbolized by the Sun, so that the fact that the Moon received its light from the Sun, meant that a king only *reflected* papal power. Writing to the nobility of Tuscany at the end of the twelfth century, Pope Innocent III declares:

> Just as the founder of the universe established two great lights in the firmament of heaven, a greater one to preside over the day and a lesser to preside over the night, so too in the firmament of the universal church, which is signified by the word heaven, he instituted two great dignities, a greater one to preside over souls, as if over day and a lesser one to preside over bodies as if over night. These are the pontifical authority and the royal power. Now just as the Moon derives its light from the sun and is indeed lower than it in quantity and quality, in position and in power, so too the royal power derives the splendour of its dignity from the pontifical authority . . .

In his *Monarchia*, Dante opposes this argument,[8]

> . . . although the Moon receives its fullness of light from the sun alone, it does not follow that the Moon is derived from the sun. Thus it should be recognized that the existence of the Moon is one thing, its powers another and its operation yet another. The Moon does not in any way depend upon the sun – as far as its existence is concerned, nor as far as its powers are concerned, nor with regard to its operation as such. Because its movement arises from its own operation, its influence from the power of its own rays; it is even the source of some of its own light, as is obvious during its eclipse. But in order to operate more effectively and powerfully it receives something from the sun, i.e. abundant light; having received this, its power is increased. Similarly I say that temporal government does not owe its existence to the spiritual government, nor its power (which constitutes its authority), nor even its operation as such – though it certainly receives from the spiritual government the energy to operate more powerfully, by the light of

[8] For the Sun-Moon dispute, see: Smith, *Architectural symbolism*, pp. 102–4; Pope Innocent III, letter to nobles of Tuscany; Ullmann, *Papacy*, p. 86, Kantorowicz, *Two-Bodies*, p. 460; Dante, *Monarchy*, pp. 70–1; Newton, *Theological manuscripts*, pp. 121–3. For Saturn as king, see Alan of Lille, *Plaint*, p. 83; for Jupiter, see fig. 50, Seznec, *Survival*, p. 138, from Mantegna's *Tarocchi*. In Roman mythology Saturn had been supreme ruler of the universe during the Golden Age, before Jupiter's coup ushered in the Age of Silver: see, e.g., Ovid, *Metamorphoses*, p. 32 [= I.113–115].

grace which God infuses into it in heaven and which is dispensed to it on earth by the Supreme Pontiff.

3 The Metaphor as Divine Rhetoric

This dispute is very instructive. For the fact that it was deemed reasonable to disagree about the symbolism, indicates that the symbolism was seen as much more than a mere figure of speech. One cannot argue about the correctness of a pure metaphor, only about its aptness. In this case however, it seems to have been something like the truth of the symbolism which was in dispute. For the metaphor was seen as a divine communication and the problem of detecting the correct symbolism was the problem of decoding God's instructions to mankind.[9] *Genesis* tells us that God placed two great lights in the sky, and this statement – like all Biblical assertions – was to be interpreted at a number of levels. It was both a description of astronomical fact *and*, at an appropriate level of allegory, a description of political fact. It was God's way of telling humanity that it was subject to a dual political authority, with one component of that authority, the secular king, subservient to the other component, the Papacy.

This allegorical approach to the physical world was much promoted by Christian Platonism. To Platonists, the sensible world is a defective illusion. What we understand about it is at best only the shadow of truth, and behind the phenomena of the sensible world lie arrays of far more important truths about ultimate reality. Christians had traditionally believed in a similar dichotomy between the everyday world and permanent reality, as well as in the notion of a 'mystery', a fact which could not be grasped through normal human cognition, but which might be partially understood through various supernormal processes. To Christian Platonists, the physical universe could readily become a guide to these mysteries. By finding the correct allegorical interpretation of facts about nature, one could increase one's understanding of matters which were far more obscure. Renaissance symbolism is thus much more than a literary device. As God's own rhetoric, it becomes a guide to the conduct of life.

A good illustration of this idea that the natural world functions as a symbol for deeper mysteries is provided by alchemy. It was, in fact, commonplace to interpret alchemical processes as parallelling other processes in the universe. While some alchemists may have been searching for material gold, many viewed themselves as seeking spiritual gold. They believed that God arranged the properties of chemicals so that by observing the reactions which took place in his laboratory, the successful adept would have various mysteries such as the resurrection or the trinity disclosed to him.[10]

[9] Gombrich, 'Icones Symbolicae'; Praz, *Studies*, pp. 18–21; Mazzeo, *Studies*, pp. 52–9, *Medieval tradition, passim.*, esp. pp. 133–73; Chenu, *Man, nature and society*, pp. 99–145; Mâle, *Gothic image*, pp. 15–22, 29–34.

[10] Eliade, *Forge and Crucible*, pp. 149–52; Needham, *Science and Civilization*, V(5), pp. 1–20.

Another illustration of this same attitude is provided by the microcosm-macrocosm doctrine. According to this widely held view, God had designed the universe so that there were close analogies between its various parts. The structure of the heavens was supposed to immitate that of the terrestial world, the interior of the human body, the distribution of the virtues, the array of metals, etc. etc. So from the relationships inside our body, and relationships in the cosmos, we can discover the relationships which do exist, or perhaps should exist, between the various elements of human society. So the study of cosmology, or physiology, became laden with political significance. If we understand how the organs relate to each other and the rest of the body, or the stars relate to each other and the rest of the cosmos, we will have decoded divine instructions for the way the components of society should inter-relate.[11]

Hence there are good reasons for suspecting that the political symbolism attached to astronomy was intended to be taken seriously. It was certainly colourful language, but it was not just colourful language. If this is accepted, then important questions arise. What did this iconography portray? What political messages were conveyed to renaissance society by the symbolism attached to the stars? What political tensions in that society did its 'decodings' of this symbolism represent?

4 *The Cosmos as a Political Symbol*

At the core of the political meanings attached to astronomy seems to be the idea, deeply entrenched by the time of the renaissance, that the Heavens were significantly different in kind from the Earth, one above and the other below. This idea had been formalised as one of the fundamental tenets of Aristotelian philosophy, but the distinction between the sky and the Earth is a much wider doctrine, common to many mythologies and philosophies. Furthermore, by this same time, the idea had also become deeply entrenched that the Heavens symbolise rulers and their authority, while the Earth below symbolises the common people. So the physical distinction between Heaven and Earth would seem to represent the social distinction between patricians and plebeians. Thus Pico della Mirandola writes, in his commentary on *Genesis*:

[11] For examples of these analogies, using variously the body and different cosmologies, see: Livy, *Early History*, pp. 125–6 (= II.32); John of Salisbury, *Polycraticus, passim.* esp. p. 245; Shakespeare, *Coriolanus, passim.*; 'Regimentsspiegel' as in note 4; Desaguliers, *System*; Newton, *Theological manuscripts*, pp. 119–126; Prince Philip's Entry, as in note 5; Lewis, *Discarded image*, p. 74; 'Sphaera civitatis' engraving from J. Case, *Sphaera civitatis*, 1588 as reprod. fig. 9c, Yates, *Astraea*; Pope Boniface VIII, *Unam sanctum*; 'Otto II in Majesty', miniature from Aachen Gospels, ca.975, reprod. as fig. 5 in Kantorowicz, *Two-Bodies*, and discussed on pp. 61–78; 'Rudolf's army swears fealty 1611', illustration from von Chadow, *Reiss-Buch*, reprod. on p. 133, Dickens, *Courts*. For the triple analogy between body, cosmos and society, see: Chalcidius, *Commentary*, CCXXXI–III; Bernardus Silvestris, *Cosmographia*, pp. 121–7; Alan of Lille, *Plaint*, pp. 118–23; Harvey, *Movement*, pp. 3–4.

> After discussing the celestial and elemental spheres, and thus the whole of the physical universe, it remained for Moses to speak of the inhabitants and citizens of this universal city, not only those in the Heavens, whom he had to discuss first, as though the senate and prefects of the city, as it were, but also those on Earth, as though the plebeians and the people.

One might suppose that this idea could not possibly survive the demise of Aristotelian philosophy. For the Copernican theory undoubtedly undermines its implicit presuppositions. Newton however provides us with clear evidence that the doctrine did not disappear promptly. For not only does he endorse this 'mystical . . . analogy', but he treats it as one of the undisputed facts about the world which can be harnessed to help solve the problem of Biblical interpretation. To Newton, the Biblical prophets were conveying political messages, and we could derive assistance in decyphering these messages from the political messages embedded in the physical cosmos. 'The whole world natural consisting of Heaven and Earth', he wrote,[12]

> signifies the whole world politic consisting of thrones and people . . . and the things in that world signify the analogous things in this. For the Heavens with the things therin signify thrones and dignities and those that enjoy them, and the Earth with all the things therein the inferior people, and the lowest parts of the Earth called Hades or Hell, the lowest and most miserable part of the people . . . Great Earthquakes and the shaking of Heaven and Earth [thus stand for] the shaking of Kingdoms so as to overthrow them.

As this example indicates, political symbolism which refers only vaguely to the fact that the sky is distinct from the ground is unlikely to have been critical to the abandonment of Ptolemaic astronomy. To perceive the political resonances of the Copernican theory, we need to advance to the *details* of the Heavens, and especially the question of the arrangement of the planets.

The fact that these were distinguished from the fixed stars in Aristotelian cosmology meant that the Heavens themselves were thought of as a composite, with two distinct regions, one located in an intermediary position between the Heavens at the top and the Earth below. The material universe was thus divided into three important components, and these were held to reflect both the tripartite division of society into royalty, nobility, and commoners, and the division of the body into head, chest, and abdomen. Just as the heart in the chest governed the desires in the abdomen on behalf of the mind in the head, so too did the planets in the sky administer the Earth below on behalf of God in Heaven, and in the same fashion, the nobles were seen as charged with the task of applying the policies of the ruler in his citadel to the lower orders of society. In his *Commentary* on the *Timaeus*, Chalcidius claims that Plato draws:

12 Mirandola, *Heptaplus*, p. 132; Newton, *Theological manuscripts*, pp. 120–1.

> a parallel between the capacities of one man and the noble ordering of a state and people . . . The principal men of the city, being the most prudent and wisest should live in the loftier areas of the city; and below them, the young military placed under arms; to these he subordinates mechanics and the common people. So the first, being wise, rule; the military act and execute; and the common people provide suitable and useful service. We see that the soul is ordered in this way. Its rational part being the wisest occupies the most eminent part, the citadel, as it were, of the whole body; its energetic part, the military youth, remains in the encampment of the heart; the common people and mechanics, which are cupidity and lust, are naturally removed to the lower hidden parts.

These

> parts of man's body follow the arrangement of the body of the universe. Therefore, the world and the world soul are so ordered that the summit is measured out for the celestial beings, and adjacent to these are the divine powers called angels and demons, and the Earth for terrestrial beings. The celestial powers command, the angelic execute, and the terrestrials are governed: the first occupy the highest place, the second the middle; and those who are governed the lowest place. Accordingly also, there is in man, something princely; there is also something in the middle position; and some third thing in the lowest position: the highest which rules; the middle which acts; the third which is governed and directed. Therefore the soul rules, the energy seated in his breast executes, and the rest as far as the genitals and below is governed and regulated.

Such notions survived right through to the renaissance. We find exactly the same ideas, extended to refer explicitly to the role of the planets, in Caxton's translation of Lull's *Book of Chivalry*:[13]

> Unto the praysynge and dyuyne glorye of God, which is lord and souerayne kynge above & ouer alle thynges celestyal and worldly, we begynne this book of the ordre of chyualry; for to shewe that to the sygnefyauance of God, the prynce almyghty which seygnoryeth above the seven planettes that make the cours celestyal, and haue power and seygnorye in gouernynge and ordeynynge the bodyes terrestre and erthely, that in lykewyse owen the kynges, prynces and grete lordes to haue puyssaunce and seygnorye vpon the Kynghtes, and the Knyghtes by symylytude oughten to have powere and dominacion ouer the moyen peple.

[13] Chalcidius, *Commentary*, pp. 246–7 [= CCXXXIII, CCXXXII], trans. A. Blake. Lull, *Chivalry*, first page of text. For other similar examples, see: Freemantle, *Town Hall*, pp. 45–6 (for planets representing the components of government); 'Otto II in Majesty', as in note 10; symbolism in Desaguliers 'System'; Heninger, *Cosmographical Glass*, p. 153.

In the imagery we have just been perusing, a loose analogy is drawn between God, King, Heaven, stars, and brain, and this group of objects is contrasted with another group of analogues, heart, angels, planets, and aristocracy. In other varieties of celestial symbolism, the *difference* between God and King is attended to by placing the King *among* the planets, as one of the intermediaries in applying divine authority to the common people. We have already met examples of this in opening discussion above of the identification of the King with the Sun, Moon, Jupiter or Saturn. In physiological terms, this idea was expressed by the very common identification of the King with the heart instead of the brain.[14]

Implicit in such symbolism is the idea that a King's relationship to his political environment parallels the heart's relationship to its organic environment, and these in turn parallel the Sun's relationship to its cosmical environment. All in particular are shown to be accompanied by a retinue. Thus Copernicus describes the Sun as 'seated on a royal throne govern[ing] his household of Stars as they circle around him'. Copernicus here echoes *The Dream of Scipio*, well-known throughout the Middle Ages, where Cicero writes:[15]

> [A]lmost midway between Heavens and Earth blazes the Sun. He is the prince, lord and ruler [*dux et princeps et moderator*] of all the other worlds, the mind and guiding principle of the entire universe [*mens mundi et temperatio*] . . . In attendance upon the Sun are [his courtiers] Venus and Mercury . . .

A good illustration of this imagery, and the way it can effect astronomical thought, is provided by the ordering of the planets in medieval astronomy. As is well-known, there were no sound theoretical grounds for placing the planetary orbits in any particular order in the Aristotelian universe. Yet the conventional setting out of the system placed the Moon, Mercury and Venus below the Sun, and Mars, Jupiter and Saturn above the Sun, so the Sun was flanked by three planets on either side. This is the sense in which Cicero says the Sun is in the *centre* of the Heavens, and for this reason the Ptolemaic system was often quite self-consciously portrayed as a *heliocentric* system.[16]

[14] For examples, see: Kepler, *Harmonies*, p. 1081; Regiomantanus, as cited Zinner, *Entstehung*, p. 160; 'zodiac man' (1500), reprod. p. 170, Harthan, *Books of Hours*; Heninger, *Cosmographical Glass*, pp. 144–158; Alan of Lille, *Plaint*, pp. 82–5; Harvey, *Movement*, pp. 3–4; Hill, 'Harvey', *passim.*; Vondel, as quoted Fremantle, *Town Hall*, p. 60.

[15] Copernicus, *De revolutionibus*, p. 50; Cicero, *Dream*, p. 347. I have added the words 'his courtiers' to emphasis the political connotations, for a later audience, of the Latin, *comites* (= counts). Cf. Koestler, *Sleepwalkers*, pp. 74–75; Du Bartas, *Devine Weeks*, pp. 133–4; L'Orange, 'Expressions', p. 317.

[16] E.g. Lucan, *Pharsalia*, pp. 4–5 (= I.58); Dio Cassio, *History*, pp. 144–5; L'Orange, 'Expressions', *passim.*; Bernardus Silvestris, *Cosmographia*, p. 92; Berchorius, as quoted in Panofsky, *Meaning*, p. 262; Ficino, as discussed Kuhn, *Cop. Revolution*, pp. 129–30; zodiac wheel from *Vatican Ptolemy*, as in note 4; Cicero, *Dream*, p. 347; view of the universe from Saturn, early 15c. Paduan illustration of Dante, *Paradiso* canto xxii as reprod. Brieger *et al.*, *illuminated manuscripts*, vol. 2, p. 491. See also the discussion below of cosmologies organised around a central 'mystic' Sun, with Earth at the periphery. A good many discussions of precedents for the Copernican

This image, of a single object of especial importance, flanked on either side by a linear array of subordinates, often three in number, is however a very common one in medieval and renaissance iconography, and it was a standard method of representing the King in his court.[17] The political symbolism thus provides an explanation for why the planets were ordered as they were.

The social ideas embedded in these analogies are fairly easy to recognise. Firstly there are the basic notions that rulers are important, and that the nobility is an intrinsic part of the world-order. Secondly, there is the idea that political power comes from God in Heaven. Thirdly there is the idea that God does not apply his powers directly, but uses intermediary agents, such as Kings, popes and angels. Fourthly, we see that a King should do likewise, and not rule directly. Instead he should surround himself by a noble court who exercises this power for him.[18]

The symbolism thus functions to limit the status of a King. For while it gives him divine backing, it does not unequivocally give him direct divine backing, for it emphasises the rarity of direct contact with divinity. It supports in particular the Papal claim that intermediaries are associated with the distribution of political authority, and the claims of the nobles to a share of this authority.

From this perspective the incoherence of the Ptolemaic system becomes of considerable significance, for it would seem to endorse this idea that political power should be decentralised, and spread throughout the aristocracy. For the orbits of the planets are not bound to each other in the Ptolemaic system, and can be modified independently of each other. They are subject to a mild constraint by the Sun, but can expand and contract at will. As circumstances require, *ad hoc* adjustments can be made to individual details of the system, and the form which the system takes at any paticular stage is the product of

---

hypothesis (e.g., Rybka, 'Influence', pp. 166–7), appear to overlook the fact that heliocentricity was not the novel feature of Copernicus' theory. What Copernicus did was insist that it was the physical Sun rather than the mystic Sun which was central, and central in the sense of being near the centre of the Earth's orbit, rather than just at the centre of the space between Earth and Heaven.

[17] The game of chess provides a good example of this imagery: cf. Richard, *Dialogus*, p. 7. See also: apse mosaics in San Marco, Venice and Sta Prassede, Rome, as photographed in Oakeshott, *Mosaics*, pl. xxiii and fig. 124; facades facing Grand Canal and the Piazza San Marco, Doge's Palace, Venice, as photographed, for example, on p. 29 of Pertegato, *Venice*; Tintoretto's ceiling, Sala del Senato, Doge's Palace, Venice (an interesting example because it reveals a willingness to re-arrange the planets for the purpose of political display); 'Saint Louis leaves for the Crusades', early 14c. French miniature from *Chroniques de France*, reprod. as fig. 38, Evans, *Flowering*, p. 30; Bohemian drinking glass, 1593 as reprod. fig. 32, Trevor-Roper, *Age*, p. 128; Rudolfine 'Ship of State' clock ca.1580 in British Museum, as phographed in Dickens (ed.), *Courts*, p. 130; 'Charles V and His Enemies' engraving by van Heemskerck, from *Divi Caroli V Victoriae*, 1556, as reprod. Yates, *Astraea*, fig. 3b. For explicit recognisable connections with the planets, see: Kepler's frontpiece to Rudolfine tables, as reprod. in fig. 139 p. 139 of Dickens (ed.), *Courts*; Swoboda, *Roemische Palaeste*, pp. 238, 255–9 ('sun' on building facades such as Doge's palace mentioned above); emblems representing division of 'candle of life' in 'Hieroglyphic' section of Quarles, *Emblemes*; Mâle, *Gothic image*, p. 11; planets on Campanile of Giotto in Florence.

[18] Richelieu, *Political testament*, as quoted Elton *Renaissance*, pp. 138–9.

accumulated tradition. These features of the Ptolemaic system were also features of medieval political organisation, and of medieval conceptualisations of political organisation. Cities and Lords jealously guarded their idiosyncratic inheritances of rights and privileges, and consistently refused to allow their independence to be compromised for the sake of any coordinated uniformity.[19] There is thus a very close parellel between the structure of medieval society and the structure of the Ptolemaic system, and between the later attacks on these structures.

This parallel can be further extended. For although the standard Ptolemaic system placed the King at the centre of the ruling elite, it did not place him at the centre of the universe. That position, of undoubted prestige in medieval iconography[20] and the most important single position in the physical universe, was accorded to the Earth, symbol of the people, and one can hardly avoid asking whether this arrangement was an expression of the political centrality of the people. For in medieval and renaissance political theory, the people were widely seen as the supreme political authority. In some theories, God dispensed power through the Pope, but in others, especially those associated with Aristotelian naturalism, God dispensed political power through the people, who then passed a portion of that power on to their King via some form of social contract. Only in the early middle ages, before the arrival of Aristotle in the West, was a King widely seen as receiving his power directly from God.[21] So once again, the Ptolemaic system can be interpreted as functioning to reduce Kingly claims to political dominance. Though placing the Earth at the centre enabled a King to be seen as particularly close to God, it also had the opposite effect, and enhanced the status of the people.

The Ptolemaic system might thus seem to be a radically democratic theory, but I doubt if this is the correct interpretation. For many medieval and renaissance political theories promoting the rights of the common people functioned in reality as apologetics for patrician claims to power. In entering the social contract which created their King, the people operated through noble representatives, whose importance in the social scene is emphasised by the 'fact' that they are the entities which pass political authority on to the King from its sources. So once again the Aristotelian cosmology can be interpreted as functioning to enhance the status of the aristocracy by comparison with royalty.

I must admit that I have no direct evidence for this hypothesis that geocentricity functioned to endorse the political sovereignty of the people,[21a] but there is some strong indirect evidence available. For in medieval Europe,

[19] Mumford, *City*, pp. 282–394.

[20] The standard cathedral rose-window is probably the best example of this. See also: L'Orange & Nordhagen, *Mosaics*, pp. 21–24.

[21] See: Gierke, *Political Theory*, pp. 39, 146 (nn. 140, 141); Copleston, *History*, III(1), p. 130; *Vindicae* as trans. Laski, *Defense*, p. 188; Bouwsma, *Venice*, p. 349; Skinner, *Foundations*, vol. 2, p. 154.

[21a] For arguments supporting a similar interpretation of *ancient* astronomy, however, see Vernant, *Myth and thought*, pp. 125–234, esp. 186–192.

it was overwhelmingly common to represent the universe in forms other than the standard Aristotelian one. Particularly widespread was the notion that the universe is like a ladder, reaching from Earth at the bottom up to Heaven at the top. In many cases such ladder cosmologies may be interpreted as short-hand representations of a portion of a genuinely Aristotelian universe, but many of them cannot be so interpreted: they often contain spheres arranged one above the other instead of concentrically, for example. This suggests a genuine interest in departing from the correct Aristotelian form.[22] Furthermore, when divine sovereignty was to be stressed, it was extremely common to represent the Aristotelian cosmos in *inverted* form with Heaven at the centre, and the Earth either absent, or at the periphery. Implicit in such a radical inversion of the physical order of the cosmos into its mystic order is the idea that the literal Aristotelian geometry gives too much prestige to the common man. Furthermore, in these inverted diagrams of the universe, God and Heaven are commonly represented by an image of the Sun, so they represent an explicitly *heliocentric* cosmology.[23] And we know that this imagery was sometimes adapted to represent the divine authority of earthly rulers – the motif indeed may well derive from ancient representations of secular power.[24] So placing the Sun at the centre was a well-accepted mode of expressing lordship by the time of the renaissance.

Such a remark brings us of course to the Copernican theory, for what Copernicus did was effectively to remove the distinction between the mystic and the material geometry of the cosmos by placing the physical Sun at the physical centre. That is, he took a symbol commonly applied to supernatural authority, and imposed it upon the natural universe. The Earth, symbol of the common people, was removed from its prestigeous position and replaced by the symbol of the ruler. This ruler now governed his subjects, the encircling planets, directly, without apparent need to share his power with

[22] For examples, see: Botticelli, sketch for the ascent from the sphere of Saturn to the sphere of the fixed stars, illustrating Dante, *Paradiso* xxii, reprod. p. 199 Clark, *Drawings of Botticelli*; Heninger, *Cosmographical glass*, p. 26. In the *Convivio*, Dante suggests that the Aristotelian spheres are often identified with the Ptolemaic epicycles. This could generate the models of the universe composed of a stack on non-concentric spheres: see his pp. 75–6, and compare the planetary illustrations to *Paradiso* in Pope-Hennessey, *Sienese codex*, figs 42–4, 47–50, 52–53, 55, 59–61, 64, 66–73.

[23] I am currently preparing a study of these alternative cosmologies. For some examples, see Lewis, *Discarded Image*, p. 58; Dante, *Paradiso*, pp. 313–319 = canto xxviii; Botticelli, drawings for *Paradiso* xxiii-xxvi, reprod. Clark, *Drawings of Botticelli*, pp. 200–207; Pintoricchio, 'Coronation of the virgin', Pinacoteca Gallery, Vatican, reprod. on p. 161, Ricci, *Pintoricchio*; Sun-Christ and stars on facade of Sta Croce, Florence; Giovanni di Paolo's mid 15c. illustration to *Paradiso* xxviii, reprod. Brieger *et al.*, *Illuminated manuscripts*, vol. 2, p. 506; Veronese, ceiling fresco from Stanza dell'Olimpo, as in note 6; Seznec, *Survival*, figs 13, 28. See also next note.

[24] For some ancient political examples of this image, see Dio Cassio, *History*, pp. 144–5; Hermocles' poem in honour of Demetrius, as trans. Barker, *Alexander to Constantine*, p. 90; Procopius, *Buildings*, pp. 83–87 [= I.x.10],; L'Orange, *Iconography*, pp. 90–109. For later examples, see: zodiac wheel from *Vatican Ptolemy*, as in note 4; Prince Philip's Entry, as in note 5; Visconti sun in rose window in Milan Cathedral; emblems on Luebeck *Rathaus*; emblem of Pope Alexander VI, ceiling, Hall of the Liberal Arts, Borgia Apartment, Vatican, as depicted on pp. 10, 12 of Ferrazza & Pignatti, *Borgia apartment*.

other components of the cosmos. Not only was the status of the Earth reduced, but so too was the status of the planets, for these became identified with the Earth, so that the former tripartite division of the cosmos into terrestrial, planetary and stellar regions was replaced by a duality. The new universe contained only active Sun-stars and passive planet-earths. In addition, the traditional freedom of the planetary orbits was completely curtailed: in the interests of communal order, such orbits became tightly coordinated, and all were subject to an overriding solar dominance. The system represents an ideological image of the powerful ruler, who is freed from any obligation to respect the rights of an independent aristocracy. Through the prevailing conception of the Sun as symbolising the monarch, changes in society appear to have operated to generate an aesthetic sense as to what is the appropriate portion of the physical cosmos in which to seat the solar power, and it is this aesthetic sense, I suggest, which was instrumental to gaining acceptance for the new cosmology.

Furthermore, without any apparent 'scientific' evidence at all, supporters of the Copernican theory soon began to *multiply* the number of Suns in the universe, and to conceive of the universe as a patchwork of separate regions, each dominated by its own Sun, rather than as a single empire loosely ruled by a single emperor. The large issues of early modern political development[25] are thus clearly reflected in the re-arrangement of the universe which Copernicus initiated.

A similar ideological interpretation can also be placed on the historical process which we might typically refer to as 'the decline of astrology in the seventeenth century'. It has sometimes puzzled me whether this is an appropriate label for the events it is applied to. For so many of the issues which were controversial in renaissance disputes over the status of astrology were to become unequivocally settled in favour of the *proponents* of astrology as a result of the events of the seventeenth century. Their belief that the motions of the planets could be accurately charted and predicted, their belief that the Heavens were causally responsible for events on Earth, and their belief that the Heavens acted on the Earth by occult forces were all affirmed in the course of the Scientific Revolution, and became relatively uncontroversial. But what did occur as part of this process was that all causal powers in the Heavens were effectively concentrated in the Sun and Moon. The influence of the planets was not so much denied as declared to be insignificant.[26] Their light, heat and gravity were judged too small to cause any notable effects on Earth. Seen through ideological spectacles, this process emerges as a *genuine* case of decline, for it marks a distinct propaganda victory for centralism. It affirms the political triviality of the aristocracy by

[25] For summaries where the relevant issues emerge clearly and concisely, see E. N. Williams, 'Absolutism' and 'Enlightened Absolutists' on pp. 13, 132–5 of *Idem., Dictionary*, and Poggi, *Development*, pp. 16–85.

[26] See, for example, the discussion by Rohault and Clarke on pp. 86–91 of Rohault, *System*, II. See also: Locke, *Essay*, pp. 313–5; Boyle, *Works*, vol. 5, pp. 638–45; Hutchison, 'Occult qualities'; L'Orange, 'Expressions', pp. 317–8.

comparison with the King. Once again we find a remarkably clear parallel between social change and changing conceptions of the cosmos.

5 The Problems of Causation

There can be little doubt that these relatively precise parallels are more than coincidences: so the existence of causal links between social changes and astronomical theory seems highly likely. Furthermore, it is hardly conceivable that the vast movement towards centralised government was itself an effect produced by the relatively late re-orientations within astronomy: though astronomy was undoubtedly harnessed to provide ideological support for social change, one can hardly doubt that these changes were being independently generated. So the causal links must be overwhelmingly of the form of influences from society as a whole onto one section, astronomers, modified by minor additions of the reverse process.

To recognise that such links exist is, however, only the beginning of the problem: we need to document the causal mechanisms behind them. Until this is done we are dependent on a relatively weak argument based on the existence of parallels, albeit parallels which participants themselves refer to.

I am, however, not yet in a position to attempt a causal analysis. So I need alternative methods to strengthen the plausibility of my interpretation. One such method of especial promise would seem to be to embed the Copernican Revolution into a yet larger change, where independent grounds exist for believing in causal links with the contemporary centralisation of political power. We might, for example, be able to uncover evidence justifying placing a similar interpretation upon the Scientific Revolution as a whole. I shall not be attempting this large task here, though I have made a slight beginning on it elsewhere, by arguing that the mechanical philosophy supported centralised political authority, because of the way it undermined papal opposition to the divine right theory of Kingship. By apparently indicating that God acted directly in the world, like an absolute monarch, rather than through a hierarchy of noble forms as Catholic Aristotelian matter theory seemed to show (and as Catholic political theory required), the mechanical philosophy functioned to enhance the prestige of Kings.[27]

6 Baroque Aesthetics

Alternatively, we might be able to find some large conglomeration of intellectual changes which are *already* widely accepted as having been in some reasonable sense 'generated' by political developments, then seek to show that the acceptance of the motion of the Earth, is an example of this species

[27] See Hutchison, 'Reformation politics'.

of change. One important possibility here is the rise of what I shall refer to as 'baroque aesthetics'. Whether this is one of the happiest applications of that awkward word 'baroque' is not important here, and 'renaissance aesthetics' might be an equally appropriate term. What I am referring to shall become clear from my examples: it is that new style of design urged by Battista Alberti, when he told architects to ensure that the parts of a building 'bear that proportion among themselves, that they may appear to be an entire and perfect Body, and not disjointed and unfinished Members'.[28]

This aesthetic image was a common one in our period of interest. It was self-consciously applied to both society, as in the emblem chosen by Grimmelshausen as the frontspiece to *Simplicissimus*, and to astronomy, as in Copernicus' well-known complaint (some fifty years after Alberti) about the 'monstrosity' of traditional astronomy.[29] Indeed, the aesthetic principles which were emphasised in baroque design were precisely those which were so instrumental in the fall of the Ptolemaic account of the Heavens: coherence, harmony, and absence of *ad hoc* deviations from a principle of organization.

A good specific illustration of the contrast between the old and new architectural styles, and the parallelism with the aesthetics of the Copernican Revolution is provided by the new town hall built in Amsterdam in the middle of the seventeenth century.[30] The old town hall had never been designed. Some sort of town hall is known to have existed in the fourteenth century, but what the seventeenth century inherited was greatly different from it. For whenever the old building had been found inadequate for its purpose, some *ad hoc* adjustment had been made. A Court of Justice was added in 1418, extensive restoration was carried out after the fires of the middle fifteenth century, a tower and a hospice were added, then neighbouring houses were taken over to increase the accommodation. In the seventeenth century however, a decision was taken to demolish the old town hall and to replace it with a new one. In the new design, all traces of ad hocery have disappeared. The parts of the building fit together as a harmonious and coherent whole, and their proportions could not be changed without disrupting the plan completely. The new town hall illustrates Alberti's principles of design as perfectly as the heliostatic theory illustrates those of Copernicus. The reconstruction of this town hall and the reconstruction of planetary theories run in parallel.[31]

If we turn from the design of buildings to the design of cities, then we observe a similarly precise parallel.[32] Once again we find that the medieval

[28] Alberti, *De re aedificatoria*, I.9, as quoted Kaufmann, *Architecture*, pp. 91–3; cf. Palladio, as quoted in Wittkower, *Principles*, p. 22.

[29] Copernicus, *De Revolutionibus*, p. 25; Grimmelshausen, *Simplicissimus*, frontspiece.

[30] Fremantle, *Town Hall, passim.*, esp. pp. 19–20, 42–6, and figures 4 and 5.

[31] It is of some interest that the decoration of the new town hall contains much planetary and cosmological imagery: for these purposes, the Earth (Cybele) is counted as a planet, but so is the Sun (Apollo). See Fremantle, *Town Hall*, pp. 42–8.

[32] For examples of the contemporary recognition of such aesthetic parallelism, see: Donne, 'Progres', as quoted Byard, 'Poetic responses', p. 123; Descartes, *Philosophical Works*, vol. 1, pp. 87–9 (= *Discours*, ii).

city was designed on the principal of *ad hoc* adaption. Its lines followed the contours of the terrain rather than any geometric pattern, and its components could be individually adapted to new functions without regard to the effect of such adaption on the City as a whole. A map of a medieval city is as incoherent as Ptolemy's map of the planets.

With the rise of the baroque, however, this began to change. New principles of design came to the fore. A city was to be conceived of as a whole, and was expected to have a balanced and harmonious pattern, with clear centres of dominance. If we look at seventeenth century city maps of Amsterdam or Nice, the contrast between the old and the new sections of the town is patent. To the historian of astronomy, it is even startling – for the geometric pattern adopted for the new portions of the town is the traditional symbol of authority the wheel of fortune, the same pattern as used by Copernicus and denied by Tycho, concentric circles, together with radial avenues. In a brand new town, such as Palma Nova founded at the end of the sixteenth century, there were no medieval residues to be incorporated, and the 'Copernican' pattern could be perfectly followed.[33] It seems then, that the change from Ptolemaic and Tychonic to Copernican astronomy is part of a much wider change in aesthetic tastes, the emergence of the baroque. It follows that whatever explanation we offer of the baroque will be *a fortiori* an explanation, or at least a partial explanation, of the Copernican Revolution. Historians of the baroque, however, are in wide agreement that the aesthetics of this movement are to a large extent a reflection of the political aspirations of late renaissance Europe. Powerful centralised governments were then attractive because they could impose order upon their realms, and stamp out the destructive anarchy which resulted from an overmighty nobility. Medieval tastes favoured incoherent design because that reflected a desire for, and belief in, regional autonomy, guaranteed by a complex network of privileges, monopolies, charters and customary rights. As one historian expresses it[34]

> The baroque towns, as they began to be planned in the sixteenth century in Italy and developed over much of Europe in the seventeenth

[33] For examples, and relevant discussions of the contrasting principles of town-design, see: Koenigsberger & Mosse, *Europe*, pp. 77–82 (includes plan of Palma Nova); Mumford, *City*, pp. 282–467 and included illustrations (contains maps of Nice and Amsterdam); Baron, *Crisis*, pp. 202–4 (long footnote discussing the vistas of Florence depicted in the illustrations between p. 180 and p. 181); Martinez, *Power and imagination*, pp. 271–6. It is of some interest that the map of Palma Nova reproduced by Koenigsberger and Mosse does not depict the centre of town accurately. Recalling Alan of Lille and Chalcidius (cf. Lewis, *Discarded Image*, pp. 57–8), it places a citadel in the centre, whereas in reality the centre is an open space surrounding a tall pole at the top of which is a golden ball and the town flag. In the light of Cook's remarks (*Zeus*, vol. 1, p. 291), this suggests that cosmological imagery, and not just defensive functionalism, was involved in the design of the town. Cf. Aristophanes, *The Birds*, pp. 97–99, [= lines 992–1010]; Eliade, *Myth*, pp. 6–21; L'Orange, *Iconography*, pp. 9–8. I do not believe, however, that the town represents a Copernican universe. It is much closer the inverted Aristotelian cosmos as discussed in the text above.

[34] Koenigsberger, *Europe*, pp. 82–33. See also the sources cited in the last footnote.

> and eighteenth, became part of the deliberately dramatic and theatrical appeal of absolutist monarchy. Just as the new Baroque style of church decoration developed a deliberate popular appeal by making the interior of the church, and especially the high alter, into a kind of stage where mass was celeberated almost as a theatrical performance for an audience-like congregation, so the Baroque city became a huge theatrical setting for the display of the court, the princes of the church, the nobility and other rich and powerful persons. It was the visual aspect of the political and social change from the city state, with its free citizens, to th capital of the absolute monarch, with its court and its subject inhabitants.

One should not take such explanations of the baroque for granted. Not everyone accepts them and they tend to be dogmatically announced rather than patiently argued: so they may well be wrong. But what they do show is that my interpretation of the iconography of astronomical systems accords with other quite independent historical beliefs.

They also provide us with one of the elusive mechanisms needed to understand how the social environment can affect the internal development of scientific theories. These influences are sometimes portrayed in terms of crude self-interest, as if Kepler were to be understood as choosing the Copernican system because he wished to further his own political preferences, *rather than* because he saw it as the more rational theory. My argument is different. I have very little idea what Kepler himself felt about the desirability of powerful and centralised government, and I am happy to accept that Kepler preferred the Copernican system because he believed it the better theory. But in choosing criteria to effect this judgment of superior, Kepler borrows from the aesthetic tastes of his community. And it is apparently these that have here been influenced by political events. What seem to be rational factors may be merely social ones in disguise. We do not have to accept that every individual scientist is making a conscious political statement when he chooses between two theories. Social influences can operate by mediation of the very canons of rationality which the scientist sincerely deploys. Indeed, in conscientiously applying these canons the scientist will probably see himself as *avoiding* social influences!

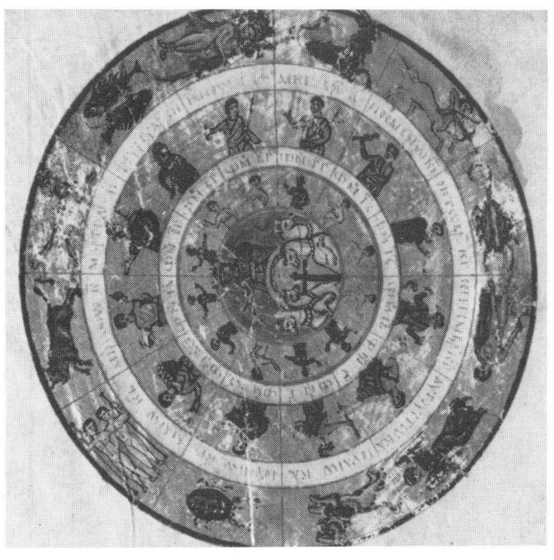

I. Celestial Imagery

Ia. The Emperor as Sun-God Helios at the centre of the zodiac, frontispiece to Vatican Ptolemy, 9th century. [From David Rice, ed., *The Dark Ages*, (London: Thames and Hudson, 1965), fig. 16.] Cf. figures Va–e below.

Ib. Aristocratic fireplace decorations, Bourges. Emblems of Moon, Sun, and fleur-de-lys (= iris = Urania = starry heavens: cf. Dioscorides, *Greek Herbal*, trans. J. Goodyer (1655), ed. R. Gunther (Oxford: Oxford University Press, 1934), p. 5). [Author's photograph.]

Ic. 'George Clifford', by Nicholas Hilliard, ca.1590. Miniature portrait in National Maritime Museum, Greenwich. Gold stars on blue costume; universes on sleeves; central Earth with Sun, Moon and Stars on shield. [From John Murdoch *et al.*, *The English Miniature* (London: Yale University Press, 1981), plate 8, p. 47].

Id. Detail of 'Ommeganck Procession, 1615' by Denis van Alsloot, Victoria and Albert Museum, London. Sun as conductor of orchestra; Moon; stellar canopies, chariot wheels. [From Hugh Trevor-Roper, ed., *The Age of Expansion* (London: Thames and Hudson, 1968), fig. 33.]

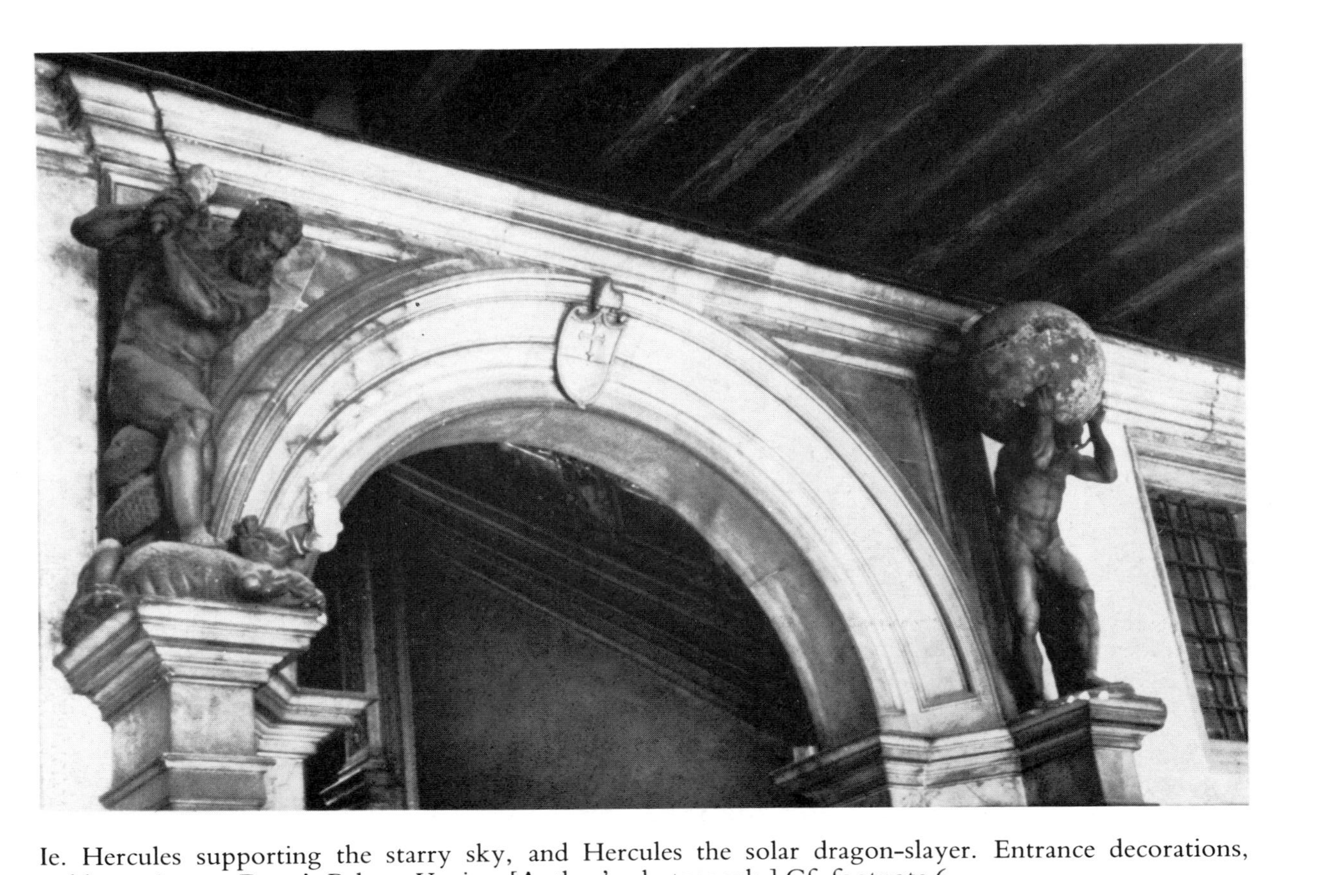

Ie. Hercules supporting the starry sky, and Hercules the solar dragon-slayer. Entrance decorations, golden stairway, Doge's Palace, Venice. [Author's photograph.] Cf. footnote 6.

If. The solar dragon-slayer. Decoration on main gateway, Versailles. For identification of the Archangel Michael with the Sun, see figure IIb. [Author's photograph.] Cf. also footnote 6.

II. The Aristotelian-Ptolemaic Universe

IIa. Clock Tower on St Mark's Square, Venice. Clock face depicts the standard medieval geocentric universe, with Earth at centre of sphere of fixed stars, and Sun and Moon (representing the planets) in orbit about the Earth. Tower decorations typify the references to the starry sky so common in monuments associated with public authority. The rear dial of this clock (not photographed here) depicts the inverted Aristotelian cosmos, as portrayed in figures Va–e below. [Author's photograph.]

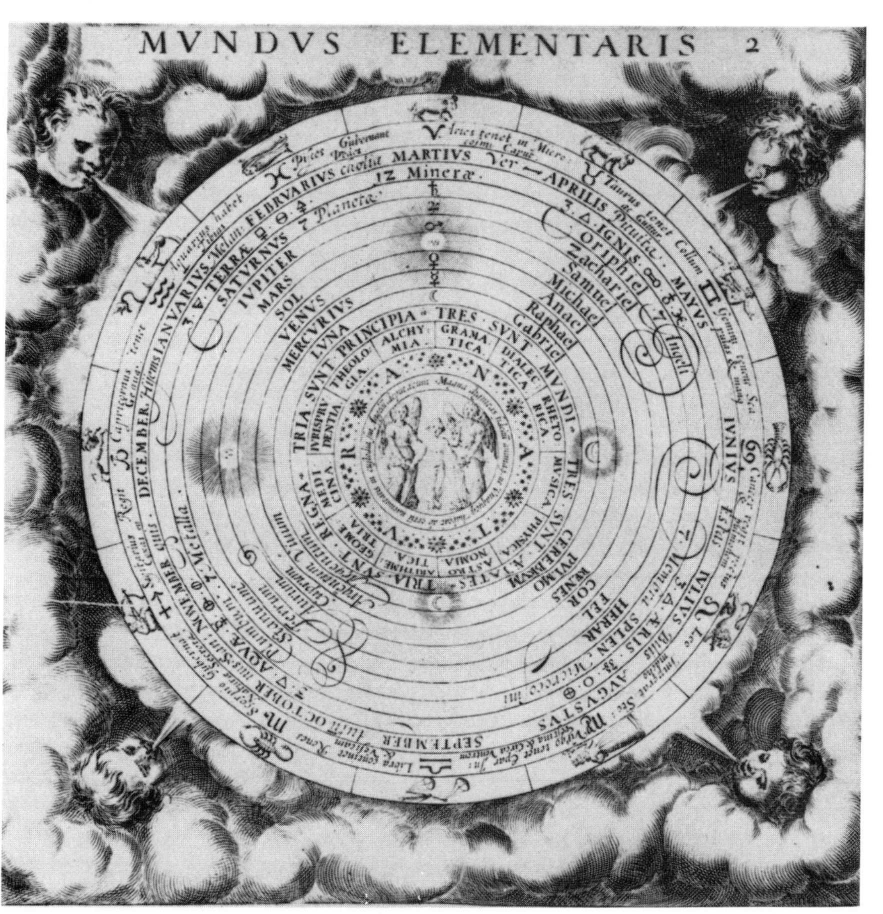

IIb. The Aristotelian pattern of the universe and its component analogies. Note the arrangement of the planets in their 'three-one-three' configuration – a 'central' Sun flanked by Moon, Mercury and Venus on one side with Mars, Jupiter and Saturn on the other. [From Warren Kenton, *Astrology* (London: Thames and Hudson, 1974), fig. 56.]

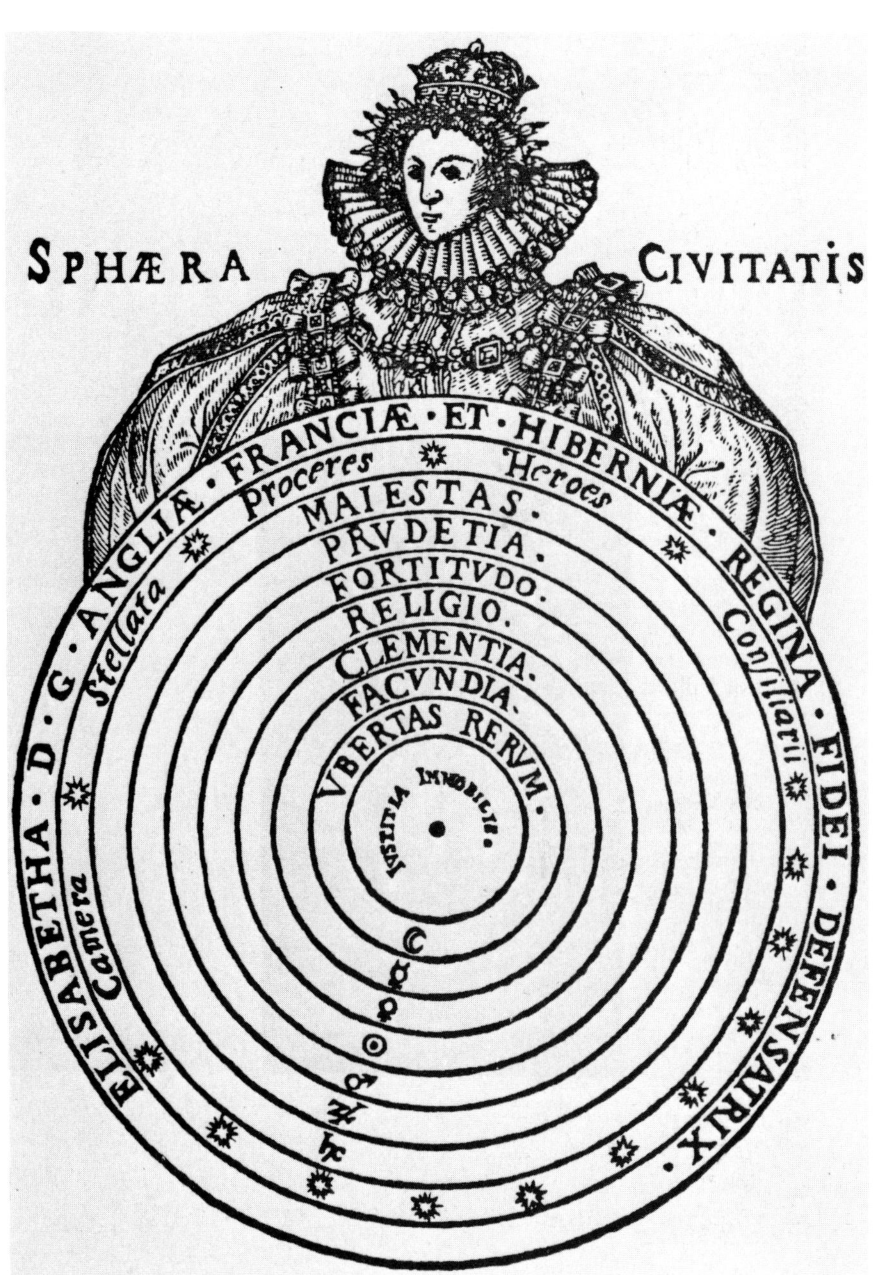

IIc. The civil sphere, from John Case, *Sphaera civitatis* (1588). [Photograph from Warren Kenton, *Astrology* (London: Thames and Hudson, 1974), fig. 49.]

III. The Three-One-Three Configuration of the Planets

IIIa: Bell-tower by Giotto, 14th century, Florence Cathedral. Depicts some of the analogies of the Universe, involving the three-one-three configuration of the planets (top left row). [Author's photograph.]

IIIb. Baroque Facade near Piazza Nicosia, Rome. The three-one-three configuration of windows on the first floor ('piano nobile') about the central balcony ('solario') is extremely common. In many cases the roof line of the building is also decorated with stars or star-substitutes. In this case, small stars (= planets?) are used to decorate the six side windows, while the central windows of the *solario* are decorated with Sun-faces. [Author's photograph.]

IIIc. The King in his Court. St Louis leaving on the 7th crusade, from *Chroniques de France ou de Saint Denis*. [Photograph from Joan Evans, ed., *The Flowering of the Middle Ages* (London: Thames and Hudson, 1966), fig. 38.]

IIId. Detail of Tintoretto ceiling, 'Triumph of Venice, Queen of the Seas', 1581–4, Senate Chamber, Doge's Palace, Venice. Planets are here re-arranged for the purpose of political display: the Moon, the watery planet associated with maritime Venice, is placed in the central position normally allocated to the Sun. [Author's photograph.]

IV. The Universe as a ladder

IVa. Fresco from the cathedral, Florence, by Domenico de Michelino, 1465. Shows portions of the planetary spheres of the canonical Aristotelian universe, truncated so as to indicate the ladder structure of the cosmos. [From Warren Kenton, *Astrology* (London: Thames and Hudson, 1974), fig. 47.]

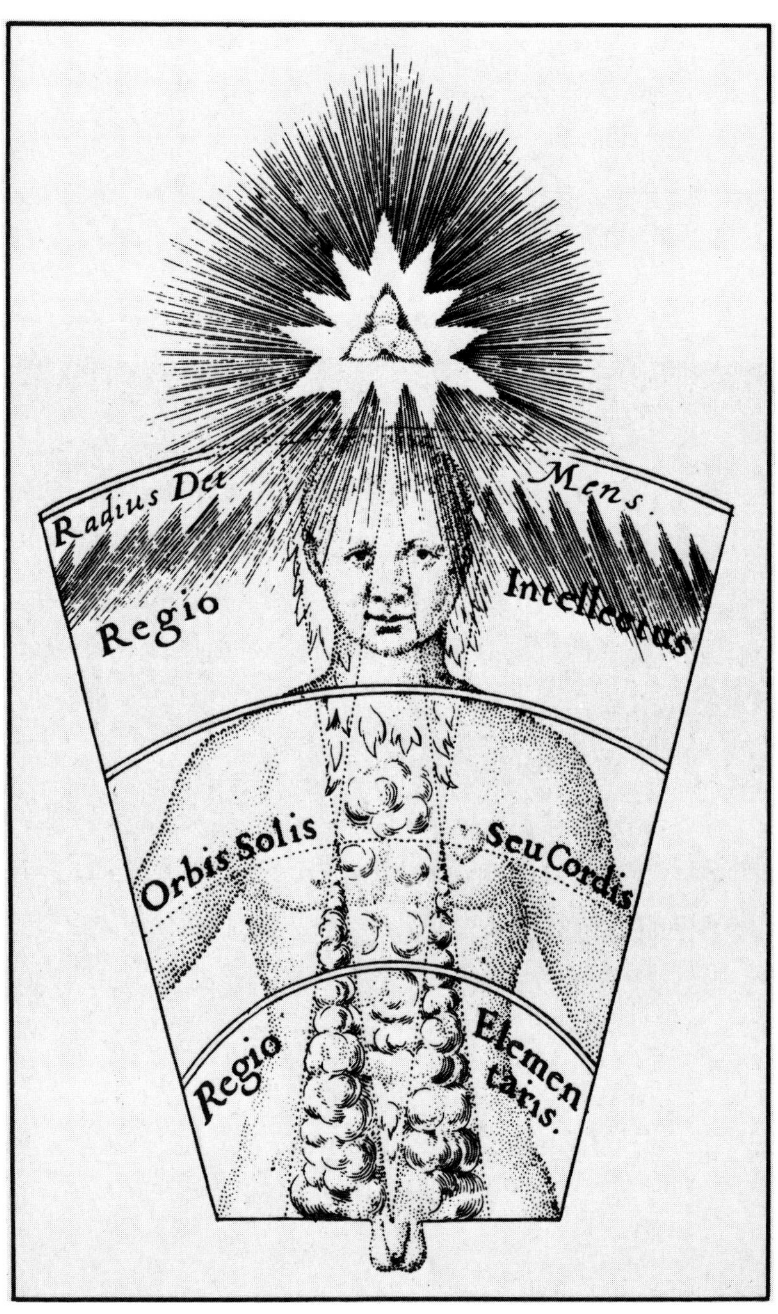

IVb. An Analogy with the Body, from Robert Fludd, *Utriusque cosmic . . . historia*, 4 vols (Oppenheim, 1617–19), III.83. The tripartite division of the universe into heaven, planets and Earth corresponds to the division of the body into head, chest and abdomen, and to the division of society into royalty, nobility, and commoners. [Photograph: S. K. Heninger, *The cosmographical glass* (San Marino, Calif.: The Huntington Library, 1977), fig. 84.]

IVc. 'The Fall of the Rebellious Angels', Louvre Museum (DL-1967-11-A), Paris. Reflects the same tripartite division of the cosmos as IVb, and especially the description of this division by Chalcidius, as quoted in the text. [Photograph: Musée de Louvre, Paris.]

IVd. 'Otto II in Majesty', miniature from the Aachen Gospel, ca.975. The emperor Otto is represented as intermediary between heaven and earth on the model of the universe as a ladder of spheres. [Photograph: Cathedral Museum, Aachen.]

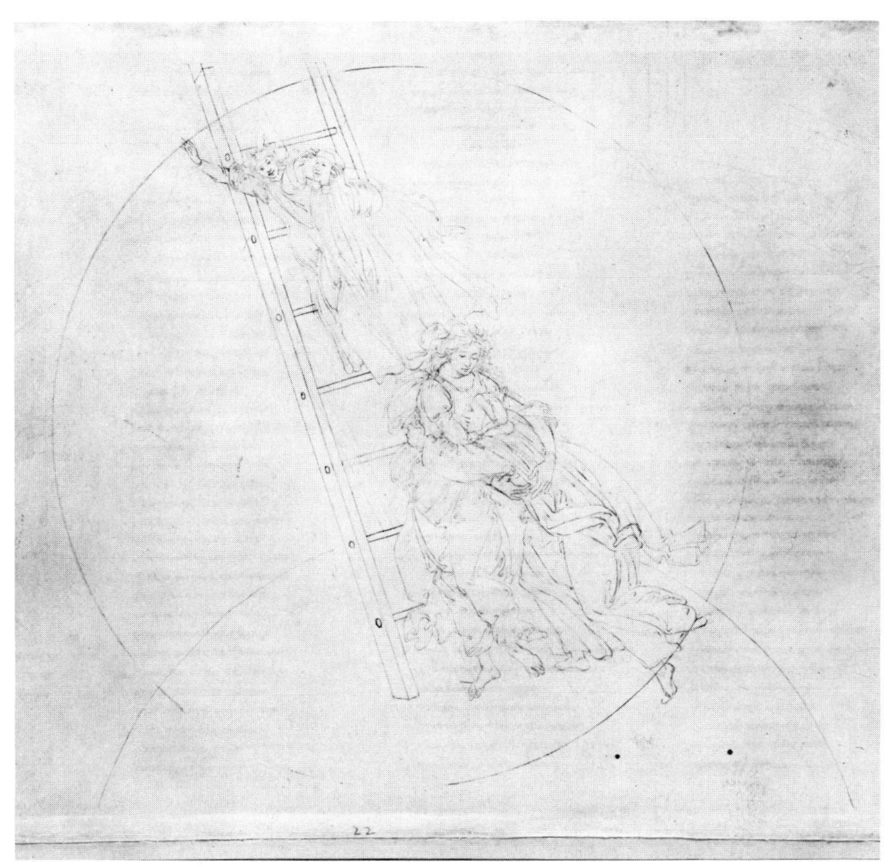

IVe. The universe as a ladder of spheres, one above the other. Sketch by Botticelli to illustrate the ascent from the sphere of Saturn to the sphere of the fixed stars in Dante's *Paradiso*. Note that this arrangement of the spheres is quite incompatible with those in the canonical Aristotelian model of the universe: but if the word 'sphere' is applied to the epicycle of the Ptolemaic model, as Dante tells us it commonly is (cf. *Convivio*, trans. P. H. Wickstead (London: Dent, 1940), pp. 75–6), and the deferents are ignored, then one can derive this ladder model out of the Aristotelian tradition. [From Kenneth Clark, *The drawings of Sandro Botticelli for Dante's Divine Comedy* (London: Thames and Hudson, 1976), p. 199.]

V. Pre-Copernican 'Heliocentric' Cosmologies, obtained by inverting the Aristotelian Universe

Va. Pinturicchio 'Coronation of the Virgin', early 16th century, Pinacoteca, Vatican Museums. The virgin is crowned by Christ in heaven, and in this context heaven is very commonly represented as the 'Sun' at the centre of the universe. [Photograph: Photo Alinari.]

Vb. 'Coronation of the Emperor': ceremonial arch for the entry of Prince Philip into Antwerp, 1549. The lamps were in motion about the central sun: for a contemporary description, see Cornille Greffier, *Les tresadmirable . . . entree du treshault et trespuissant Prince Philipes . . .* (Anwers, 1549). The model for this arch is clearly the standard 'Coronation of the Virgin': cf. fig. Va. [From Roy Strong, *Splendour at Court* (London: Weidenfeld and Nicolson, 1973), fig. 84, p. 85.]

Vd. Botticelli, sketch illustrating the arrival of Dante and Beatrice in Heaven, *Paradiso* 23–26. Dante sees the universe inside-out as he arrives in heaven. Botticelli represents the new arrangement of universe as 'heliocentric', replacing Dante's bright point of light with a 'sun' figure. [From Kenneth Clark, *The drawings of Sandro Botticelli for Dante's Divine Comedy* (London: Thames and Hudson, 1976), p. 207.]

Ve. Raphael's cupola, Chiggi tomb, Santa Maria del Popola, Rome, ca.1516. God in the centre is surrounded by the seven planets plus the sphere of the fixed stars. Further out but not encompassed by this photograph are the four elements. [Author's photograph.]

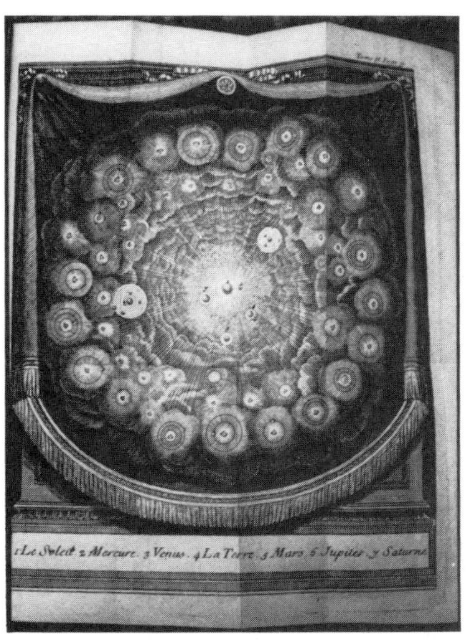

VI. The New Universe

VIa. The Copernican Universe. Frontispiece to Bernard Le Bovier de Fontenelle, *Entretiens sur la pluralité des mondes* (Paris, 1762). [Author's photograph.]

VIb. 'Regimentsspiegel' ('mirror of government'), 17th century, triptych from Swiss National Museum, Zürich. The components of city government in Zürich are arranged on the model of the Copernican system, with the mayors at the centre. [Photograph: Schweizerisches Landesmuseum, Zürich.]

VII. Baroque Aesthetics

VIIa. Medieval Valenciennes. [From Lewis Mumford, *The City in History* (Penguin, 1979), Plate 32.]

VIIb. Baroque Palmanova. [From H. G. Koenigsberger and George L. Mosse, *Europe in the Sixteenth Century* (London: Longman, 1977), p. 80.]

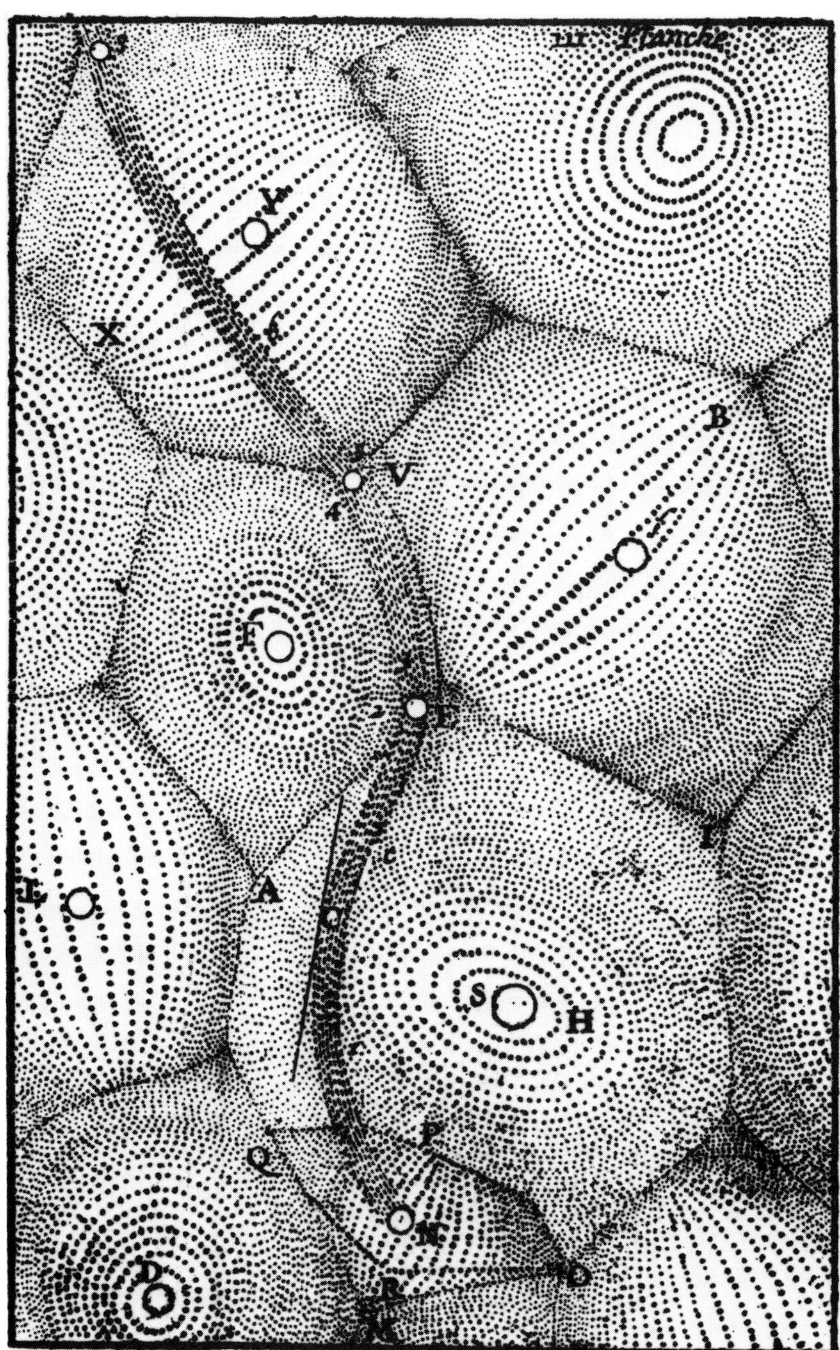

VIc. The Infinite Cosmos of Descartes, from Rene Descartes, *Oeuvres*, vol. IX, ed. C. Adam and P. Tannery (Paris, 1904), plate iii. The new universe now looks like a map of 17th-century Europe. [Photograph: CSHE, University of Melbourne.]

VIIc. The Medieval Town-Hall of Amsterdam. [From Katharine Fremantle, *The Baroque Town-Hall of Amsterdam* (Utrecht: Decker and Gumbert, 1959), fig. 4.]

VIId. The Baroque Town-Hall of Amsterdam. [*Ibid.*, fig. 5.]

BOOKS AND ARTICLES CITED

ALAN of Lille, *The plaint of nature*, trans. J. Sheridan (Toronto: Pontifical Institute of Medieval Studies, 1980).

ANGLO, Sydney, *Spectacle, Pageantry, and Early Tudor Policy* (Oxford, 1969).

ARISTOPHANES, *The Birds*, trans. G. Murray (London: George Allen & Unwin, 1950).

BAKER, Ernest, (ed., trans.), *From Alexander to Constantine: Passages and documents, illustrating the history of social and political ideas 336 B.C. – A.D. 337* (Oxford: Clarendon Press, 1956).

BARON, Hans, *The Crisis of the Early Italian Renaissance*, rev. edn (Princeton, 1967).

BERNARD SILVESTRIS, *The Cosmographia of Bernardus Silvestris*, trans. W. Wetherbee (New York: Columbia UP, 1973).

BOYLE, Robert, *The Works of the Honourable Robert Boyle*, ed. T. Birch, 6 vols (London, 1772).

BOUWSMA, William J., *Venice and the Defense of Republican Liberty* (Berkeley: Univ. Calif. Press, 1968).

BRAHE, Tycho, *Recent phaenomena* (excerpts) trans. in Marie BOAS and A. Rupert HALL, 'Tycho Brahe's system of the world', *Occas. Notes R. Astron. Soc.*, 3 (1959), pp. 253–63.

BRIEGER, Peter, M. MEISS & C. SINGLETON, *Illuminated manuscripts of Dante's Divine Comedy*, 2 vols (London: RKP, 1969/70).

BROWN, Jonathon & J. H. ELLIOT, *A palace for a king: the Buen Retiro and the court of Philip IV* (New Haven: Yale Univ. Press, 1980).

BURTT, Edwin, *The Metaphysical Foundations of Modern Physical Science*, 2nd rev. edn (London, 1964).

BYARD, Margaret, 'Poetic responses to the Copernican Revolution', *Scientific American*, 236 (1977), pp. 120–127.

CHALCIDIUS, *Commentary* on the *Timaeus* of Plato, in *Plato Latinus*, ed. R. Klibansky, Vol. 4: *Timaeus a Calcidio translatus commentarioque instructus*, ed. J. Waszink (London: Warburg Institute, 1962).

CHENU, Marie Dominique, *Man, nature and society in the twelfth century*, ed., trans., J. Taylor & L. Little (Chicago: Univ. of Chicago Press, 1968).

CICERO, *The Dream of Scipio* (from *The Republic*) in *idem.*, *On the Good Life*, trans. M. Grant (Penguin, 1971), pp. 337–55.

CLARK, Kenneth, *The drawings of Sandro Botticelli for Dante's Divine Comedy* (London: Thames & Hudson, 1976).

COOK, Arthur, *Zeus: a study in ancient religion*, 3 vols (Cambridge: Univ. Press, 1914–1940).

COPERNICUS, N., *On the Revolutions of the Heavenly Spheres*, trans. A. M. Duncan (New York: Barnes & Noble, 1976).

COPLESTON, Frederick, *A History of Philosophy*, Vol. 3, Part 1 (New York: Image, 1963).
DANTE, *The convivio*, trans. P. H. Wicksteed (London: Dent, 1940).
——, *Monarchy*, trans. D. Nicholl (London: Weidenfeld & Nicolson, 1954).
——, *Paradiso*, trans. C. Singleton (Princeton: Univ. Press, 1975).
DESAGULIERS, J. T., *The Newtonian System, an Allegorical Poem* (London, 1728).
DESCARTES, René, *The Philosophical Works of Descartes*, trans. E. S. Haldane & G. R. T. Ross, 2 vols (Cambridge: Camb. Univ. Press, 1931).
DE TOLNAY, Charles, *Michelangelo*, vol. 3 (Princeton: Univ. Press, 1948).
DICKENS, A. G., ed., *The Courts of Europe* (London: Thames and Hudson, 1977).
DIO CASSIO, *Roman history*, trans. E. Cary, vol. 9 (London: Heinemann Loeb series, 1925).
DU BARTAS, Guillaume, *Bartas: His Devine Weeks and Works* (freely) trans. J. Sylvester (London, 1605).
ELIADE, Mircea, *The myth of the eternal return*, trans. W. Trask (New York: Bollingen, 1954).
——, *The Forge and the Crucible*, trans. S. Corrin (New York: Harper & Row, 1971).
ELTON, G. R., *Renaissance and reformation* (New York: Macmillan, 1963).
EVANS, Joan, ed., *The Flowering of the Middle Ages* (London: Thames and Hudson, 1966).
FERRAZZA, Mario and Patrizia PIGNATTI, *The Borgia apartment and contemporary art in the Vatican* (Vatican City: Monumenti, Musei e Gallerie Pontificie, 1971).
FOWLER, Kenneth, *The Age of Plantagenet and Valois* (London: Ferndale, 1980).
FRANKFORT, Henri, *et al.*, *Before philosophy* (Penguin, 1949).
FREMANTLE, Katharine, *The Baroque Town Hall of Amsterdam* (Utrecht: Decker and Gumbert, 1959).
GIERKE, Otto, *Political Theory of the Middle Age*, trans. F. W. Maitland (Boston: Beacon, 1958; 1900).
GOMBRICH, 'Icones symbolicae', on pp. 123–195 of *idem, Symbolic images* (Oxford: Phaidon, 1978).
GRAVES, Robert, *The Greek Myths*, 2 Vols, rev. edn (Penguin, 1966).
GRIMMELSHAUSEN, *Der Abentheurliche Simplicissimus Teutsch* (1669), reprint ed. R. Tarot (Tubingen: Niemeyer, 1967).
GYR, Salomon, *Zuercher Zunfthistorien* (Zurich, 1929).
HARTHAN, John, *Books of Hours and Their Owners* (London: Thames and Hudson, 1977).
HARVEY, William, *An Anatomical Disputation Concerning the Movement of the Heart and Blood in Living Creatures*, trans. G. Whitteridge (Oxford: Blackwell, 1976).
HAWKES, Jacquetta, *Man and the Sun* (London: Cresset, 1962).
HENINGER, S. K., *The Cosmographical Glass* (San Marino, Calif.: Huntingdon Library, 1977).
HILL, Christopher, 'William Harvey and the Idea of Monarchy', paper and

debate, in C. Webster, ed., *The Intellectual Revolution of the Seventeenth Century* (London: RKP, 1974), pp. 160–196.

HUTCHISON, Keith, 'What happened to occult qualities in the Scientific Revolution?', *Isis*, 73 (1982), pp. 233–53.

——, 'Reformation politics and the new philosophy', *Metascience* (Occasional Papers of the Australasian Association for the History Philosophy and Social Studies of Science), 1/2 (1984), pp. 4–14.

JOHN OF SALISBURY, *Policraticus* selections in *idem.*, *The Statesman's Book of John of Salisbury*, trans. J. Dickinson (New York: Russell and Russell, 1963).

KANTOROWICZ, Ernst, *The King's Two Bodies: A Study in Mediaeval Political Theology* (Princeton, 1957).

KAUFMANN, Emil, *Architecture in the age of reason* (Cambridge, Mass.: Harvard Univ. Press, 1955).

KEPLER, Johannes, *The Harmonies of the World*, Bk. V, trans. C. Wallis in *Great Books of the Western World*, vol. 16 (Chicago: Encyclopaedia Britannica, 1952), pp. 1009–1085.

——, *Epitome of Copernican Astronomy*, Bks IV and V, trans. C. Wallis in *Great Books of the Western World*, vol. 16 (Chicago: Encyclopaedia Britannica, 1952), pp. 845–1004.

KOENIGSBERGER, H. G. and G. L. MOSSE, *Europe in the Sixteenth Century* (London: Longman, 1968).

KOESTLER, Arthur, *The Sleepwalkers* (Harmondsworth: Penguin, 1964).

KUHN, Thomas, *The Copernican Revolution* (Harvard, 1957).

LASKI, Harold J., *A Defence of Liberty Against Tyrants* (New York: Burt Franklin, 1972). Reprints a 16c. trans. of the anon. *Vindiciae Contra Tyrannos* (1579).

LEWIS, C. S., *The discarded image* (Cambridge: Univ. Press, 1964).

LIVY, *The Early History of Rome*, [= *History*, books 1–5], trans. A. de Selincourt (Penguin, 1973).

LOCKE, John, *An Essay Concerning Human Understanding*, abridged, ed. J. Yolton (London: Dent, 1977).

L'ORANGE, Hans, *Studies on the Iconography of Cosmic Kingship in the Ancient World* (Oslo: Ashehoug, 1953).

——, 'Expressions of cosmic kingship in the ancient world', on pp. 313–4, *idem.*, *Lileness and icon* (Odense: Univ. Press, 1973).

——, and P. Nordhagen, *Mosaics*, trans. A. Keep (London: Methuen, 1966).

LUCAN, *The Pharsalia*, trans. H. Riley (London, 1853).

LULL, Ramon, *The Order of Chivalry*, trans. W. Caxton, ed. F. S. Ellis (Hammersmith, London: William Morris at Kelmscott Press, 1896).

MÂLE, Emile, *The gothic image*, trans. D. Nussey (London: Fontana, 1961; orig. 1910).

MARTINEZ, Lauro, *Power and imagination: City-states in Renaissance Italy* (New York: Vintage, 1980).

MAZZEO, Joseph, *Rennaissance and Seventeenth-Century Studies* (New York: Columbia Univ. Press, 1964).

——, *Medieval cultural tradition in Dante's Comedy* (Ithaca, NY: Cornell Univ. Press, 1960).

MIRANDOLA, Pico della, *Heptaplus* in *idem.*, *On the Dignity of Man and Other Works*, trans. C. Wallis, P. Miller, D. Carmichael (New York: Bobbs-Merrill, 1965).

MOSSAKOWSKI, Stanislaw, 'The symbolic meaning of Copernicus' seal', *J. Hist. Ideas*, 34 (1973), pp. 451–9.

MOUSNIER, Roland, 'Le Roi-Soleil' on pp. 84–127 of Jacques Goimard, ed., *La France au Temps de Louis XIV* (Paris: Hachette, 1965).

MUMFORD, Lewis, *The City in History* (Penguin, 1979).

NEEDHAM, Joseph, *Science and Civilisation in China*, Vol. 5, Part 5 (Cambridge: Cambridge Univ. Press, 1983).

NEWTON, Isaac, *Theological manuscripts*, ed. H. McLachlan (Liverpool: Univ. Press, 1950).

OAKESHOT, Walter, *The mosaics of Rome* (London: Thames and Hudson, 1967).

OVID, Publius, *Metamorphoses*, trans. M. Innes (Harmondsworth: Penguin, 1955).

PANOFSKY, Erwin, *Hercules am Scheideweg* (Leipzig: Teubner, 1930).

——, *Meaning in the Visual Arts* (New York: Doubleday, 1955).

PERTEGATO, Francesco, *Venice*, trans. D. Mills (Milan: Graf, 1983).

PIGNATTI, Terisio, *Paolo Veronese a Maser* (Milan: Fabbri, 1965).

POGGI, Gianfranco, *The Development of the Modern State* (London: Hutchinson, 1978).

POPE BONIFACE VIII, *Unam Sanctum*, Papal Bull, 1302, trans. in Elton *Renaissance*.

POPE INNOCENT III, Letter to Nobles of Tuscany, 1198, excerpt trans. on p. 217 of Brian TIERNEY ed., *The middle ages*, vol. 1, 2nd edn (New York: Knopf, 1973).

POPE-HENNESSEY, John (ed.), *A Sienese codex of the Divine Comedy* (Oxford: Phaidon, 1947).

PRAZ, Mario, *Studies in Seventeenth-Century Imagery*, 2nd rev. edn (Rome, 1964).

PROCOPIUS OF CAESAREA, *Buildings*, trans. H. Dewing, [= *Works*, Vol. 7] (London: Heinemann Loeb Series, 1940).

QUARLES, Francis, *Emblemes divine and moral; together with hieroglyphicks of the life of man* (London, 1736; 1634 & 1658).

RICCI, Corrado, *Pintoricchio*, trans. F. Simmonds (London: Heinemann, 1902).

RICE, David, ed., *The Dark Ages* (London: Thames and Hudson, 1965).

RICHARD, Son of Nigel, *Dialogues de Scaccario*, trans. C. Johnson (London: Nelson, 1950).

ROHAULT, Jacques, *A System of Natural Philosophy* 2 vols, trans. and extensively annotated by Samuel Clarke (1723; reprinted New York: Johnson, 1969).

RYBKA, Eugeniusz, 'The influence of the Cracow intellectual climate at the end of the fifteenth century upon the origins of the heliocentric system', in Arthur Beer, ed., *Vistas in Astronomy* (Oxford: Pergamon, 1967), pp. 165–9.

SEZNEC, Jean, *The survival of the pagan gods*, trans. B. Sessions (New York: Pantheon, 1953).

SHAKESPEARE, *The Collected Works*, ed. W. J. Craig (London: OUP, 1930).
SKINNER, Quentin, *The Foundations of Modern Political Thought*, 2 vols (Cambridge: Cambridge Univ. Press, 1978).
SMITH, Earl Baldwin, *The Dome* (Princeton: Univ. Press, 1950).
——, *Architectural symbolism of Imperial Rome and the Middle Ages* (Princeton: Univ. Press, 1956).
STRONG, Roy, *Splendour at court* (London: Weidenfeld & Nicolson, 1973).
——, 'Homage to the Queen', *The V. & A. Album*, 1 (1982), pp. 70–77.
SWOBODA, Karl, *Roemische und Romanische Palaeste* (Wien: Bohlaus, 1969).
TACITUS, *The Annals of Imperial Rome*, trans. M. Grant, rev. edn (Penguin, 1972).
TREVOR-ROPER, Hugh, ed., *The Age of Expansion* (London: Thames and Hudson, 1968).
ULLMANN, Walter, *A Short History of the Papacy in the Middle Ages*, rev. edn (London, 1977).
UTZ, Hildegard, 'The *Labors of Hercules* and other works by Vicenzo de' Rossi', *Art Bulletin*, 53 (1971), pp. 344–66.
VERNANT, Jean-Pierre, *Myth and thought among the Greeks* (London, RKP, 1983).
VIALE, Mercedes, *Tapestries*, trans. H. St. Clair-Erskine and A. Rhodes (London: Hamlyn, 1969).
VISCONTI, *The Hours of Giangaleazzo Visconti, Duke of Milan* (early 15c. illum. MS), reprod. with introd. etc. by M. Meiss and E. Kirsch (London: Thames and Hudson, 1972).
VIVIANI, Corrado, 'Henri IV, the Gallic Hercules', *J. Warburg Court. Inst.*, 30 (1967), pp. 176–97.
WILLIAMS, E. N., *The Penguin Dictionary of English and European History* (Penguin, 1980).
WINTER, Carl, *Elizabethian Miniatures* (London: Penguin, 1943).
WITTKOWER, Rudolf, *Architectural principles in the age of humanism*, 3rd edn (London: Tiranti, 1962).
YATES, Frances, *Astraea* (London: RKP, 1975).
ZINNER, Ernst, *Die Entstehung und Ausbreitung der Copernicanischen Lehre* (Erlangen, 1943).

ASTROLOGY IN KEPLER'S COSMOLOGY

J. V. Field

Historians have tended to class astrology with the occult sciences. For example, it finds its place in such works as Thomas's *Religion and the Decline of Magic* (London, 1971) and Shumaker's *The Occult Sciences in the Renaissance* (University of California Press, 1972). Indeed, in our own day astrology is occult, in the sense that astrologers wilfully ignore the results obtained in other fields (postulating instead forces of types unknown and otherwise unexampled) and use techniques associated with mainstream science – such as statistical analysis – in a manner which clearly shows their incomprehension or rejection of the actual methods of science. We are seeing here another manifestation of that 'fear of freedom' which E. R. Dodds described so well as the force behind the occultism of Late Hellenistic philosophies, and whose sinister resurgence in the earlier years of our own century gave modern relevance to his study.[1]

In Kepler's day, astrology was about to undergo a period of relatively rapid decline in its intellectual and social respectability, but was undoubtedly still taken seriously in its own right, rather than being regarded as primarily 'occult'. Indeed, the meaning of the word 'occult' seems itself to have changed quite considerably over the period between, say, 1580 and 1680, as well as apparently varying from writer to writer.[2] It is thus anything but clear whether it is legitimate, in this period, to regard astrology as inevitably

ACKNOWLEDGEMENTS
I am grateful to Dr John Henry for helpful discussions while this essay was taking shape and for his critical reading of the final draft.

ABBREVIATIONS
KGW *Johannes Kepler Gesammelte Werke*, ed. W. von Dyck, M. Caspar *et al.*, Munich, 1938–.
KOF *Johannis Kepleri Opera Omnia*, ed. C. Frisch, Frankfurt, 1865.

1 E. R. Dodds, *The Greeks and the Irrational*, University of California Press, 1951. On the question of contemporary relevance see especially pp. 253ff. It is interesting to compare Dodds' robust rejection of occultism with Shumaker's quasi-scientific argumentation against it in the introductory chapter of *The Occult Sciences in the Renaissance*, University of California Press, 1972. For a more effective method of dealing with what he calls the 'mystery mongers' of this generation see C. Sagan, *Broca's Brain. The Romance of Science*, London (Hodder and Stoughton) 1974, especially pp. 43–136.

2 For a somewhat different view of this matter see K. Hutchison, 'What happened to occult qualities in the Scientific Revolution?', *Isis*, 73, 1982, 233–253. There is an illuminating discussion of the problems of deciding the type of cause to be assigned to 'occult' phenomena in S. Clark, 'The scientific status of demonology', in B. Vickers (ed.) *Occult and Scientific Mentalities in the Renaissance*, CUP, 1984, 351–374.

allied with the other sciences that later became regarded as exclusively 'occult', such as magic. Unlike magic, palmistry, geomancy etc., astrology was taught in universities as part of the *quadrivium*. For instance, the list of texts to be studied in the astronomy course set up at the university of Krakow in the 1520s included several astrological works,[3] and the students to whom Galileo taught elementary Ptolemaic astronomy at Padua from 1592 until 1610 certainly included many who would apply what they had learned to the practice of medical astrology.[4]

The rapid decline in the intellectual standing of astrology in the course of the seventeenth century is roughly contemporary with a rise in the respectability of Copernicanism. While the causes of the latter phenomenon are not by any means completely understood, it is nevertheless tempting to see the two phenomena as linked with one another. It is not, however, the case that all Copernicans rejected astrology. The most prominent counter-example is Kepler – a Copernican since his student days at Tübingen in the late 1580s and a lifelong believer in astrology. Thus particular interest attaches to Kepler's astrological work as providing, apparently, an exception that will allow us to test our rule. There are, after all, good reasons why heliocentric astronomy should have caused a decline in astrological belief. For instance, if all the planets, including the Earth, were believed to be moving round the Sun, the fact that a particular planet was 'in' a particular constellation merely told one something about its position relative to the Earth rather than, as in a geocentric cosmology, its actual position in regard to the Universe as a whole. While this does not refute the ascription of particular 'houses' to each planet, it does somewhat decrease their cosmic significance. Moreover, a similar weakening will be found in all the reasoning which depends upon Zodiac signs: in a heliocentric system we no longer have absolute properties of the macrocosm exerting their influence upon Man, the microcosm. The ancient symmetries can no longer be maintained and the ancient arguments from analogy thus lose something of their force. Nonetheless, in a heliocentric Universe as in a geocentric one, there is still good empirical evidence for the influence of heavenly bodies upon the Earth. The changing position of the Sun in the Zodiac defines the changing seasons of the year and the Moon is widely recognised as a cause of the tides. Both these phenomena were regarded as astrological by most writers, including Kepler, who considers them in some detail in his short treatise on astrology *De Fundamentis Astrologiae Certioribus*, published in Prague at the end of 1601 (but dated 1602).[5]

[3] The relevant document is dated 15 November 1525, though the course seems to have been instituted in 1522. The astrological works included Ptolemy's *Tetrabiblos* and *Centiloquium* as well as treatises by Albohazen, Albumasar and Guido Bonatti, all available in printed form.

[4] For Galileo's own attitude to astrology see G. Ernst, 'Aspetti dell'astrologia e della profesia nel pensiero di Galileo e di Campanella', in P. Galluzzi (ed.) *Novità celesti e crisi del sapere*, Florence, 1983, and G. Righini, 'L'Oroscopo galileano di Cosimo II de' Medici', *Annali dell'Istituto e Museo di Storia della Scienza di Firenze*, 1, 1976, 29–36.

[5] Theses V to XVIII, reprinted in KGW 4, pp. 12–15. An English translation of this treatise forms the second part of J. V. Field, 'A Lutheran Astrologer: Johannes Kepler', *Archive for History of Exact Sciences*, 31, 1984, 189–272.

The second half of *De Fundamentis Astrologiae Certioribus* consists of month-by-month weather predictions for the year 1602. During the period when he was District Mathematician at Graz (1594–1600) it had been part of Kepler's duties to write annual calendars of this kind.[6] Such calendars are, in fact, a prominent feature of astrological writing in the Middle Ages and in the Renaissance.[7] (Weather prediction was, of course, of great importance to societies whose bases were very largely agrarian.) It was usual for calendars to include not only predictions of the weather but also specific predictions about the size of particular harvests, such as wheat or oil, and about the quality of that year's wine. Kepler ridicules these latter predictions (which are derived from considerations of 'houses') and says that they in fact follow from predictions of the weather.[8] However, apart from such specific prognostications, it is clear that most of the predictions in calendars are based on the idea, put forward in Aristotle's *Meteorologica*, that the roughly cyclic changes in the sublunary world are consequences of the perfect cyclic movements in the celestial world.[9] Thus the intellectual context of this kind of prediction is that of Aristotelian cosmology, which also provides the background for university teaching in astronomy and other subjects. Weather prediction, like medical astrology, clearly belongs to a learned tradition. This marks it off from the astrology of soothsaying, recovery of stolen goods etc. which is discussed by Thomas in *Religion and the Decline of Magic*. The latter types of astrology show more affinity with popular beliefs, and thus belong to folklore rather than to the mainstream of intellectual development. One cannot, of course, make a sharp division between learned ('high') and popular ('low') astrology, but it is clear that Kepler's astrology essentially belongs to the 'high' tradition and, like the astrology taught in the *quadrivium*, is an integral part of a complete system of the world.

This is apparent in Kepler's first published work, *Mysterium Cosmographicum* (Tübingen, 1596), where he includes among the observational verifications of his proposed cosmological archetype – the system of nested orbs and polyhedra (see Figure 1) – its ability to account for known astrological phenomena such as the individual characters of the planets and the existence of certain configurations which exert astrological influence (aspects).

The explanation of the characters of the planets is straightforward: Kepler derives them from the geometrical characteristics of the polyhedra, each planet being associated with the solid which defines the space between that planet's sphere and the sphere of the planet which lies next to it in the direction towards the Earth.[10] This gives us the correspondences

[6] Those he wrote for 1598 and 1599 are reprinted in KOF 1.

[7] See S. Jenks, 'Astrometeorology in the Middle Ages', *Isis*, 74, 1983, 185–210.

[8] *De Fundamentis Astrologiae Certioribus*, Thesis LXV, KGW 4, p. 31.

[9] Aristotle, *Meteorologica* I, iii (340b4–341a31) and iv (342a16–33).

[10] Since astrology is concerned with the effects of celestial bodies upon the Earth the astrology even of so convinced a Copernican as Kepler inevitably contains a geocentric element.

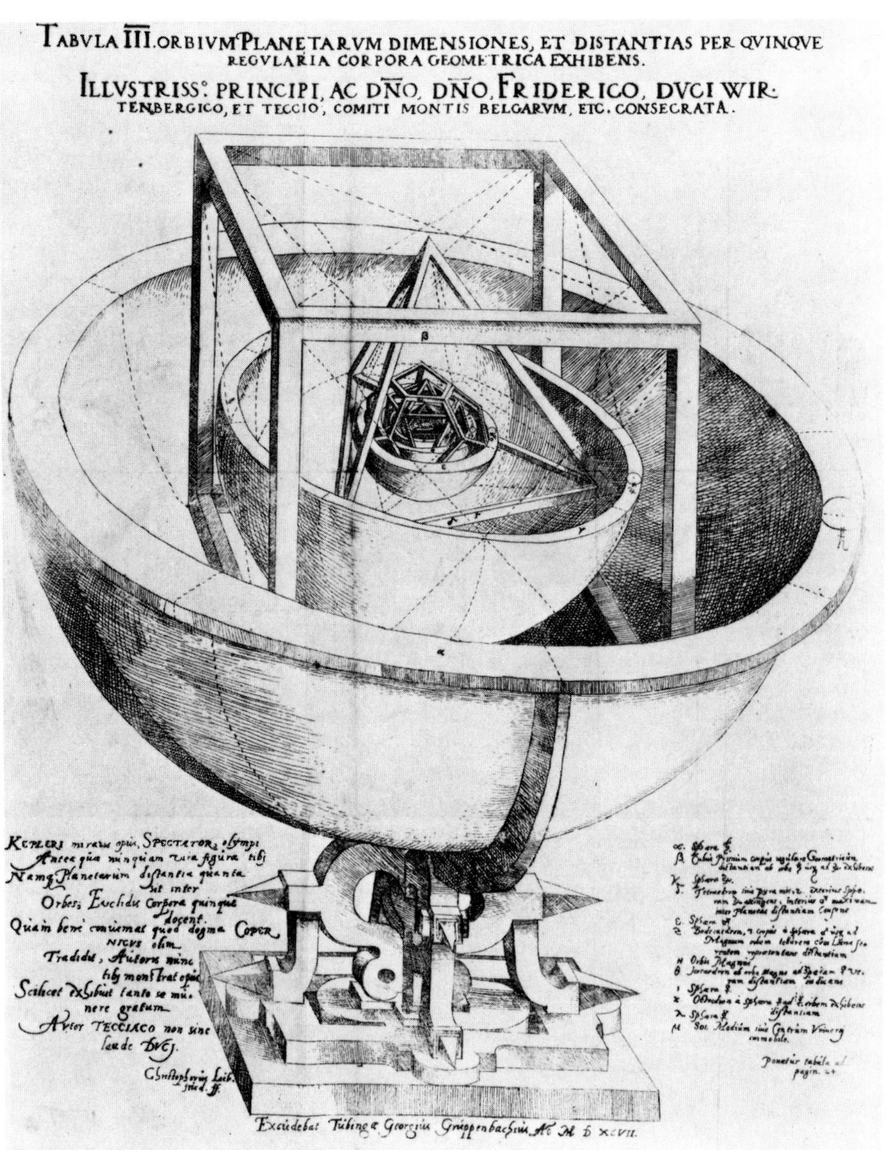

Fig. 1 Planetary orbs and regular polyhedra, from Kepler, *Mysterium Cosmographicum* (Tübingen, 1596). Photograph courtesy of the Trustees of the Science Museum, London.

Saturn – cube
Jupiter – tetrahedron
Mars – dodecahedron
Venus – icosahedron
Mercury – octahedron.[11]

The mathematical properties of the polyhedra, already discussed at some length in earlier chapters of the *Mysterium Cosmographicum*, are then used to explain the known astrological characters of the planets. For instance, we are told that

> In the case of Jupiter first, of Saturn next, and lastly of Mercury, their calm and the steadiness of their character are the result of the fewness of their faces [i.e. the faces of the polyhedra associated with the planets]; in the case of Venus and Mars their turbulence and changeability are due to their large number of faces. Woman is always fickle and capricious; and the shape of Venus is the most capricious and variable of all.[12]

The relations between planets are explained similarly. For example, the 'friendship' between Jupiter, Venus and Mercury is said to arise from the fact that the polyhedra associated with them all have triangular faces.[13] The style of argument is exactly that of Plato's distribution of the polyhedral shapes among the five elements in *Timaeus* (55E ff) – though Kepler's explanations, which deal with rather vaguer properties, seem less appropriate to his subject than Plato's did to his. Kepler's arguments nonetheless clearly echo Plato's; and we may note also that, like the mathematical forms in *Timaeus*, Kepler's polyhedra, though shown in his diagram as lying between the orbs of the planets, have no actual existence in the physical world.

The explanation of astrological aspects by means of the polyhedral archetype[14] is much more complicated, and considerably more tentative, than the explanation of the characters of the planets. Kepler begins by discussing musical consonances, that is, the ratios of lengths of vibrating strings which will give combinations of notes that are pleasing to the ear. Particular 'consonant' ratios are then associated with individual regular polyhedra – usually by appealing to the divisions set up in a circle by the inscription in it

[11] *Mysterium Cosmographicum*, Chapter IX, KGW 1, pp. 34–36. Latin text with English translation in *Johannes Kepler. Mysterium Cosmographicum. The Secret of the Universe*, translation by A. M. Duncan, introduction and commentary by E. J. Aiton, with a preface by I. B. Cohen, New York (Abaris Books) 1981, pp. 114–119. French translation in Jean Kepler, *Le Secret du Monde*, introduction, traduction et notes de A. P. Segonds, à partir d'un essai initial de L.-P. Cousin, Paris ('Les Belles Lettres') 1984, pp. 66–68.

[12] Kepler, *Mysterium Cosmographicum*, Chapter IX, KGW 1, p. 36, l. 16, trans. Duncan, *op. cit.* in note 11, p. 117, l. 42ff. Aiton's note points out that the remark about woman's fickleness, 'varium et mutabile semper fœmina', is a quotation from Vergil (*Aeneid* IV, 569).

[13] *Mysterium Cosmographicum*, Chapter IX, KGW 1, p. 35, l. 22, trans. Duncan, *op. cit.* in note 11 above, p. 117, l. 7ff.

[14] *Mysterium Cosmographicum*, Chapter XII, KGW 1, pp. 39–43, trans. Duncan, *op. cit.* in note 11 above, pp. 130–137.

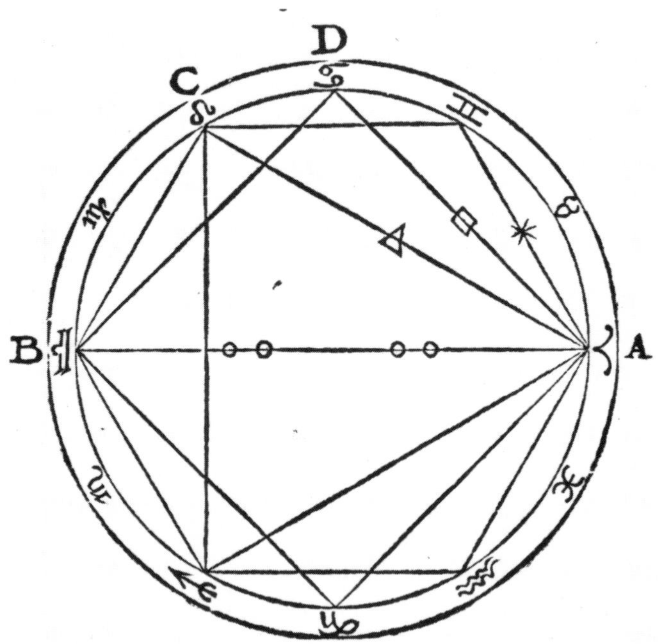

Diapaſon ſeu dupla ratio tripliciter reſpondet: Vel
- Totius circuli ad dimidium
- A B C ad A C, nēpe 8 ad 4.
- A C B ad A D, ſex ad tria.

Diapente ſeu ſeſquialtera, item tripliciter: Vel
- Totius circuli, ſeu 12.8
- D A B, ideſt, 9 ad 6
- A B, ideſt, 6 ad A C, ideſt 4.

Diateſſaron ſeu ſeſquitertia, item tripliciter: Vel
- Totius circuli ad A B C D, ſeu 12 ad 4 in A B D.
- A B C ad A B, ideſt 8 ad 6.
- A C ad A D, 4 ad 3.

Diapaſon & Diapente, item tripliciter, Biſdiapaſon uero dupliciter Tonus ſemel.

Fig. 2 Musical consonances and astrological aspects, from Ptolemy, *Harmonica*, trans. Gogava (Venice, 1562). Photograph courtesy of the British Library Board.

of a polygon associated with the solid in question. Kepler points out that the divisions of a string to give consonant ratios are like the divisions set up in the circle of the Zodiac by planets that are at aspect to one another.[15] He has thus established a rather complicated series of relations between polyhedra, aspects and some (but not quite all) of the musical ratios, and the chapter ends with some speculations as to why there should be some musical ratios with no corresponding astrological ratio.

The relation between astrology and music theory that was to become important in Kepler's later work thus makes its appeareance in his very earliest attempt to explain aspects. The origin of Kepler's theory is (at this time still indirectly)[16] the *Harmonica* of Ptolemy, a work to which Kepler refers in the closing lines of his chapter.[17] Ptolemy's scheme of equating musical and astrological ratios is shown in Figure 2. The scheme Kepler puts forward in the *Mysterium Cosmographicum* is much less neat, and even at the time, Kepler himself clearly found it somewhat unsatisfactory – to judge by the remarks at the end of his chapter. One of the remarks is, however, decidedly misleading: Kepler says he will leave it to the ingenuity of others to make a further study of the relationship between consonances and aspects.[18] It was not Kepler's habit to abandon problems half-solved, and his correspondence shows that he lost little time in returning to this one, both by attempting – eventually successfully – to procure a copy of Ptolemy's *Harmonica*[19] and by developing some further ideas of his own. The scheme he favoured (tentatively) in May 1599 is shown in Figure 3 (see over). Here, the circle of the Zodiac has been opened out to show aspects as divisions of a straight line, in the same way that consonant ratios are shown as divisions of the string of the monochord. Kepler appealed to observation to confirm the existence of the new aspects that he showed in this series of diagrams. In fact, he used the new aspects to make weather predictions in the calendar part of *De Fundamentis Astrologiae Certioribus* and called his readers' especial attention to these particular predictions for the purpose of collecting further evidence for or against his innovations.[20]

There are obvious problems associated with appealing to observational confirmation of the effects of aspects upon the weather. Not only will the data tend to be somewhat subjective (how to distinguish freshening of the wind from a fall in temperature?) but aspects are in fact so numerous that for any given change in the weather one could almost certainly find an appropriate recent aspect. Moreover, changes in the weather are also so

[15] *Mysterium Cosmographicum*, Chapter XII, KGW 1, p. 41, l. 22ff, trans. Duncan, *op. cit.* in note 11 above, p. 135, l. 30ff.

[16] See U. Klein, 'Johannes Keplers Bemühungen um die Harmonieschriften des Ptolemaios und Porphyrios', in *Johannes Kepler Werk und Leistung*, G. Maar (ed.), Linz, 1971, pp. 51–60.

[17] *Mysterium Cosmographicum*, Chapter XII, KGW 1, p. 43, l. 27, trans. Duncan, *op. cit.* in note 11 above, p. 137.

[18] *Mysterium Cosmographicum*, Chapter XII, KGW 1, p. 43, l. 26, trans. Duncan, *op. cit.* in note 11 above, p. 137, l. 29.

[19] See Klein, *op. cit.* in note 16 above.

[20] E.g. *De Fundamentis Astrologiae Certioribus*, Thesis LII, KGW 4, p. 27, l. 32ff.

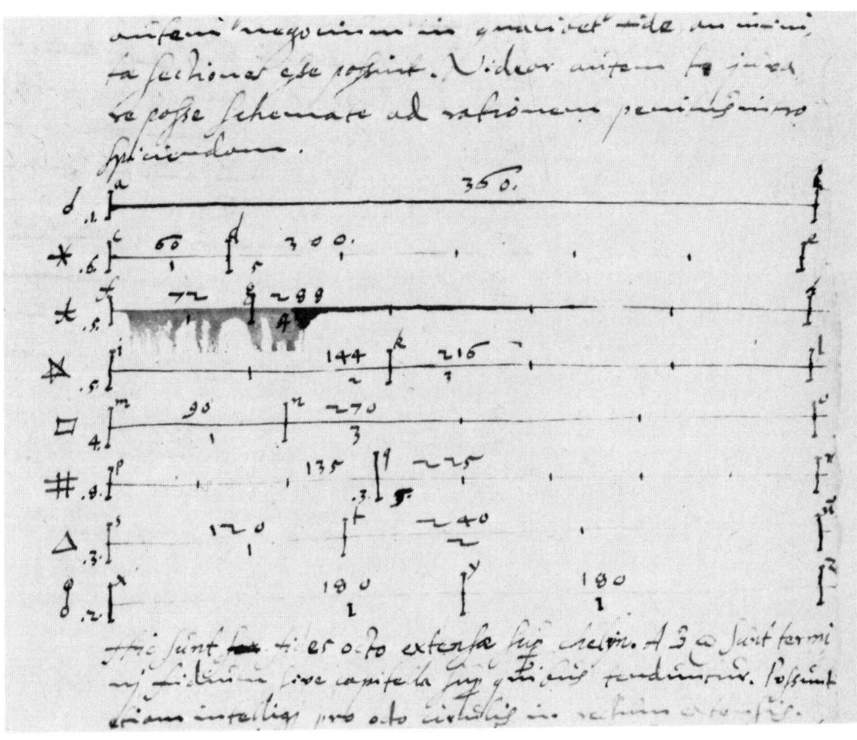

Fig. 3 Correspondence of astrological aspects and harmonic ratios, from a letter written by Kepler to Herwart von Hohenburg (30 May 1599).

Ptolemaic aspects: AB, conjuction (0°, undivided string) – used as an aspect by Ptolemy, though he does not define it as such;
CDE, sextile (60°, 1:5);
MNO, quadrature (90°, 1:3);
STU, trine (120°, 1:2);
XYZ, opposition (180°, 1:1);

New aspects: FGH, quintile (72°, 1:4);
IKL, biquintile (144°, 2:3);
PQR, sesquiquadrature (135°, 3:5).

Photograph courtesy of the Bayerische Staatsbibliothek, Munich.

numerous as to make it easy to find an effect to correspond to any aspect.[21] At least one of these objections to observational testing of aspects occurred to one of Kepler's regular correspondents, the physician Johann Georg Brengger, who noted, in a letter to Kepler dated 7 March 1608:

> You say you have confirmed by meteorological experience that there exist the additional aspects quintile, biquintile and sesquiquadrature [corresponding to angles between celestial bodies of 72°, 144° and 135° – see Figure 3]. I myself should like to see an example of this observational material, for with such a number and variety of aspects always occurring, so that one is unsure to which of them one should ascribe a change in the atmosphere, I do not know how I should make an observational test, or even whether I should find it possible to do so.[22]

Kepler replied on 5 April 1608, and provided an example that meets Brengger's objections:

> In 1600, when from 23 April until 2 May, New Style, there were no primary aspects [i.e. the standard aspects used by Ptolemy – see Figure 2], and Magini's tables showed Saturn and Jupiter to be at quintile, on 1 May there was a very heavy fall of snow both in Prague and in Styria for Ferdinand's wedding, and the jousting had to be cancelled (*hastiludia impedita fuerunt*). From observing the heavens, it was found that during these same days Saturn and Jupiter were 72° apart [i.e. at quintile, see Figure 3]. Tycho's students made the check on my behalf with Tycho's quadrant.[23]

This is, indeed, an exemplary piece of observational confirmation: other aspects have been avoided[24] and, although he took his cue from Magini's tables, Kepler had made use of the most accurate observing instrument then available to check the actual angular separation of the planets. His reason for not making the measurements himself was presumably his bad eyesight.[25]

Kepler's attitude to the relationship between theory and observation in astrology thoroughly confirms Simon's finding that Kepler's astrological work in general shows him applying the same standards in his astrology as in

21 This, at least, was my own experience in 1974 and 1975 when attempting to find *post factum* astrological explanations for the weather in London. As I remember, the cold spell in May 1975 corresponded, roughly, with Saturn's being at aspect to the Sun, and the very hot dry Summer that followed was marked by aspects between the Sun, Mars and Jupiter in June and July. (I was giving daily lectures in a planetarium at the time so I was familiar with the positions of the planets.)

22 Brengger to Kepler, 7 March 1608, letter 480, ll. 6–11, KGW 16, p. 114.

23 Kepler to Brengger, 5 April 1608, letter 488, ll. 9–15, KGW 16, pp. 137–138.

24 They are avoided also in the observations proposed in *De Fundamentis Astrologiae Certioribus*, e.g. in the passage referred to in note 20 above.

25 See M. Caspar, *Kepler*, ed. and trans. D. Hellman, London and New York 1959. (Original German edition Stuttgart, 1938.)

his astronomy.[26] However, one must not push the parallel too far, since it is clear that Kepler regarded astrology as inherently imprecise in its predictions and there is therefore no exact astrological equivalent to the astronomical demand that a model orbit of Mars should yield planetary positions as accurate as the observations that were used to construct it.[27] Nevertheless, as we have seen, Kepler did believe that the effects of aspects were sufficiently pronounced for it to be possible to use them to confirm the existence of aspects not considered by Ptolemy.

Although, as Simon has shown, many of Kepler's attempted reforms of astrology were very radical,[28] this reform of the number of aspects is not. Astrology had always been regarded as fairly unreliable – Kepler himself compares it to another hit-and-miss art, that of the physician[29] – and suggestions for minor reforms are a feature of the astrology of all periods.[30] It is presumably a tribute to the unsettled nature of the weather in Central Europe in the years from 1593 to about 1609 that by the latter date Kepler had, apparently, decided that there were more astrological aspects than musical consonances. This made it impossible for him to adopt a simple scheme like that used by Ptolemy, in which there is an exact correspondence between aspects and consonances (see Figure 2). However, he continued to follow Ptolemy's example in regarding the consonances as logically prior to the aspects, so that his attempts to explain aspects are, in effect, suggestions for how the underlying harmony of the Universe expresses itself in a slightly different way in the slightly lower form of aspects. Thus geometrical causes are given for consonances and then the same causes are shown to act in a modified manner to produce aspects.[31]

26 G. Simon, *Kepler astronome astrologue*, Paris, 1979.

27 We may note in passing that this astronomical demand – made in a crucial passage of the *Astronomia Nova* (detailed reference in note 37 below) – is, in general, untenable. Some loss of accuracy should be expected as a result of carrying out calculations, as is proved in any university course on Numerical Analysis. (At a more elementary level one finds the rule of thumb whereby the last figure of an answer is rounded off if mathematical tables have been used.) On Kepler's belief that astrological prediction could never be precise see my 'A Lutheran Astrologer' (full reference in note 5 above).

28 See Simon, *op. cit.* in note 26 above; and G. Simon, 'Kepler's Astrology: the Direction of a Reform', *Vistas in Astronomy*, 18, 1975, 439–448, where we are told (p. 446) 'the remedy [i.e. the reform designed to cure the faults of astrology] was clearly one of those that in the end kill the patient'.

29 Kepler, *Tertius Interveniens*, Frankfurt, 1610, Thesis XII, KGW 4, p. 163.

30 See A. Bouché-Leclercq, *L'Astrologie grecque*, Paris, 1899 (reprint Brussels, 1963); A. A. Long, 'Astrology: arguments pro and contra', in J. Barnes *et al.* (eds), *Science and Speculation. Studies in Hellenistic theory and practice*, CUP, 1982, 165–192; E. Garin, 'Magic and Astrology in the Civilisation of the Renaissance', in *Science and Civic Life in the Italian Renaissance*, trans. P. Munz, New York, 1969 (first edition in *Medioevo e Rinascimento*, Rome, 1966); W. Shumaker, *The Occult Sciences in the Renaissance*, University of California Press, 1972; M. E. Bowden, 'The Scientific Revolution in Astrology: the English Reformers, 1558–1686', *unpublished PhD thesis* (Yale University), 1974; C. Webster, *From Paracelsus to Newton*, CUP 1982; S. Jenks, 'Astrometeorology in the Middle Ages', *Isis*, 74, 1983, 185–210.

31 For Kepler's geometrical explanation of musical ratios see J. V. Field, 'Kepler's Rejection of Numerology', in B. Vickers (ed.), *Occult and Scientific Mentalities in the Renaissance*, CUP 1984, 273–296. A more detailed account of the evolution of Kepler's explanation of aspects is given in the paper referred to in note 5 above.

It is not clear exactly when Kepler decided that separate geometrical causes should be given for aspects. There seems to be nothing in his published works or in his extensive surviving correspondence (now also published) to mark the exact date of the change of mind which must have taken place between 1610, when the aspects are explained as subsidiary to the consonances,[32] and 1618, when they are given a separate geometrical cause as part of Kepler's scheme of *musica mundana* in *Harmonices Mundi Libri V* (Linz, 1619).[33]

Kepler had intended to write a book about world harmony for many years. We find sketches of the proposed work, with slightly differing titles, in two letters written in 1599.[34] Moreover, many of the mathematical results used in the *Harmonice Mundi* seem to have been worked out at this same time.[35] Nonetheless, the work cannot have taken its final form until Kepler had decided to give a separate mathematical account of musical 'harmonies' (consonances) and astrological 'harmonies' (aspects), for the structure of the published work corresponds closely with the intellectual structure of Kepler's theory.

The bases of the theory are laid in the first two books, which are entirely concerned with geometry. The first considers problems of inscribing regular polygons in a circle. This is a re-working of Euclid's classification of surds in Book X of the *Elements*. In contrast, the work of *Harmonices Mundi* Book II is highly original: Kepler considers how regular polygons can be fitted together to form a 'congruence', that is, a flat pattern which will cover the plane (a tessellation) or a polyhedron whose vertices are all alike (a uniform polyhedron).[36] Both these two geometrical books of the *Harmonice Mundi* are organised in the manner of the *Elements*, as orderly series of definitions, axioms and propositions. Both end with sections marked 'Conclusion' – Book II in fact having two such sections. There is thus no room to doubt the function of the two books: the sections of conclusions show that each is designed to establish a hierarchy among the regular polygons. The hierarchy established in Book I depends upon the degree of 'demonstrability' (or 'constructibility') of the side of the regular polygon, that is the number of geometrical operations required to construct it in the circle. The hierarchy

[32] Kepler, *Tertius Interveniens*, Frankfurt, 1610, Thesis LIX, KGW 4, p. 203.

[33] The title of this work has been the subject of much scholarly niggling. The first word, though written in Latin script, is Greek: ἁρμονικῆς, which is the genitive singular form of ἁρμονική (the feminine form of the adjective ἁρμονικός, -ή, -όν, meaning 'skilled in music'). Standing alone, ἁρμονική (the noun ἐπιστήμη 'learning' being understood) signifies the theory of music. Liddell and Scott give an example of this usage from Aristotle's *Metaphysics*, where τὰ ἁρμονικά (the title of Ptolemy's treatise on music) is also used to convey the same sense. Since the word 'harmonic' – formed by analogy with 'arithmetic' – already has a quite different meaning in musical theory, the best translation of Kepler's title would seem to be 'Five Books of the Harmony of the World'.

[34] Kepler to Maestlin, 29 August 1599, letter 132, l. 139, KGW 14, p. 46; Kepler to Herwart, 14 December 1599, letter 148, l. 12, KGW 14, p. 100.

[35] See Kepler to Herwart, 6 August 1599, letter 130, ll. 220–651, KGW 14, pp. 27–39.

[36] *Harmonices Mundi* Book II is translated in J. V. Field, 'Kepler's Star Polyhedra', *Vistas in Astronomy*, 23, 1979, 109–141.

established in Book II depends upon the degree of 'sociability' of the polygon, that is, the number of 'congruences' in which it can participate. The second conclusion of Book II compares and contrasts the two hierarchies. Since the hierarchy of 'demonstrability' will be used to explain consonances (in *Harmonices Mundi* Book III) and the hierarchy of 'sociability' will be used to explain aspects (in *Harmonices Mundi* Book IV), Kepler's comparison between the two hierarchies is an indication that he is still mindful of the Ptolemaic musical theory which had influenced his earlier work. Here, as elsewhere, Kepler seems to prefer to keep track of his departures from Ptolemy. The spirit is like that in which, at the very point in the *Astronomia Nova* when he rejects orbits compounded of circles, he remarks that, for his part, Ptolemy had had no reason to reject them, since they gave results accurate enough to fit the observations available at the time.[37]

The influence of the *Almagest* upon the *Astronomia Nova* is no more than an indication of the all-pervasive influence of Ptolemy's treatise upon succeeding generations of astronomers. The influence of Ptolemy's *Harmonica* upon Kepler's *Harmonice Mundi* is, however, rather more precisely definable. For instance, having established the geometrical bases of world harmony, Kepler proceeds to find harmonies in exactly the same relationships as those in which Ptolemy had found them, namely in the ratios that define musical consonances, in the angles that define astrological aspects and in the motions of the planets. The consonances, together with a complete musical theory derived from them, are explained and set out in *Harmonices Mundi* Book III.[38] Book IV is concerned with astrological harmonies (aspects). Kepler's final book, Book V, shows that the musical consonances are found among the extreme angular velocities of the planets (as seen from the Sun) and that there exists an observable celestial counterpart to the polyphonic human music whose theory was set out in Book III.

Kepler's harmonic version of the analogy between Man (the microcosm) and the Universe (the macrocosm) is a natural consequence of the fact that each is made in the image of God and thus embodies the geometrical archetype that guided the Creator. The fusion between Christianity and Platonism is complete. Kepler had described it in characteristic terms, though more briefly than elsewhere, in his letter to Galileo about the *Sidereus Nuncius* (Venice, 1610):

> Geometry is one and eternal, shining in the mind of God. That share in it accorded to men is one of the reasons that Man is the image of God.[39]

As we have seen, astrology does not play a very important part in the *Mysterium Cosmographicum* – and the notes Kepler added for the second

37 *Astronomia Nova*, Heidelberg, 1609, Chapter XIX, KGW 3, p. 177, l. 37ff.

38 See D. P. Walker, *Studies in Musical Science in the Late Renaissance*, London, 1978, and J. V. Field, 'Kepler's Rejection of Numerology' (full reference in note 31 above). The technical details of Kepler's musical theory are discussed at length in M. Dickreiter, *Der Musiktheoretiker Johannes Kepler*, Berne and Munich, 1973, and in H. F. Cohen, *Quantifying Music*, Dordrecht and Boston, 1984.

39 Kepler, *Dissertatio cum Nuncio Sidereo*, Prague, 1610, KGW 4, p. 308, ll. 9–10.

edition of the work in 1621 suggest that by then he regarded most of the astrological content as irrelevant to the main purpose of the book.[40] He did not, however, reject his earlier explanation of the characters of the planets by reference to the corresponding regular polyhedra. His note merely invites the reader to compare the reasons with those given by Ptolemy in the *Tetrabiblos* and the *Harmonica*.[41] Unlike that of the *Mysterium Cosmographicum*, the astrology of *Harmonices Mundi Libri V* is, as I have already suggested, an integral part of the cosmological model which the work describes, namely a scheme of universal harmonies (of which a slightly modified version of the theory described in the *Mysterium Cosmographicum* is also a part). The astrology of *Harmonices Mundi* Book IV has much the same status as the astronomy of Book V. Astrological aspects, like the eccentricities of the planetary orbits, are seen as the physical consequences of the mathematical truths described in the earlier part of the work.

We have already noted that, like other writers on *musica mundana*, Kepler is concerned with harmony in a wider sense than the purely musical one. It is in *Harmonices Mundi* Book IV that this wider sense first becomes relevant, and the first three chapters are accordingly concerned with the problem of the essence of harmonies (in sensible and abstract things), the soul's faculty for perceiving harmonies, and in what things harmonies may be perceived (by God or by Man). The fourth chapter discusses the distinctions to be made between the musical harmonies (consonances) considered in Book III and the astrological ones to be considered in Kepler's present book. These distinctions are all 'physical' in the wide sense in which Kepler uses the word, that is in something close to its etymological sense, to mean 'pertaining to Nature or to the way the natural world works' (the natural world being taken to include celestial as well as terrestrial phenomena).[42]

The marginal notes provide a succinct summary of Kepler's reasoning: 'The harmonies in this book are narrower' (i.e. they do not interact as musical harmonies do), 'These harmonies concern angles', 'In the form of arcs of the Zodiac', 'They are not truly celestial', 'but terrestrial' (i.e. they are perceived from the Earth, being angles made at the Earth, whereas musical harmonies do not depend upon the Earth), and so on.[43] Harmonies, of any kind, are perceived by the human soul, and since the Earth responds to harmonies (i.e. to astrological aspects) Kepler argues that the Earth too must have a soul.[44]

Having established the nature of the harmonies with which we are to be concerned, Kepler proceeds to his geometrical explanation of their nature and

[40] General note on Chapter IX and first note on Chapter XI, KGW 8, pp. 59 and 62, trans. by Duncan in *op. cit.* in note 11 above, pp. 119 and 125.

[41] General note on Chapter IX, full reference in note 40 above.

[42] See N. Jardine, 'The Forging of Modern Realism: Clavius and Kepler against the Sceptics', *Studies in the History and Philosophy of Science*, 10, 1979, 141–173.

[43] *Harmonices Mundi* Book IV, Chapter IV, KGW 6, pp. 234–235.

[44] *Harmonices Mundi* Book IV, Chapter IV, KGW 6, pp. 236–237. He had made this same suggestion, on the same grounds, many years earlier in *De Fundamentis Astrologiae Certioribus*, Prague, 1602, Theses XL to XLIII (KGW 4, pp. 23–24).

number, setting out his work as a series of definitions, axioms and propositions, with only a very small amount of linking text.[45] The style is close to that of the geometrical work of *Harmonices Mundi* Books I and II – or, indeed, Kepler's *Dioptrice* (Augsburg, 1611) – and is clearly intended to emphasise the mathematical nature of Kepler's reasoning.

After definitions of what constitutes an astrological configuration, and what we mean by describing it as powerful (*efficax*), that is an 'aspect', Kepler states two axioms, upon which, he tells us, the whole of his discussion will depend. They are:

> Axiom I
> The arc of the Zodiac cut off by the side of a convex or star polygon which forms congruences and is knowable [i.e. 'demonstrable'] measures the angle of a powerful configuration.
>
> Axiom II
> The angle of a convex or star polygon which forms congruences and is knowable is the measure of the angle of a powerful configuration.[46]

These axioms are probably more easily understood if we move a little ahead of ourselves and look at the diagrams of aspects which accompany later sections of Kepler's chapter. These diagrams are shown in our Figure 4. To illustrate the axioms we may take the aspect (sextile) shown in the centre of Figure 4c. The positions of the two heavenly bodies are each marked with a small star. In accordance with Axiom I, the arc between the bodies is that cut off from the circle by the inscription of the side of a regular hexagon, a polygon which will form congruences (as Kepler has proved in Book II)[47] and is knowable, that is, can be constructed in the circle by means of straight edge and compasses (as Kepler has proved in Book I).[48] In accordance with Axiom II, the angle of the configuration (the angle between the lines joining the bodies to the centre) is the angle of an equilateral triangle, a polygon which will form congruences (as Kepler has proved in Book II)[49] and which is knowable (as Kepler has proved in Book I).[50]

The paragraphs immediately following the statement of the two axioms are concerned with expanding their meaning. In particular, Kepler points out that the two axioms in fact lead to the same set of configurations. Again, this is most easily illustrated by reference to the diagrams of aspects given later in the chapter, shown in our Figure 4, from which we can see that the angle subtended by the side of a congruence-forming and knowable polygon at the centre of its circumcircle is always equal to the angle of another

45 *Harmonices Mundi* Book IV, Chapter V, KGW 6, pp. 239–256.
46 *Harmonices Mundi* Book IV, Chapter V, KGW 6, p. 241.
47 Sections XVIII, XX (for tessellations), XXVIII (for polyhedra), KGW 6, pp. 71–77, 84–87. The hexagon is of the third rank in congruence.
48 Section XXXVIII, KGW 6, p. 39. The hexagon is of the second rank in demonstrability.
49 Sections XVIII, XIX, XX (for tessellations), and XXV and XXVIII (for polyhedra), KGW 6, pp. 71–77, 78–80, 84–87. The triangle is of the first rank in congruence.
50 Section XXXVIII, KGW 6, p. 39. The triangle is of the third rank in demonstrability.

Fig. 4a–f Aspects as given in Kepler, *Harmonices Mundi* Book IV, Chapter V. Photographs courtesy of the Trustees of the Science Museum, London.

CAP. V.

Si duo prima axiomata ſunt conſentanea vero, erit & hoc: quia propter quod unumquodque eſt tale; illo intenſo, iſtud etiam magis erit tale, Sic autem intellige; quòd in figura circumferentiali prior ſit comparatio graduum Congruentiæ, in centrali prior graduum Scibilitatis, denique potiores partes circumferentialis figuræ.

Propoſitio IX.

Configurationes efficaces ſunt, quæ intercipiunt Arcus circuli Zodiaci iſtos:

Gr. 180. Oppoſitio ☍, ex Diametro circuli: ut in Fig. I.
Gr. 90. Quadratus □, ex Tetragono: ut in figura II.
Gr. 120. Trinus △. & 60. Sextilis ⚹, ex Trigono & Hexagono, ut in figura III. IV.
Gr. 45. Octilis vel Sequadri, & 135. Trioctilis vel Seſquadri ⚼ ex Octogono & Stella ejus: ut in fig. V. VI.
Gr. 30. Semiſexti ⚺, & 150 Quinquuncis, Ex Dodecagono & Stella ejus: ut in fig. VII. VIII.
Gr. 72. Quintilis ±, & 108. Tridecilis ſeu Seſquintilis: ex Pentagono & Stella Decagonica: ut in fig. IX. X.
Gr. 144. Biquintilis ✡, & 36. Semiquintilis ſeu Decilis: ex Stella Pentagonica & Decagono, ut in figura XI. XII.

Quòd hæ figuræ ſint Scibiles & demonſtrabiles, oſtenſum eſt Libro I. quòd & Congruæ, libro II. Quòd verò configurationes expreſſorum à talibus arcuum ſint efficaces, id habent axiomata I. II. præmiſſa.

Propoſitio X.

Efficacitatis Aſpectuum gradus primus & fortiſſimus, eſt Conjunctionis ☌ & Oppoſitionis ☍.

Characteres uſitati.

☍

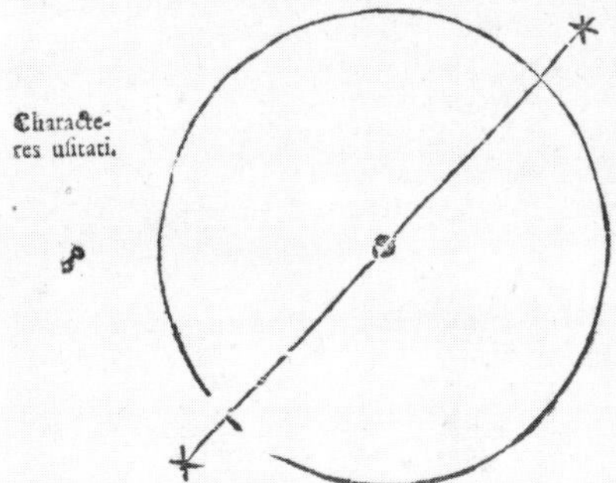

Nam in Conjunctione congruunt radij duo in eandem lineam, & ab eadem plaga deſcendunt; in Oppoſitione ☍ à plagis quidem diverſis deſcendentes, nihilò tamen minus fiunt partes unius continuæ lineæ. Hæc verò perfectiſſima eſt Congruentia & principium quoddam omnis congruentiæ. Sic cùm conjunctionem repræſentet punctum ſignatum in circumferentiâ circuli; oppoſitionem verò, Diameter; hæc certè ſunt principia, illa & menſura omnis in hoc genere ſcientiæ; cùm omnis in circulo lineæ rectæ ſcientia contineatur determinatione demonſtrativâ per Diametri vel longitudinem vel potentiam: ut libro primo

Fig. 4a From Kepler, *Harmonices Mundi* Book IV, Chapter V.

HARMONICIS LIB. IV. 145

CAP. V.

primo patuit. Ergò per Axioma III. principium etiam Efficacitatis in his est Aspectibus.

Propositio XI.

SEcundus in Aspectuum Efficacitate gradus est Quadrati □.

In Quadrato enim concurrunt prærogativæ multæ, quarum prima, quod similis est centralis figura, circumferentiali: quare quoscunque illa gradus obtinet in Congruentia & Scibilitate, iij quodammodo duplicati intelliguntur, respectu cæterorum Aspectuum. Sicut enim Quadratus* primus post Oppositum ab exilitate lineæ explicatur in aliquam latitudinem seu amplitudinem superficialem areæ Tetragonicæ: sic cæteri Aspectus ab identitate figurarum Aspectus Quadrati, discedunt in aliquam figurarum alteritatem. Cùm igitur aliàs in physicis unita virtus sit fortior, erit etiam in hac ideali & objectivâ impressione, major gradus fortitudinis, ubi figuræ locis distinctæ, altera sc. centralis, altera circumferentialis, specie eædem fuerint.

□

Deinde quantum ad Congruentiam, illa in Tetragono perfectissima est & omnivaria, nam secum ipsa congruit hæc figura in solido ad cubum formandum, qui mensura est omnis soliditatis, & congruit simplicissimè, ternis tantum angulis ascitis: congruit & in plano secum ipsa, quaternis angulis: congruit rursum in solido cum Trigono, Pentagono, Hexagono, Octogono, Decagono variè, ad formandas figuras solidas, congruit cum iisdem omnibus, insuperque & cum Dodecagono & Icosigono quadamtenus, ad planitiem sternendam: qua in proprietate illa à nullâ aliâ superatur.

Tertiò area Tetragoni est effabilis, quod principium est singularis alicujus & eximiæ Congruentiæ in plano; ut certus arearum hujus figuræ numerus absumat certum quadratorum diametri numerum, & sic figuræ non tantùm ipsæ inter se angulis & lateribus congruant, sed quodammodò, certis sc. suis lineis, etiam cum quadrati diametri lateribus. In hac proprietate Quadratus aspectus solum Semisextum habet ex parte socium. Vide lib. II.

Quartò, nec ignobilis est gradus scientiæ lateris, quod est effabile potentiâ; quo gradu præcellit cæteris figuris omnibus, excepto Sexangulo: neque tamen illi propterea loco cedit; cùm Scibilitas non sit comparanda Congruentiæ, ut explicatum est suprà; & verò valeat accumulatio prærogativarum, ad augendam Efficaciam, per Axioma III. hujus.

Propositio XII.

TErtius Efficacitatis Gradus est Trini △, Sextilis ⚹ & Semisexti ⚹.

Quod Trinum, Sextilem, & Semisextum, in eodem gradu colloco; T facit

Fig. 4b From Kepler, *Harmonices Mundi* Book IV, Chapter V.

146 DE CONFIGURATIONIBUS

CAP. V. *facit proprietatum non identitas, sed æquipollentia. Primùm eorum figuræ*

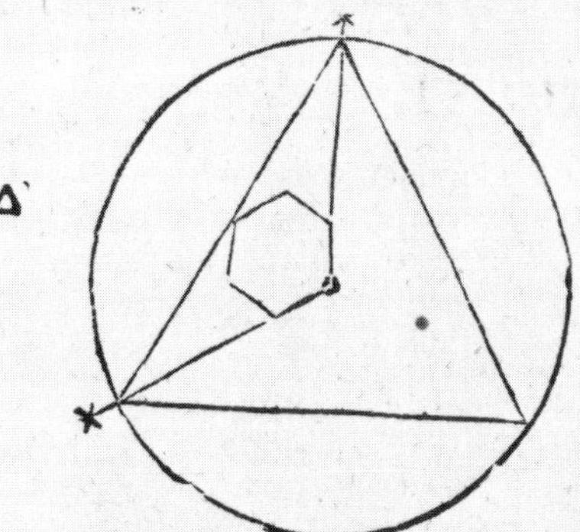

principales, in congruentia plana tradunt mutuas operas: coëunt n. & inter se variè & cum alijs, ut quadrato. Præcellunt quidem hic Trigonus & Hexagonus, quia etiam secum ipsæ singulæ species congruunt; præcellit Trigono Hexagonus, quia perfectissimam obtinet in plano congruentiam, solis sc. ternis angulis: præcellunt ambo Dodecagono, quia etiam in solido congruunt illi cum figuris alijs, quod nō potest Dodecagon' At vicissim præcellit reliquis Dodecagon' effabilitate, cùm illorū areæ sint mediales & sic ignobiliores: quæ arearū differentia, ut jam modo dictum, redundat in congruentiæ perfectionem.

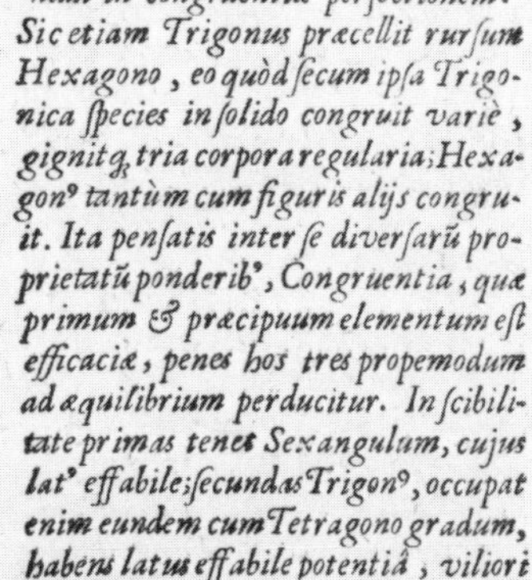

Sic etiam Trigonus præcellit rursum Hexagono, eo quòd secum ipsa Trigonica species in solido congruit variè, gignitque tria corpora regularia; Hexagon' tantùm cum figuris alijs congruit. Ita pensatis inter se diversarū proprietatū ponderib', Congruentia, quæ primum & præcipuum elementum est efficaciæ, penes hos tres propemodum ad æquilibrium perducitur. In scibilitate primas tenet Sexangulum, cujus lat' effabile; secundas Trigon', occupat enim eundem cum Tetragono gradum, habens latus effabile potentiâ, viliori tamen proportione: ultimus hic est Dodecagonus, habens latus ineffabile. Verùm scibilitas nec præcipuum est ad Efficaciam argumentum, nec in figura præcipuâ, hoc est circumferentiali, consideratur; sed tantum in centrali min' præcipuâ. Quæ si quid potest, Trinum paulò reddit efficaciorem Sextili, quia Trinum format angul' Hexagoni in centro; paulò min' utrisque efficacem Semisextū, quem metitur angul' Stellæ Dodecagonicæ in centro. Est tamē cæteris sequentib' nobilior scientia Semisexti, quia lat' centralis figuræ, in Ineffabilib' nobilissimæ est speciei, sc. Binomiū, & in earū subdivisione duplici, semp priores tēet, adeò ut cum sociâ suâ, latere circumferentialis figuræ, rectangulū effabile formet, quod est nota perfectionis penè absolutæ; adeòque etiam cum Trigono & Hexagono hanc figuram scibilitate facit contendere, propter hanc pensationem ineffabilitatis suæ, ponderosam admodum.

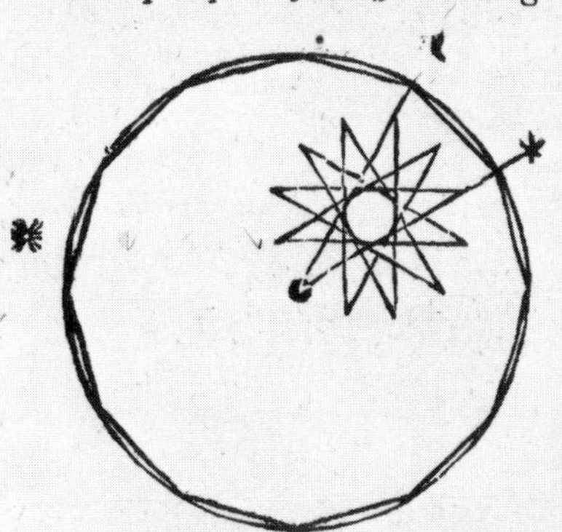

Propo-

Fig. 4c From Kepler, *Harmonices Mundi* Book IV, Chapter V.

HARMONICIS LIB. IV. 147

CAP. V.

Propositio XIII.

QUartus in Efficacitate Configurationum Gradus est Quintilis, Biquintilis, & Quincuncis.

His enim communis est congruentia figurarum primariarum totarũ in plano, non tamen singularum specierum secum ipsis; sed primarum duarum inter se mutuò, ultimæ cum alijs sibi cognatis. Præcellunt duo priores aspectus eo, quòd congruunt figuræ, Pentagonus & stella ejus, etiam in solido, faciuntque duas figuras solidas regulares; qua nobilitate penè associant aspectus suos Trino & Quadrato; stella Dodecagonica in solido non congruit. At vicissim præcellit & Dodecagonica, congruentiâ planâ; quam habet continuabilem in infinitum; cùm illæ non longè continuari possint sine mixturâ irregulari. Vide hæc omnia libr. II.

Quod scibilitatem attinet laterum in figuris centralibus; hîc etiam medio loco consistunt latera Decagoni & Tridecilis & Dodecagoni, quæ sunt hac in classe centrales, inter latus Trigoni præcedentis, & latera Pentagoni, stellæque Pentagonicæ, centralium figurarum in classe sequenti, Nam libro I. demonstratum est, prius esse in scientia, Decagonicum latus Pentagonico, Tridecile stellari Pentagonico. Itaque & Scibilitas eodem ducit, quo & Congruentia, per pr. VII: quæ hujus potissimùm demonstrationis causâ fuit præmittenda, ne Decilis vel Tridecilis præferrentur Quintili & Biquintili. Si verò quis missâ figurâ centrali, Scibilitatem potiùs in circumferentiali quærere velit, non minus quàm Congruentiam: etsi fatendum est, hoc pacto prælatum iri Decilem Quintili, Tridecilem Biquintili; at meminerit is, præcipuas esse partes Congruentiæ, ut

Fig. 4d From Kepler, *Harmonices Mundi* Book IV, Chapter V.

148 DE CONFIGURATIONIBUS

CAP. V.

ostendimus pr. IV. majus igitur est, & plus ad efficacitatem potest, creare figuram solidam (quæ est veluti idea quædam mathematica efficacitatis physicæ) quàm latus habere perfectiori gradu scibile. Latus quidem Dodecagoni hoc pacto confert aspectum suum in eandem classem cum subtensis decimæ parti circuli, tribusque decimis: quia contendunt inter se præstantiâ scibilitatis. Nam sicut sociantur inter se duæ illæ subtensæ, fitque minor majoris pars, in proportione divinâ sectionis secundùm extrema & medium: sic etiam latus Dodecagoni & latus ejus stellæ sociantur, & hoc etiam respectu sectionis & compositionis alicujus, non tamen proportionalis. Et hæc quidem biga posterior cadit in primam speciem Ineffabilium, quæ complectitur Binomines & Apotomas; at vicissim illa prior biga acquirit novam proprietatem sectionis secundùm extrema & medium: ut videre est lib. I. Quare non tantùm pensantur hi gradus, sed etiam præcellit nonnihil Decanguli latus. Rectè igitur factum, quod Aspectum Quincuncem seu Gr. 150. cum Quintili Gr. 72. & Biquintili 144. eodem gradu locavi, primâ tamen sede his datâ.

Propositio XIV.

QUintus, ultimus & imbecillissimus aspectuum Gradus est Decilis & Tridecilis, Octilis & Trioctilis.

Quintum locum feci Decili & Tridecili (Mæstlinus Semiquintilem, & Sesquiquintilem appellat) quos in Ephemeridibus hactenus omisi: quibus associavi Octilem & Trioctilem seu Sequadrum, & Sesquadrum; quos Calendariographi ex mea quidem suggestione & nonnullâ Ptolemæi authoritate, sed nimis calidè & & inconsideratè arripuerunt. Probandum igitur est utrumque, primùm imbecilliores esse hos quatuor, Quintili & Biquintili; deinde, Decilem & Tridecilem, fortiores Octili & Trioctili, parùm admodùm.

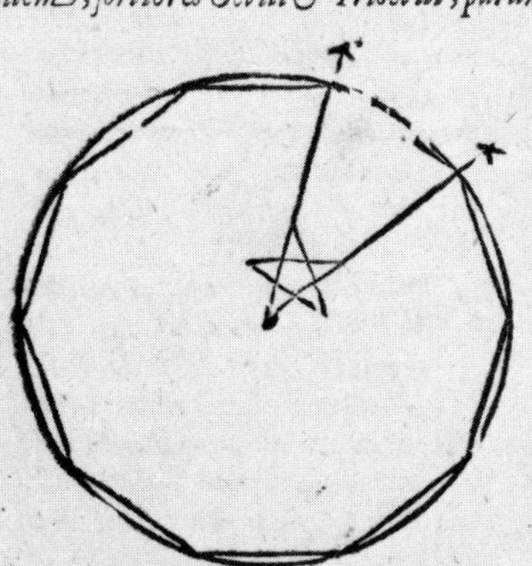

Cùm igitur Propositiones nostræ præcipuum ad Efficaciam momentum collocent in congruentia figuræ præcipuæ, hoc est circumferentialis: manifestum est, Pentagonum & stellam ejus, congruere cum suæ quamque speciei figuris, ad solidum perfectum formandum, ut jam modo dictum; congruere etiam inter se pulchrè ad planum sternendum. Decagonus vicissim & Octogonus cum stellis suis, quæque cum suæ speciei aliis, in solido congruere non possunt. Congruunt quidem, Decagonus & Octogonus, sed cum aliis non omnibus sui generis; stellæ verò inchoant aliquam congruentiam in solido, at non absolvunt: etiam in plano ignobilior est earum congruentia, quia nec mutuas tradunt operas, quævis figura cum suâ stellâ solitariæ, ut Pentagonus cum suâ: sed cum illis suis stellis, & Octogonus cum Tetragono, in societatem veniunt, congruentiæ alie-

Fig. 4e From Kepler, *Harmonices Mundi* Book IV, Chapter V.

HARMONICIS LIB. IV. 149

CAP. V.

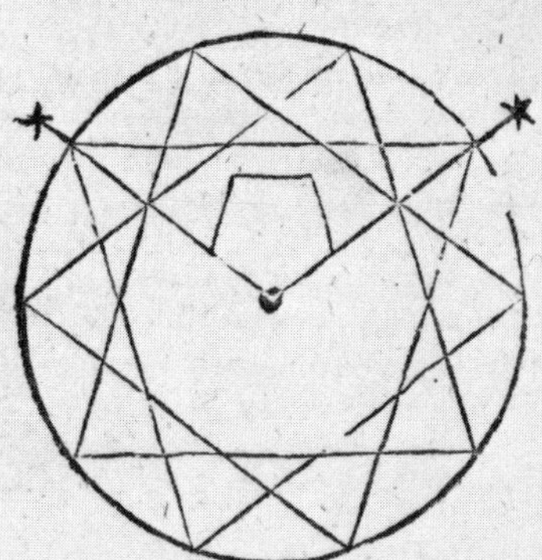

tiæ alienæ, illamque, quo minus continuari possit, ipse Decagonus impedit: Stella ejus etiam hiulcam in mediis interceptis spacijs facit congruentiam. Octogonus verò & stella ejus, alternis juvant continuationem congruentiæ, admixtis Quadratis; Congruentia fit diversiformis. Ita penè pares sunt hæ quatuor in congruentiâ planâ; præsertim cùm areas utræque figuræ habeant ineffabiles. At in scibilitate multùm præcellit secta Pentagonica. Primùm si centrales figuras consideremus, quæ sunt hic jam Pentagonus & Stella ejus, illarum quidem latera sub eãdem speciem ineffabilium cadunt, cum lateribus Octogoni & stellæ; existentia Elasson & Mizon: sin circumferentiales, quæ hic sunt latera Decagoni & stellæ ejus: illa non tantum sunt ex specie nobiliori Binominum & Apotomarum, cùm Octogonicæ lineæ sint ex quarta specie, quæ est Mizonum & Elassonum: sed acquirunt etiam omnia latera Pentagonicæ sectæ, nobilissimam proprietatem sectionis secundùm extrema & medium; quæ planè nihil attinet lineas Octogonicas. Quòd si Octogonica secta nonnihil præpollere visa est in Congruentiâ; hic jam vicissim, multò fortiùs deprimitur à Pentagonicâ. Rectè igitur utrasque, ut de præstantia contendentes, in unam classem redegi, præmissâ tamen Pentagonicâ. Consulatur de his identidem liber I.

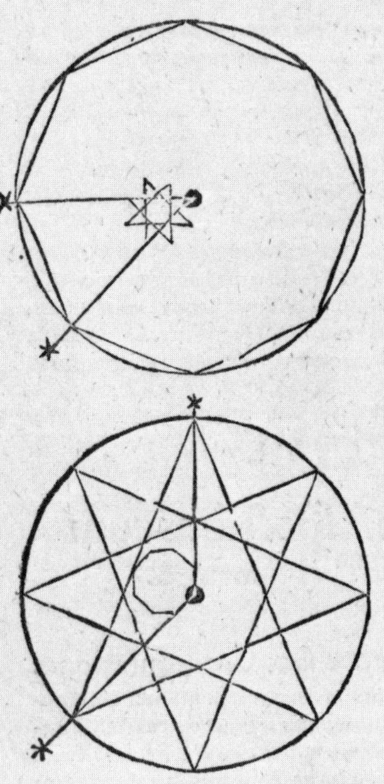

Est & peculiaris prærogativa Biquintilis, præ Tridecili & Trioctili, etiamque Quincunce, quòd stella Pentagonica, primaria sc. illius figura, aptissimum & Trigonici æmulum habet angulum: quia ut tres anguli Trigoni, sic etiam quinque anguli stellæ Pentagonicæ, junctim utrinque æquantur duobus rectis, ut sic latera angulos formantia, circuli arcus',

T 3 quæ-

Fig. 4f From Kepler, *Harmonices Mundi* Book IV, Chapter V.

congruence-forming and knowable polygon (for the square it is in fact an angle of the same polygon). Since both axioms give the same set of aspects, it would appear that Kepler has committed the mathematical solecism of employing two axioms where one would have sufficed. However, the use of both axioms has the advantage of allowing him to relate each aspect to two polygons: the 'central' one and the 'circumferential' one. Perhaps a decision between the two seemed unnecessarily arbitrary at this stage of Kepler's reasoning, for as we shall see, he does not seem to regard his axioms as being true by definition, but rather as being reasonable assumptions that might be susceptible of proof.

The series of propositions which follows this discussion of the axioms is concerned with establishing the relative importance of the 'central' and 'circumferential' polygons for their corresponding aspect, and deciding which properties of the polygons determine the properties of the aspects. The first proposition looks back to the consonances:

> Aspects are more closely connected with the circle and its arcs than the consonances are.[51]

This is proved simply by considering the relationship of each to its associated circle:

> the consonances do not depend immediately on the circle and its arcs on account of their being circular, but on account of the length of the parts, that is, their proportion one to another, which would be the same if the circle were straightened out into a line.[52] Whereas the Aspects, by definition I, are angles, which the circle measures with its arcs, and in no other way except by remaining what I have called it, that is, by continuing to have a circular shape and to remain complete.[53]

Kepler adds that the consonances do not always involve the whole circle, but only ratios of parts of it, whereas aspects do always concern the whole circle. This mixture of physical and mathematical reasoning is characteristic of most of the propositions, though the relative importance of the two components varies from proposition to proposition. For example, Proposition IV that

> Congruence of figures is more influential than Knowability in making a configuration powerful,[54]

is clearly mainly a matter of physics; whereas the following proposition, that

[51] KGW 6, p. 242, l. 28.
[52] Compare Ptolemy's system, shown in Figure 2 above.
[53] KGW 6, p. 242, l. 35ff.
[54] KGW 6, p. 245.

> Congruence is a property of the Circumferential rather than the central figure,

is entirely mathematical. By way of proof of this latter result, Kepler merely makes the (entirely reasonable) assertion that the capacity to form congruences is a property of the figure as a whole and the circumferential polygon is employed as a whole (since the circle goes through all its vertices) whereas the central polygon has only one of its angles at the centre.[55] This simple mathematical distinction justifies using the hierarchy of polygons defined by considering congruences to determine the relative status of aspects – once Kepler has shown (Proposition VI) that aspects are determined more by their circumferential than their central polygon, and (Proposition VII) that for the circumferential polygon congruence-forming is more important than knowability of the side (and vice versa for the central polygon).[56]

The following proposition (VIII) is a partial converse of Axiom I, namely that the arc of the circle cut off by a figure which does not form congruences does not correspond to an aspect. It is not clear why Kepler did not make this proposition an axiom. Possibly he thought it too important to be asserted without proof, for he comments

> Behold the cause why, although the knowable figures are infinite in number, though of various rank, yet aspects are few.[57]

However, it is of the nature of axioms to have just this importance, as Kepler had acknowledged after stating his two axioms earlier in this same chapter. Moreover, although the two axioms quoted are called axioms, the series of propositions which follows them does seem to have been at least partly designed to justify their acceptance, and Kepler's comment on his third axiom (which is introduced immediately after proposition VIII and states that the arcs of a circle corresponding to polygons of higher rank in Congruence and Knowability will give more powerful configurations) begins 'If the first two axioms are acceptable (*consentanea*) . . .'.[58] It therefore appears that the word 'axiom' does not mean quite the same to Kepler as it does to a modern mathematician – and seems to have done to Archimedes. Kepler's use of the word in this chapter is rather closer to Copernicus' usage in the *Commentariolus*, where 'axiom' apparently means something we shall use as if it were true but would like to prove one day if we can.[59]

Having stated this third, and final, axiom Kepler proceeds to list the twelve aspects whose existence can be deduced from the axioms, referring to

55 KGW 6, p. 245, ll. 35–38.
56 KGW 6, pp. 246–250.
57 KGW 6, p. 250, ll. 17–18.
58 KGW 6, p. 250, l. 23.
59 See N. Swerdlow, 'The Derivation and First Draft of Copernicus' Planetary Theory: A Translation of the Commentariolus with Commentary', *Proceedings of the American Philosophical Society*, 117, 1973, 423–512. Like Kepler, Copernicus also shows no anxiety to reduce his axioms to the minimum number or to ensure their mutual independence.

the results of Books I and II for proof that the polygons involved are congruence-forming and knowable.[60] The next group of propositions (X to XV) is concerned with establishing the relative degrees of power of the aspects, on the basis of the nobility of the corresponding polygons, as established in Books I and II, that is, using the criteria set out in Axiom III. Kepler works from the strongest aspects, opposition and conjunction, which correspond to the diameter of the circle[61] (see Figure 4a) to the weakest:

> configurations which hesitate between power and powerlessness, namely the 24° arc from the pentekaedecagon and the 18° arc from the icosigon.[62]

The lowest grade of configurations which are definitely accepted as aspects is that of the decile, tridecile, octile and trioctile aspect[63] (see Figure 4f). This enumeration of the effective configurations in order of decreasing power – which resembles the standard classification of consonances in music theory – brings Kepler's chapter to an end.

The next chapter (VI) returns to the comparison between astrological and musical harmonies, and discusses the fact that there is not the same number of aspects as of consonances. As we have seen, Kepler had already discussed the physical distinctions to be made between the two types of harmony (in Chapter IV). The new discussion is concerned with differences in the mathematical formulation (dictated by the physical differences) which have led to aspects which do not correspond exactly with the consonances described in *Harmonices Mundi* Book III. Chapter VI thus provides philosophical justification for the mathematical reasoning of Chapter V. Presumably Kepler thought he should first show his mathematical method would give the desired result and leave philosophical arguments until later.

The final chapter of *Harmonices Mundi* Book IV, Chapter VII, discusses in some detail the nature and activities of the soul of the Earth (whose existence was suggested in Chapter IV).[64] In neither of these two final chapters of Book IV is there any discussion of the practical business of astrology.[65] Indeed, in the first sentence of Chapter VI Kepler explicitly rules out such practical considerations, referring the reader instead to *De Stella Nova* (Prague, 1606).[66] Thus although the main title page of *Harmonices Mundi Libri V* describes the fourth book as 'metaphysical, psychological and astrological', and the title page of Book IV reads

[60] Proposition IX, KGW 6, pp. 250–251.
[61] Proposition X, KGW 6, p. 251.
[62] Proposition XV, KGW 6, p. 256. Kepler gives no diagrams for these aspects.
[63] Proposition XIV, KGW 6, p. 254.
[64] See passage referred to in note 44 above for Chapter IV. KGW 6, p. 264–286.
[65] Similarly, the astronomy of *Harmonices Mundi* Book V is unrelated to the usual tasks of an astronomer.
[66] KGW 6, p. 257, l. 4ff. The chapters to which Kepler refers are in fact concerned with arguing for the power of aspects and the significance of points of Great Conjunctions, see below.

> On the harmonic configurations of stellar rays at the Earth, and their effect on the weather and other natural phenomena,

the contents on the book are very little more than a discussion of the theoretical foundations of astrology.

It would seem that this restriction of the scope of Book IV indicates what Kepler takes to be the intrinsic limitations of his mathematical theory. In his notes for the second edition of the *Mysterium Cosmographicum* (Frankfurt, 1621) he says that when he first wrote the work he hoped to extend the use of mathematical archetypes to other things, but his attempts to do so had taught him that

> the heavens, the first of God's works, were laid out much more beautifully than the remaining small and common things.[67]

Kepler's exemplar in constructing his geometrical archetype was certainly Plato's *Timaeus*,[68] but, whereas Plato had given a mathematical description of the sublunary world as well as the celestial, Kepler found himself compelled to apply an Aristotelian distinction and confine his mathematical cosmology to the heavens. We must suppose he did so with regret, since he does not seem to have believed that celestial physics should be distinguished from terrestrial. However, *Harmonices Mundi* Book IV succeeds at least in giving an *a priori* account of the means by which celestial bodies exercise influence upon terrestrial ones. It was Kepler's sole contribution to the mathematisation of terrestrial physics – apart from his description of the close packing of equal spheres to explain the shape of the honeycomb (but not the snowflake) in *De Nive Sexangula* (Prague, 1611).

The earlier form of Kepler's geometrical model for the Universe, the polyhedral archetype described in the *Mysterium Cosmographicum*, does indeed invite an immediate comparison with *Timaeus*. Like Plato, Kepler is concerned with polyhedra and, as we have already noted, some of his arguments echo those used by Timaeus in assigning a polyhedral form to each element. Nonetheless, in connection with astrological aspects we also find an indication that Kepler is already interested in Ptolemy's *Harmonica*. Ptolemy's cosmological theorising differs from Timaeus' in being more closely concerned with technical matters that were considered the province of the mathematician in Kepler's own day (as they no doubt had been in Ptolemy's) namely astronomy, astrology and music.

As we have already noted, the plan for Kepler's own book on world harmony was conceived in or before 1599. It seems always to have resembled that of Ptolemy's *Harmonica*. Moreover, the resemblance was still clear when

67 *Mysterium Cosmographicum*, Frankfurt, 1621, note on the titlepage, KGW 8, p. 15, ll. 14–18, trans. by Duncan in *op. cit.* in note 11 above, p. 51.

68 I have argued this in more detail in 'Kepler's Geometrical Cosmology', *unpublished PhD thesis* (London University), 1981, and 'Cosmology in the work of Kepler and Galileo', P. Galluzzi (ed.), *Novità celesti e crisi del sapere*, Florence, 1983, 207–215.

the 'short book' planned in 1599 was eventually published twenty years later as a folio of some 320 pages. This is not to say that the details of Kepler's *Harmonice Mundi* resemble those of Ptolemy's *Harmonica* any more than the details of the astronomy of the *Astronomia Nova* resemble those of the *Almagest.* From the first, Kepler had rejected the numerological explanation of consonances that formed the basis of Ptolemy's work.[69] In time he came also to reject Ptolemy's direct equation of astrological aspects with musical consonances (as shown in Figure 2 above). Moreover, as we shall see, he also rejected almost all of the astrological lore catalogued and codified in Ptolemy's *Tetrabiblos.* He also, of course, rejected Ptolemy's geocentric description of the planetary system. However, to Kepler the most important difference of all was certainly that, whereas Ptolemy's description of world harmony was merely cosmological, Kepler regarded his own model as cosmogonic, that is as the archetype according to which God had created the Universe. As Kepler put it in a letter to Christopher Heydon written in October 1605:

> Ptolemy had not realised that there was a creator of the world: so it was not for him to consider the world's archetype, which lies in Geometry and expressly in the work of Euclid, the thrice-greatest philosopher.[70]

Nevertheless, despite these very important points of difference, it is clear that Kepler regarded his *Harmonice Mundi* as closely related to Ptolemy's *Harmonica.* This can be seen, for instance, in the introduction to Kepler's fifth book, which makes repeated references to the *Harmonica* – the first being to remark upon the fact that Kepler had not yet read the work when he wrote the *Mysterium Cosmographicum.*[71] Having expressed admiration for Ptolemy's work, he ascribes its shortcomings to the deficiencies of the astronomy of Ptolemy's day and notes that Ptolemy's failure may well have deterred others from attempting a similar task.[72] He then, somewhat elliptically, refers to his own very recent discovery of the law relating the period of each planet to the mean radius of its orbit – the law is important because it allows him to deduce the sizes of the orbits from the harmonic archetype[73] – and acclaims his own success:

> . . . nothing holds me back, I can give myself up to the sacred frenzy, I can have the insolence to make a full confession to mortal men that I

[69] I have argued this in detail in 'Kepler's Rejection of Numerology' (full reference in note 31 above).

[70] . . . et nominatim in Euclide philosopho ter maximo. Kepler to Heydon, October 1605, letter 357, ll. 164–167, KGW 15, p. 235.

[71] KGW 6, p. 289, l. 4.

[72] KGW 6, p. 289, ll. 24–29.

[73] Kepler carries out this calculation in *Harmonices Mundi* Book V, Chapter IX, section XLVIII, KGW 6, pp. 356–359. I have given an outline of his working in 'Kepler's Cosmological Theories: their agreement with observation', *Quarterly Journal of the Royal Astronomical Society*, 23, 1982, 556–568.

> have stolen the golden vessels of the Egyptians to make from them a Tabernacle for my God far from the confines of the land of Egypt. If you forgive me I shall rejoice; if you are angry, I shall bear it; I am indeed throwing the die and writing the book, either for my contemporaries or for posterity to read, it matters not which: let the book await its reader for a hundred years; if God himself has waited six thousand years for his work to be seen.[74]

Comparison of himself with Moses and Julius Caesar is not part of Kepler's usual stock-in-trade, though the reference to God is highly characteristic.[75] The interpretation usually put upon Kepler's exhilaration is that he had only three days earlier discovered the period-radius relationship that he had been seeking for more than twenty years. However, it seems to me to be clear from the context of this outburst that, unless Kepler is so happy that he would write in this style about anything related to the cause of his happiness, this cause is in the success of the harmonic archetype, which now, thanks to the period-radius law, can give a full account of the observed structure of the planetary system. Although this passage occurs not in the introduction to the whole work but in the introduction to Book V, I think it is intended to refer to the whole work. Its not being put in the general introduction could well be explained by the fact that Kepler's discovery only affected the astronomical part of his archetype, which is dealt with in Book V. Thus in this rhetorical passage, Kepler seems to be proclaiming his own success in the task at which Ptolemy had failed, namely to give a true account of the harmony of the world, expressed in musical consonances, astrological aspects and the motions of the planets. Astrological harmony is an integral part of Kepler's work as it is of Ptolemy's.[76]

Now, *musica mundana*, whether astronomical or astrological, and as described by either Ptolemy or Kepler, has rather little to do with the practical concerns of an astronomer or an astrologer. We have already noted that Kepler refers his readers elsewhere for instruction in the practice of astrology.[77] This reference, to Chapters VIII, IX and X of *De Stella Nova* (Prague, 1606), turns out to be decidedly unhelpful. The chapters concerned contain only arguments for the power of aspects and for the significance of the points at which Great Conjunctions take place (against Pico della Mirandola). In fact, much of *De Stella Nova* is taken up with arguments not for traditional astrological tenets, as in these particular chapters, but with arguments against them – against almost all of them. Kepler gives a fair summary of the astrological part of the work in a letter to Herwart von

74 *Harmonices Mundi* Book IV, Proemium, KGW 6, p. 290, ll. 3–9.

75 One feels Kepler would have insisted upon His being included in the Personenregister of the KGW.

76 It seems, moreover, to have been a dissatisfaction with the astrological-musical part of the theory described in the *Mysterium Cosmographicum* which set Kepler upon the particular direction he took in seeking means to improve upon it. See above and my 'A Lutheran astrologer' (full reference in note 5 above).

77 See passage referred to in note 66 above.

Hohenburg written about a year after the publication of *De Stella Nova*:

> And what is the whole book if not a solemn execution (ἀποτυμπάνισις) of almost the whole of judicial Astrology, only Aspects being taken over as pertaining to nature?[78]

The form of execution to which Kepler's Greek word seems to refer is death by means of many blows.[79] Several of Kepler's are aimed at vital areas of traditional astrology. For instance, the fourth chapter discusses whether the division of the Zodiac into twelve equal parts is natural – and concludes it is not natural but merely a matter of human convention.[80] Kepler's argument is mainly historical. The following chapter discusses whether the names of the signs are arbitrary or natural and significant. They are considered arbitrary. As before, much of Kepler's argument is historical. As his comment to Herwart indicates, these chapters are in fact typical of the astrological part of Kepler's book.[81] Thus, apart from the influence of the Sun upon the atmosphere and the Moon upon the sea, Astrology is reduced to the study of aspects. These are, of course, angles between heavenly bodies as perceived from the Earth and their nature is thus unaffected by whether one happens to believe that the planetary system is geocentric or heliocentric. Unlike that of his astronomy, the nature of Kepler's astrology is not, in itself, an argument for Copernicanism.

His astrology nonetheless forms part of a cosmological system that is conceived as an argument for Copernicanism, though it was one which seems to have attracted little favourable attention from the astronomers of succeeding generations. The reasons for this are not far to seek. Kepler's cosmological model makes the Sun unique, or at least gives it a privileged position in the Universe: the harmony refers only to the Solar system. The theory thus necessarily seemed unsatisfactory to his successors, who were more and more inclined to believe the Sun to be only one star in an infinite Universe.[82]

Thus Kepler's defence of astrology, being embedded in an unacceptable cosmological theory, seems to have been entirely without effect upon later generations. Indeed, as the prestige of astrology waned among serious students of Natural Philosophys the fact that Kepler had both practised and defended astrology no doubt helped to give him the reputation for 'mysticism' which has clung to him for so long. Galileo, as so often, spoke

[78] Kepler to Herwart [April 1607], letter 424, ll. 172–174, KGW 15, pp. 453–454.

[79] See Liddell and Scott *Lexicon*, 1968, article ἀποτυμπανισμός, in main text and in Supplement (the exact form Kepler uses is not listed).

[80] *De Stella Nova*, Chapter IV, KGW 1, pp. 168–172.

[81] For a detailed discussion of Kepler's reasons for rejecting conventionaal astrological beliefs, see the works by G. Simon referred to in notes 26 and 28 above.

[82] This change in the nature of cosmological thinking is caught perfectly by the title of Koyré's classic study *From the Closed World to the Infinite Universe* (Baltimore, 1957). In this, Kepler stands out as a paradoxical figure, though Koyré notes that he gives good observational arguments against believing the Universe to be infinite. See also my 'Kepler's Geometrical Cosmology', Chapter II (full reference in note 68 above).

for the future in his wonderment (perhaps not entirely ingenuous?) that such an otherwise independent-minded man should have lent his ear to astrological nonsense.[83]

Kepler's cosmological model cannot easily be adapted to an infinite Universe of many Suns. The cause is in its very perfection: it finds all possible harmonies in our one Solar system, which is thus a complete expression of the nature of the Creator.[84] Later generations were to see Kepler as the Copernican *par excellence* because he believed the new astronomy made it possible to write a more accurate version of Ptolemy's *Almagest*. Kepler's cosmological work, on the other hand, shows his conviction that the new astronomy also made it possible to write a more accurate version of Ptolemy's *Harmonica*, in which astrological aspects were included among the manifestations of the harmony of a finite world. Kepler was by no means defending traditional astrology and he rejected (indeed often ridiculed) those parts of it based upon the signs of the Zodiac.[85] His reasons for this rejection were indeed connected with his Copernicanism (as Simon has shown), and his case suggests not that Copernicanism did not contribute to the decline of astrology, but rather that the element in seventeenth-century Copernicanism which was most influential against astrology was the fact that most Copernicans went on to draw the consequence that the Universe contained an infinite (and uniform) array of stars. Constellations thus became no more than patterns of stars, determined merely by the position of the particular Sun around which we move. This clearly represented a very important weakening of traditional astrology, which was much concerned with constellations, particularly the signs of the Zodiac. Kepler, as we have seen, accepted this weakening (though for rather different reasons), and thus does not provide a counter-example to the thesis that the rise in respectability of Copernicanism was, to some extent, a cause of the decline in respectability of astrology. The astrology he did believe in was an astrological form of *musica mundana* which later generations inevitably rejected because it made sense only in a finite Universe.

[83] It was mere bad luck that the piece of nonsense in question should have been Kepler's belief that the Moon caused the tides. Galileo Galilei *Dialogo sopra le due massimi sistemi del mondo*, Fourth Day, *Edizione Nazionale*, 7, p. 486. For Galileo's own astrological beliefs and practices see the works by Ernst and Righini referred to in note 4 above.

[84] It is, however, not philosophical conviction but observational evidence that Kepler urges as an argument for others against the suggestion that the Universe is infinite. See the works referred to in note 82 above.

[85] See, for instance, the works by Simon referred to in notes 26 and 28 above and E. Rosen, 'Kepler's Attitude toward Astrology and Mysticism', in B. Vickers (ed.), *Occult and Scientific Mentalities in the Renaissance*, CUP 1984, 253–272.

METOPOSCOPY: AN ART TO FIND THE MIND'S CONSTRUCTION IN THE FOREHEAD

Angus G. Clarke

'Liues there anie such slowe yce-brained beefe-witted gull, who by the riveld barke or outwarde rynde of a tree will take vpon him to forespeak how long it shall stand, what mischances of wormes, caterpillers, boughs breaking, frost bitings, cattells rubbing against, it shall have? As absurd is it, by the external branched seames or furrowed wrinckles in a mans face or hand, in particular or general to coniecture and foredoom of his fate.'
Thomas Nashe *The Terrors of the Night* (London, 1594) (sigs E4r–F1r, cited by C. Camden (1941) 411. See note 58 below).

Metoposcopy is a means of discerning an individual's character, or destiny, from the lines and marks on the forehead. It resembles chiromancy – which attempts to do the same with the lines on the hand – in that it has an astrological rationale. It is, therefore, a branch of physiognomics which, in its widest sense, seeks to relate all the individual's exterior lineaments to his or her character, disposition or destiny.

Though only a branch of physiognomics, metoposcopy raises in a more manageable fashion many of the issues and problems posed by physiognomics in general. Also, amongst the parts of the body studied by the physiognomist, the forehead enjoys a pre-eminence which is rivalled only by that of the eyes – the 'windows of the soul'.

Before looking more closely at the theoretical techniques of metoposcopy I shall briefly consider its origins.

Although there is a modest priority dispute about the matter, the first printed work on metoposcopy seems to be Tadeas Hajek's *Aphorismorum metoposcopicorum* (1561/2), an extremely rare book better known from a translation into French in 1565 and from its second Latin edition in 1584.[1]

[1] Tadeáš Hájek z Hájku *Aphorismorum metoposcopicorum libellus unus* (Prague, 1562). Thorndike, L., *A History of Magic and Experimental Science* (New York, 1923–58), VI, 504–5, and Müller-Jahncke, W.-D., 'Zum Prioitätenstreit um die Metoposkopie: Hájek contra Cardano', *Sudhoffs Archiv* 66, 1982, 79–84 passim, both give 1561 as the date of publication. Müller-Jahncke describes it as 'extrem rar und weder in den einschlägigen Bibliographien noch Bibliothekskatalogen nachzuweisen' (p. 80, n. 16). For the date 1562 I have followed Schüling, H., *Bibliographie der psychologischen Literatur des 16. Jahrhunderts* (Hildesheim, 1967) (s.v. Hájek) who describes the

Before Hajek went into print, however, he knew that Girolamo Cardano had projected a book on the same topic. Cardano's *Metoposcopia* did not, in fact, appear in print until 1658, when it was published simultaneously in French and Latin.[2] The fact that this book was not included in Cardano's collected works (published five years later) raises a small question-mark against its authenticity. There is no doubt, however, that a metoposcopy attributed to Cardano circulated in manuscript before that date.[3] Cardano did not claim to be original, but says he was taught the art of metoposcopy by one Girolamo Visconti, whose identity remains obscure.[4] Well before this, in any case, we find metoposcopy listed amongst the divinatory arts by Heinrich Cornelius Agrippa in his *De incertitudine et vanitate scientiarum* (1530). He tells us that:

> Metoposcopie doth avaunte that she can foretel al mens beginninges, proceedinges and endinges, with a very sharpe wit, and learned experience, by the onely beholding of the foreheade: she naming herself also a nourished member of Astrologie.[5]

---

book fully and notes that the dedicatory letter to Kaiser Ferdinand is dated 1561 – which is possibly the source of the confusion. The later date is confirmed by Urbánková, E. and Horský, Z. in their catalogue for the exhibition *Tadeáš Hájek . . . and his Time* (Prague, 1975, with notes in Czech), pp. 23 and 47. It is translated as *Nouvelle invention pour incontinent juger du naturel de chacun par l'inspection du front et de ses parties, dicte en grec Métoposcopie* (Paris, 1565 and 1567). The second Latin edition was published at Frankfurt a. Main in 1584. The priority dispute is first noted by Thorndike and is discussed in more detail by Müller-Jahncke.

2 Cardano, G., . . . *Metoposcopia libris tredecim, et octingentis faciei humanae eiconibus complexia . . . interprete Claudio Martino Lavrenderio* (Paris, 1568). It was translated by de Lavrendière into French as *La metoposcopie de H. Cardan . . .* (Paris, 1658).

3 G. Naudé, the collector of Cardano's collected works thought the Metoposcopy was 'plane intercidisse' and did not seem to know of the Paris editions. Kristeller, P. O., *Iter italicum* (London & Leiden, 1963–) records ms metoposcopies attributed to Cardano in Biblioteca nazionale Braidense (Milan), Bib. Palatina (Parma), Bib. Casanatense (Rome) and in the Bib. Apostolica Vaticana. The latter also has an ms entitled *Tadei Aggecii aphorismorum metoposcoporum* at Vat.Lat. 12706 Misc.XVI (Kristeller, II, 348). Müller-Jahncke (1982) 79, n. 4, cites Samuel Fuchs mentioning Cardano's metoposcopy in 1615. A work on metoposcopy (presumably, but not necessarily, ms, and the same as the printed work) by Cardano is mentioned occasionally in the Florence metoposcopy ms, dated 1605 (see note 27 below).

4 Cardano, *De vita propria* (written 1574–76) *Opera omnia* (Lyons, 1663) I, 42: 'Metoposcopiae libros e xiii in septem contraxi, pars est Physiognomiae. Accepi ex Hieronymo Vicecomite'. De Lavrendière, in his introduction to the 1658 edition(s) mentions Christophe Visconti (p. 27) and Hierosme Visconti (p. 45) – presumably a confusion. Ferrari, L., *Onomasticon – repertorio biobibliografico degli scrittori italiani* (Milano, 1947) offers one possible candidate of this name. This single individual turns out to be a conflation of (a) an otherwise unknown Milanese physician who died in 1572 only two years after joining the college of physicians – and who was probably too young to have taught Cardano anything in the 1550s, and (b) a cleric who dedicated a poetic composition to Francesco Sforza in 1490 – and was therefore probably too early to have taught Cardano who was only born eleven years later.

5 This quotation from the London, 1569 English translation (Repr. Northridge, Cal., 1974, ed. Dunn, C.M.), p. 111, ch. 34. The original (Cologne, 1531) reads: 'Metoposcopia autem ex solius frontis inspectione omnia hominum initia, progressus et fines . . . issimo ingegno ac docta experientia se praesentire jactat, ipsa se etiam Astrologiae alumnam faciens.'

Although there seems to be no text-book on the subject, the notion of metoposcopy had clearly been in circulation, in some form or other, for a considerable time.[6] To take an example – almost at random – of the wide-spread use of metoposcopical motifs, we find that Petrarch talks occasionally of knowing Laura's emotions from her forehead. In *Canzoniere* ccxxii, for instance, he says:

> Spesso ne la fronte il cor si legge.
> (Often the heart can be read from the forehead).

And, not surprisingly, there are solid classical precedents for this sentiment.[7] One of the most influential – and one which was to become a commonplace in seventeenth-century metoposcopies – is an episode in the life of the Emperor Titus, as it is narrated by Suetonius (active early in the second century A.D.):

> At that time, they say, a metoposcopus was brought . . . to examine Britannicus. He stated most insistently that Britannicus would never rule, and that Titus, who was standing nearby at the time, would certainly become emperor.[8]

Whether the incident is true, and what kind of metoposcopy was involved we cannot know, but the commentators generally agree that it was astrological and divinatory.

Having regressed this far in search of the origins of metoposcopy we might as well go right back to classical Athens and trace the development of physiognomics more broadly defined.

Physiognomics flourished with the classical Greeks, if not before, and continues to do so today in some form or other. The earliest known text is a physiognomics attributed to Aristotle. It is a work of considerable interest and intelligence.[9] As we shall see, it also had a colossal influence right down to the seventeenth century. There is no space to discuss it here, but one point should be made about it. The second section – which differs somewhat in tone from the first – is devoted to the comparison of various human physical

[6] Scholem, G., for instance states: 'Über Metoposkopie scheint es überhaupt keine Texte vor dem Spätmittelalter zu geben', in 'Ein Fragment zur Physiognomie und Chiromantik der spätantiken jüdischen Esoterik', *Liber amicorum: Studies in hon. of Prof. Dr C. J. Bleeker* (Leiden, 1969), p. 175.

[7] Cf. *Canzoniere* xxxv, 8 and lxxvi, 11. See also *Concordanza delle rime di Francesco Petrarca* (Oxford, 1912) compiled by McKenzie, K., s.v. 'fronte'. Cf. also Ovid, *Amores* I, 11, ll. 17–18.

[8] Suetonius *Opera omnia* (London, 1826) II, p. 941: 'Quo quidem tempore, aiunt, Metoposcopum . . . adhibitum, ut Britannicum inspiceret, constantissime affirmasse, illum quidem nullo modo, ceterum Titum, qui tunc prope astabat, utique imperaturum'.

[9] The most authoritative Greek text, plus the translation into Latin by Barthomew of Messina (made at the Sicilian court of King Manfred between 1258–66), are given by Förster, R., *Scriptores physiognomonici graeci et latini* (Leipzig, 1893), I, pp. 1–90. For English translation see following note.

types with animals and the consequent assessment of various human types on the basis of the characteristics ascribed to the most similar animal. For instance,

> A small forehead means stupidity, as in swine; too large a forehead, lethargy, as in cattle. A round forehead means dullness of sense, as in the ass; a somewhat long and flat forehead, quickness of sense, as in a dog. A square and well-proportioned forehead is a sign of a proud soul, as in the lion. A cloudy brow signifies self-will, as in the lion and the bull.[10]

The comparison of human types with animals and judgement thereby develop into a major physiognomical tradition which we will have to disregard entirely here. Like the astrological tradition, this 'bestiarum humanum' tradition is reluctant to make unsupported statements about the significance of human appearance without using some kind of external criterion, however unreliable it might seem to us.[11]

There are other classical physiognomies. An important one, which develops the ps-Aristotle material especially in its detailed treatment of the eyes, is by Polemon (or Polemo) of Laodicea (d.144 A.D.).[12]

There is a solitary and anonymous Latin physiognomics, which has been attributed to various hands, including Apuleius and (naturally) Aristotle.[13] This and the ps-Aristotle work were the first to be known to the Latin Middle Ages. Since the anonymous Latin text opens with the claim that it is based on three authorities – Aristotle, Polemon and Loxus – it is usually possible to infer that medieval physiognomy writers mentioning these three were basing themselves on the anonymous text (not least because nothing by Loxus was, or is, extant). One such is Peter of Abano.[14]

[10] The *Physiognomonica*, p. 811b of *The Works of Aristotle*, vol. VI (Oxford, 1913) translated by Loveday, T. and Forster, E. S.

[11] For an introduction to this topic see Baur, O., *Bestiarum humanum: Mensch-Tier-Vergleich in Kunst und Karikatur* (München, 1974) and Baltrušaitis, J., *Aberrations. Quatre essais sur la legende des formes* (Paris, 1957), ch. 1, 'Physiognomonie animale' (pp. 7–46).

[12] Förster (1893) I, 94–294 for Latin and Arabic texts. Less important texts include one by Adamantius (Förster, II, 296–426 for Greek, and Latin translation by Janus Cornarius (Basle, 1544)), which is regarded as a paraphrase of Polemon by Pack, R. A., 'Artemidorus and the Physiognomists', *Transactions & Proceedings of the American Philological Assoc*, 72, 1941, p. 333, n. 60.

[13] Förster, II, 3–145 for the text. Denieul-Cormier, A., 'La très ancienne physiognomonie et Michel Savonarole', *Biologie medicale*, 45, 1956 (avril, numéro hors serie) dates its authorship to ca.3rd/4th century A.D. and gives a useful description of the contents. A translation into French, ed. and trans. André, J. under the title *Anonyme Latin Traité de Physiognomonie* was published in Paris, 1981. Secondary studies of classical physignomics include: various essays by Evans, E. C., the latest of which is 'Physiognomics in the Ancient World' in *Transact. Amer. Philos. Soc.*, ns 59, 1969; Megow, R., 'Antike Physiognomielehre' in *Altertum*, 9, 1963, 213–221; and various papers in *Transactions & Proceedings Amer. Philol. Ass.* (*TAPhA*) in the 1940s, including Armstrong, A., 'The Methods of the Greek Physiognomists', *TAPhA*, 72, 1941, 321ff.

[14] See note 17 below.

The classical corpus of physiognomical literature passed into Islam and Arabic. The Islamic tradition is extrememly complex, is by no means fully elucidated yet and its details do not concern us here.[15] During this passage through Islamic culture physiognomy is given considerable impetus. It was certainly regarded as a necessary ancillary to such activities as the appointment of subordinates to positions of responsibility, the purchase of slaves, the selection of spouses, servants and prostitutes and in assessing the truthfulness of witnesses. For our present purpose an interesting issue is the emergence of an astrological physiognomy. According to Yusouf Mourad, the first to achieve this convergence was the Syrian scientist and cosmographer known as ad-Dimashqi who died in 1327.[16] Though Mourad's suggestion has been adopted by subsequent scholars, the situation is not entirely clear. Before any work by ad-Dimashqi could have been known in the Latin West there already were astrological physiognomies circulating in Latin. Two celebrated examples are those by Michael Scot,who died in 1235, nearly a century before ad-Dimashqi, and by Peter of Abano, an almost exact contemporary of ad-Dimashqi.[17]

In fact the notion that the planets and zodiacal signs were associated with certain distinct physical and psychological types was current nearly a millennium earlier in Ptolemy's *Tetrabiblos*.[18]

The answer to the questions when and by whom was physiognomy first astrologised is not really crucial. Once Galen had produced the humoral model of human physiology which was to remain standard wisdom for the enusing millennium and a half, both astrology and physiognomy were furnished with a common ground. We may take it that any medieval savant in search of a more 'scientific' and rigorous physiognomy not principally

[15] For preliminary indications see *Enzyklopädie des Islām* (Leiden/Leipzig, 1913–38) and *The Encyclopaedia of Islam* (New edition, Leiden/London, 1960–) s.v. Firāsa. The fullest accessible work on the subject seems to be Mourad, Y., *La physiognomie arabe et le Kitāb al-Firāsa de Fakhr ad-Dīn ar-Rāzī* (Paris, 1939). A Spanish translation of Fakhr ad-Dīn ar-Rāzī – who is not to be confused with the better known (and variously spelt) Rhases (Abū Bakr Muḥammad ibn Zakariyyā ar-Rāzī) whose *Liber almansoris* also contains a section on physiognomy – is edited by Viguera, M., *Dos cartillas de fisiognómica* (Madrid, 1977). Something can be gleaned from Grignaschi, M., 'La "Physiognomie" traduite par Hunayn ibn Isḥāq', *Arabica*, 21, 1974, 285–91, who shows that the work in question is a translation of the ps-Aristotelian physiognomics.

[16] Mourad (1939), 8–9. See *Encyclopaedia of Islam*, s.v. Dimashki for his full name and activities. He was known to the West as Damascenus and should not be confused with various other bearers of this toponymic.

[17] Thorndike, II, 308 reports that 18 editions of Scot's *Phisionomia* are said to have appeared between 1477 and 1660. Thorndike, II, 910–18 describes the content of Peter of Abano's *Phisionomia* and gives details of mss and editions. I have used the one edited by Michelangelo Biondo, *Decisiones physionomiae . . . petri de Abbano* (Venice, 1548).

[18] See *Tetrabiblos*, III, 11 and 13. Cf. J. Firmicus Maternus *Matheseos libri VIII*, IV, 19. Pack, R. A., 'Auctoris incerti de physiognomonia libellus', *Arch. hist. doct. litt. Moyen Âge*, 41, 1974, 113–138 discusses, and gives the text of, a short physiognomical work which gives examples of the character and outward features associated with each planet (see chs 3–10). The conflation of planetary and humoral types 'seems' to have been 'definitely' established by the Arabs in the ninth century, according to Saxl/Panofsky/Klibansky *Saturn and Melancholy* (London, 1964), 127.

dependent on circular analogies with animals would find the astrologisation of physiognomy both easy and commonsensical.

Evidence for the considerable impetus given to physiognomics by Islam, and for the subject's widespread popularity in the Middle Ages and Renaissance, is afforded by the numerous manuscript and printed works about it. Very schematically, the situation is this: there are seven major texts originating either in antiquity (ps-Aristotle, Polemon, Adamantius and the anonymous Latin work) or with the Arabs (the section on physiognomics in the *Liber almansoris* by Rasis, the work by Fakhr ad-Din ar-Rasi and the relevant section of the *Secret of Secrets*).[19] All these derive from, or are to some extent indebted to, the ps-Aristotelian compilation. At various times and in various ways all are absorbed by the Latin West and emerge as translations, translations-plus-commentaries and as 'original' works. Their authors include Albertus Magnus, Aldobrandino da Siena, Roger Bacon, Jean Buridan and Cecco d'Ascoli, as well as Peter of Abano and Michael Scot who have already been mentioned.[20]

The fifteenth and sixteenth centuries produce a great deal of physiognomical literature. The better known authors of the period include Alessandro Achillini, Michelangelo Biondo, Cocles (Bartolomeo della Rocca), Geronimo Cortes, Johannes Glogoviensis, Johannes ab Indagine (von Hagen), Girolamo Manfredi, Agostino Nifo, Giambattista della Porta, Giuliano Ristori, Giorgio

[19] In addition to these 7 major texts there are, of course, others such as the Latin translation from Arabic of a work by ps-Polemon (Förster, II, 149–160) as well as a variety of fragments. Any discussion of these complicates rather than advances the argument. The *Secret of Secrets* (in Arabic: *Sirr al-Asrar*), also known as the *Letter to Alexander*, was first translated by John of Seville in the first half of the 12th century (Thorndike, II, 269) and became the prototype for the Salernitan *Regimen sanitatis*. A more influential translation by Philip of Tripoli ca.1243 (Thorndike, II, 270) spawned numerous mss and printed versions in Latin and the vernaculars. Bower, M. R., 'Relations between Art and Physiognomics, 1400–1550' (M.Phil.Diss, University of London, 1973) ch. 3 and notes passim, reports over 200 extant Latin mss of the *Secret of Secrets*. Its Greek and Arabic roots are elucidated by Braekman, W. L., '"Den mensche te bekennen bi vele tekenen": Het mnl. prozatraktaatje over fysiognomie en zijn bron', *Scientiarum historia (Antwerp)* 12(3), 1970, pp. 113–42. For further details see the edition of Roger Bacon's version of the *Secret of Secrets, Opera hactenus inedita* V, ed. Steele, R. (Oxford, 1920), pp. 164–72 and 218–24. I have used the Italian translation by Giovanni Manente, *Col nome de dio il segreto de segreti . . . et la phisionomia d'Aristotile* (Venice, 1538).

[20] For Albertus see the comment on his *De animalibus* by Siraisi, N., 'The medical learning of Albertus' in *Albertus Magnus and the Sciences: Commemorative Essays*, ed. Weisheipl, J. A. (Toronto, 1980), pp. 396–8, and also Albertus' *De secretis mulierum*. Whether or not this is by Albertus is not crucial since the period in question ascribed it to him. Its popularity is attested by more than 50 incunable editions (Bower, 1973, p. 25). I used the 1648 Amsterdam edition (Warburg Institute FBH 425) which includes Michael Scot's *De secretis naturae*. Aldobrandino of Siena (d.1287) wrote a brief physiognomics based on the *Liber almansoris* (Bower, 1973, p. 17). I used *Philosomia de gli huomini facta p. Aldrobaldio philosopho . . . Traducta per Baptista Caracino* (no place, 1495, British Library, IA 27903). For Bacon, see the work mentioned in Note 19 above. For Buridan, see Thorndike, L., 'Buridan's questions on the physiognomics ascribed to Aristotle' in *Speculum* 18, 1943, 99–103. Cecco d'Ascoli (né Francesco degli Stabili, and burnt for heretical and astrological views in 1327) gives some physiognomical material in his poem *L'acerba*, and he wrote a brief prose treatise on the subject: see Boffito, G., 'Il *De principiis astrologiae* di Cecco d'Ascoli novamente scoperto e illustrato', *Gior. stor. lett. ital.*, Suppl. n.6, Torino, 1906, pp. 65–73.

Rizzacasa and Michele Savonarola.[21] My impression – and this is subject to a more extensive analysis of the texts – is that despite a degree of originality the shadow of the ps-Aristotelian physiognomics never entirely disappears. The earliest contribution which is in some degree original is that of Savonarola. The novelty is not in the content but in the approach – he is concerned with how, not whether, broad shoulders, say, may indicate a fine intellect.[22]

Physiognomics flourishes even more rampantly in the seventeenth century, as a glance at Heinrich Laehr's bibliography reveals.[23] Unfortunately space does not permit us to go into this important and almost entirely neglected

[21] See Achillini's *Quaestio de subiecto physionomiae et chiromantiae* (Bologna, 1503) which was reprinted in 1517, 1518, 1520, 1523 and 1568; and more generally Zambelli, P., 'Aut diabolus aut Achillinus. Fisionomia, astrologia e demonologia nel metodo di un Aristotelico' *Rinascimento*, 18, 1978, 59–86. See Biondo's *De cognitione hominis per aspectum* (Rome, 1544), which is non-astrological and non-divinatory, and his edition of *Decisiones physionomiae . . . petri de Abbano* (Venice, 1548), which is astrological. The bibliography of Cocles is complex, due to numerous translations, similarly titled works and the problem of whether they are by, or compiled by, Cocles; see in any case his *Physionomiae ac chiromantiae anastasis* (Bologna, 1504) which was reprinted a dozen times in various languages before 1550. The chiromantic section is by Andrea Corvi of Mirandola. For Cocles' relations with Achillini see the article by Zambelli mentioned above, and more generally, Thorndike, V., ch. 4 passim. For Cortes, see *Libro de phisonomia natural, y varios secretos de naturaleza* (Madrid, 1598) (published with his celebrated *Lunario nuevo, perpetuo, y general, y Pronostico de los tiempos*). Some general information is to found in Thorndike, VI, 166f and VII, ch. 35 passim. Though Cortes' work is a treasury of 'mirabilia naturae' it is strictly non-divinatory and barely astrological. John of Glogau's *Phisionomia* (Cracow, 1518) is a non-astrological compilation. It is briefly described by Chojecka, E., 'Theorie und Praxis des Porträts der Frührenaissance: Die *Phisionomia* des Johann von Glogau (1518)', in *Wissenschaftliche Zeitschrift der F.-Schiller-Universität Jena* (Gesellsch. u. Sprach-wissensch. Reihe) 18, 1969, 177–80. See also Thorndike, IV, 702–3. For John of Indagine, see *Introductiones apotelesmaticae in physiognomiam, astrologiam naturalem, complexiones hominum, naturas planetarum cum periaxiomatibus de faciebus signorum et canonibus de aegritudinibus hominum* (Strasburg, 1522). The edition I used (Ursellis, 1603, BL: C.79.a.2.(2)) included a short work on improving the memory by Guglielmo Gratarolo and part of the *De sculptura* by Pomponio Gaurico (see note 55 below). What little is known about John of Indagine is to be found in Hermann, F., 'Johannes Indagine', *Archiv für hessische Geschichte u. Altertumkunde* 18, 1934, 274–91. For Girolamo Manfredi – a noted Bolognese astrologer – see his *Liber de homine* (Bologna, 1474), which was frequently reprinted. Nifo's commentaries on Aristotle included one on the (ps-)Aristotelian physiognomics, included in the *Parva naturalia* (Venice, 1523, repr. 1550). More generally see Zambelli, P., 'I problemi metodologici del necromante Agostino Nifo', *Medioevo*, 1, 1975, 127–71. For Della Porta, see note 31 below. For Ristori, see the *Trattato di fisionomia e chiromanzia* (Firenze, Bibl. Riccardiana, ms. Ricc. 1221[6]), mentioned by Thorndike, V, 326–7. Ristori's *Pronostico sopra la genitura di Cosimo I* (de' Medici), (Firenze, Bibl. Medicea Laurenziana, Pluteo 89 sup. 34) dated 1537, also contains physiognomic material (see Thorndike, V, 326, and the catalogue *Astrologia, magia e alchimia nel Rinascimento fiorentino ed europeo* (Firenze, 1980, ed. Zambelli, P. in the series of exhibitions *Firenze e la Toscana dei Medici nell'Europa del Cinquecento* (1980)), pp. 379 and 390. For Rizzacasa, see *La fisionomia del Rizzacasa* (Carmagnola, 1588), which, though non-divinatory and non-astrological, was denied publication permission by the Inquisition, and was dedicated to Elizabeth I of England, Ireland and 'France'. For Savonarola see note 22.

[22] Savonarola (1384–1466), grandfather of Girolamo Savonarola, composed his *Speculum physiognomiae* before 1450. A good synopsis is given by Denieul-Cormier (1956) 41f. The fact of Savonarola's reliance on previous work is attested by his incorporating into his work the second book of Peter of Abano's physiognomics (ibid, p. 53) – which is strongly astrological.

[23] Laehr, H., *Die Literatur der Psychiatrie, Neurologie und Psychologie von 1459 bis zum 1799* (Berlin, 1900, 3 vols).

field. However, as was suggested above, many of the key issues in physiognomy are also raised by metoposcopy. At this point therefore, it would be helpful to have a clearer idea of how the metoposcopist thought he should operate, and of the principles of his art. The best way to achieve this is by looking more closely at a work on the subject. Since most metoposcopies share a broadly similar approach, one text can fairly introduce the rest.

One of the most popular works on metoposcopy was published in 1626 by Giambattista Spontoni who claimed that the book was the labour of his late father Ciro Spontoni (c.1552–c.1610). This work was reprinted seven times in the following three decades and once more, in a somewhat debased form, in 1789.[24]

A further motive for examining this particular book is that it was not in fact by Spontoni but was plagiarised from the celebrated mathematician, astronomer, cartographer and astrologer Giovanni Antonio Magini (1555–1617). Magini was the professor of astrology/astronomy/mathematics (all three were effectively synonymous at the time) at the Studio of Bologna from 1588 until his death. His astronomical ephemerides for the period 1580 to 1630, furnished with excellent isagoges to all aspects of astrological techniques, were justly regarded throughout Europe as the best and most accurate available.[25] No less a person than Kepler tried to enlist Magini's collaboration in the production of a new set of tables.[26]

Apart from the printed editions under the name of Spontoni, Magini's work on metoposcopy exists in several manuscripts (none autograph). One is in Florence and, though undated, seems to be the earliest, or at least the roughest and most amateurishly copied. There are two others in Venice and are dated 1605 and 1628 respectively.[27] The story of the plagiarism is told by the copyist of the 1628 Venice manuscript – a certain G.F.T. – in his

[24] *La metoposcopia overo commensuratione delle linee della fronte trattato del Signor Cavalier Ciro Spontoni* . . . (Venice, 1626). The *National Union Catalogue* gives subsequent editions as: Venice, 1629, 1637, 1642, 1645 and 1651. To these may be added an edition in Venice, 1636 (recorded in the Catalogo magliabecchiano of the Bibl. naz. centrale di Firenze, but not seen by me because it was destroyed in the 1966 Arno flood) and an edition in Venice, 1654 (mentioned by Niceron, J.-P., *Memoires pour servir a l'histoire des hommes illustres dans la republique des lettres* (Paris, 1732) vol. 27, p. 317f) and finally an edition in Venice, 1789 (in Bibl. universitaria di Bologna). BL has editions for 1629, 1637 and 1642. With the exception of the 1789 (and 1636?) edition, all these editions are identical.

[25] For Magini, see Favaro, A., *Carteggio inedito di Ticone Brahe, Giovanni Keplero e di altri celebri astronomi e matematici dei secoli XVI e XVII con Giovanni Antonio Magini* (Bologna, 1886), which gives a wealth of biographical data, the bulk of Magini's extant correspondence with the major – and minor – savants of his day, and an almost complete bibliography of his prolific printed works. Clarke, A. G., 'Giovanni Antonio Magini (1555–1617) and Late Renaissance Astrology', PhD thesis, University of London, 1985, examines Magini's astrological writings in detail. Almagià, R., *'L'Italia' di G.A. Magini e la cartografia dell'Italia nei secoli XVI e XVII* (Napoli, 1922) does justice to Magini's dominance of Italian cartography for most of the seventeenth century. As a curiosity, Magini seems to have been the first to use the decimal separatrix – see Ginsburg, J., 'On the early history of the decimal point', *Scripta mathematica*, 1 (1), 1932, 84–85 and idem., 'Predecessors of Magini' in idem., 1 (2), 1932, 168–9..

[26] Favaro, (1886), 96–100.

[27] Florence, Bibl. naz. centrale, Fondo palatino, Serie Targioni-Tozzetti, 80 (101) s.xvii. Venice, Bibl. naz. marciana, Cod. marc. ital. II, 5043 and 4994, dated 1605 and 1628 respectively. I am grateful to Dr Charles Schmitt for drawing my attention to these other mss of Magini's

dedicatory letter.[28] The fact of the plagiarism is confirmed beyond any doubt by a comparison of the Florence manuscript with the printed text. After certain initial dissimilarities in the first two pages, where however the gist of the argument is identical, the bulk of the text is identical.[29] Finally, and ironically, the printed version is incomplete. It lacks the brief third chapter which survives only in the 1628 Venice manuscript and, catastrophically, it lacks the hundreds of illustrations which adorn the manuscript.[30]

Regarding date of composition and motive, Magini had certainly written something before the summer of 1594 and he seems to have been busy with the subject again around 1601. The work was composed at the request of a prince of the Gonzaga family in Mantua, where Magini was court-astrologer and mathematics tutor.[31]

---

metoposcopy: Vatican, Biblioteca Apostolica, Lat. 12770 and Chantilly, Museé Conde 423 (No. 887). I was unable to see these two mss but I would be very surprised if they add anything material to the present argument. The Chantilly ms is short – only 59 folios, of which 46 contain illustrations – and is unlikely to contain material not in the very much larger Florence ms. The Vatican ms mentions in its title a 1652 edition of Spontoni's metoposcopy, so it must postdate 1652 and is too late to be closely connected with Magini (d.1617).

28 'Morto il Maggini gl'anni passati, il/ signor Giovan Battista Spontoni . . . stampò il primo, et il secondo libro di questa metocospia (sic) sotto il nome del Cavalier Cyro suo padre, affermando haverla attrata dalle cenere dell' oblivione delle fatiche di suo padre . . . Questa uscita alla luce fu tosto conferita con questa del Maggini, ritrovata esser una copia, et non opera nuova'. (ff. 5r–5v).

29 The first six pages of the printed version (pp. 7–12) are slightly longer than the Florentine ms version, but the additions are purely stylistic and not substantial. After this point the two versions are identical, except for a handful of trivial deviations, such as the occasional reversal of the noun/adjective order or the replacement of a noun by a pronoun and vice versa. The only effect of the initial padding is that 'Spontoni' has 10 chapters in the first book where Magini has only nine. To this extent, Frati, C. and Segarizzi, A., *Catalogo dei codici marciani italiani* (Modena, 1909), I, 259, are correct but slightly misleading when they state that ms 4994 is not identical to the 'Spontoni' version.

30 G.F.T., in the preface to ms 4994, notes that Spontoni does not include the third book 'ch'è il più bello . . . contenendo 357 osservationi, nelle quali vien praticata tutta questa scienza'. (f. 6r). Later, however, he (or she) refers to the brief section on measurement to discover which lines are present on the forehead, as the third book. The Florence ms has the same illustrations and describes them as the third book. Flood damage has rendered the table of contents of this book partly illegible in the latter ms. Unlike the rest of the text it is in Latin and claims to follow the order used by Cardano.

31 In a letter to Giambattista della Porta, dated 27 July 1594, Magini writes that : 'Alli giorni passati ad istanza d'un principe mio padrone ho messo insieme un trattato di Metoposcopia al meglio ch'io ho saputo'. (The letter is in Milan, Bibl. Ambrosiana, ms S.94. sup. misc. XVI, f. 225r–v. It is not autograph and came from the ill-fated library of Vincenzo Pinelli in Padua – see Rivolta, A., *Catalogo dei codici pinelliani dell'Ambrosiana* (Milano, 1933) passim. Pinelli, Magini and della Porta had a number of common friends or correspondents whether in Paduan intellectual circles or in the Accademia dei Lincei in Rome.) The letter continues with a list of the contents of Magini's metoposcopy (ff. 26r–27v). Della Porta finished his *De humana physiognomia* (Napoli, 1586) in 1583, translated it himself under the pseudonym Giovanni de Rosa as *Della fisionomia dell'huomo libri sei* (Napoli, 1598), issued an expanded version of the original work in 1600 and then produced his *Coelestis physiognomia libri sex* (Napoli, 1603) which is not so much a work of physiognomy as a rather confused, and probably simulated, refutation of astrological ideas.

Magini's interest in metoposcopy is suggested by a letter to him, dated 4 October 1601, from Ilario Altobelli (an enthusiast for astrology who became the 'chronologus' of the Franciscans, his order) where Altobelli says he will try out Magini's metoposcopic theories and report on the

What exactly does metoposcopy consist of? First, Magini tells us what, and where, the forehead is, and he derives the word (Italian *fronte*, Latin *frons/frontis*) from the Latin *fero* (I carry), because it is that part of the body which is 'carried in front'. Then he sketches out the underlying rationale of the art:

> The forehead is an index which reveals the passions of the soul: sadness and happiness, clemency and severity. When someone is sad, he wrinkles up his forehead; when happy, his forehead becomes smooth. So it is with good reason that Aristotle declared the forehead to be the seat of shame and honour in man, because it is so close to the imaginative virtue which, together with common sense, is to be found in the anterior part of the brain.[32]

The skin of the forehead is variable (*variabile*) according to the individual's will by means of certain nerves and muscles. Thus, opening and closing one's eyes causes the forehead to contract and expand. These contractions and expansions of the skin, when repeated, cause certain lines or wrinkles of the skin. However – and it is a big however – not everyone has the same lines, and not all lines are caused by these muscular contractions. Many lines

> are caused by nature in order to show man the variety of accidents which life may bring.[33]

We should note in passing that Magini does not clearly make the distinction, which we make, between moods – happiness and sadness – and character – severity, mercifulness. We should also note the apparently unsupported conceptual leap from the commonsensical observation that lines are caused by repeated movements of the subcutaneous musculature to the notion that nature causes certain lines in order to show us what upsets life may have in store for us. That is to say, not only is there no consistently coherent distinction between underlying personality and transient responses, there is no clear distinction between the empirical detection of personality and the divination of future events. The latter non-distinction reflects the blurring of diagnosis and prognosis which is familiar to us from the theoretical expositions of astrological medicine.[34]

---

outcome. This letter is reprinted by Favaro (1886), Lettera XX, pp. 244–5. For the knowledge that the princely patron was Francesco Gonzaga we are indebted to G.F.T. in the preface to the Venice 1628 ms. Francesco Gonzaga was only nine in 1594.

32 All quotations, unless otherwise indicated, are from the Florentine ms (see note 21 above). 'La Fronte dimostra le noti, o passioni dell'animo essendo ella uno indice, o dimostrazione della tristezza, o allegrezza, clemenza o severità dell'uomo, e perciò che vediamo chiaramente che quando l'uomo s'atrista la Fronte s'increspa, e quando si rallegra la Fronte si dilata, onde non senza ragione Arist: pone la Fronte come sede della vergogna, ed onore dell'uomo, e ciò per la vicinanza della virtu immaginativa. La qual insieme col senso comune è posta nell'anteriore parte del Cervello'. (f. 80r).

33 '. . . che le dette linee, o segni sono state causate dalla natura per dimostrare all'uomo la varieta delli accidenti che nella vita occorrono'. (f. 80v).

34 Astrological medicine is a vast subject. Useful recent surveys include: Müller-Jahncke, W.-D., *Astrologisch-magische Theorie und Praxis in der Heilkunde der frühen Neuzeit* (Habilta-

Magini now goes on to look at details. He tells us to consider how many lines there are. The number varies from individual to individual and it is rare to find less than three. No lines at all signifies a short life, death at the hands of thieves, crippling accidents, death from dropsy in youth; many lines is positive in that it denotes intelligence and ability, but negative because it indicates a multiplicity of intractable worldly affairs, heavy responsibilities and problems throughout life.[35]

In addition to their quantity, we should consider also the position of the lines, their length, depth and width, their colour, whether or not they are straight or otherwise, whether continuous or not. We must also consider the random variations (*accidenti*) which befall lines, such as moles, warts, dimples, scars, birthmarks and the like. All this is best done when the subject has just got up in the morning, before he has had feelings and emotions which might cause superficial variations.[36]

Regarding the location of the lines, the basic idea is that the forehead is divided into a number of zones, each of which is governed by a planet. Lines and other marks which fall within that area are indicative of the relative planet's influence. The kind of line indicates the quality of the planet's influence – whether weak or strong, favourable or malign, and so on. The longer the line, the longer the planet's influence lasts. The width of the line

> reveals the greatness of the effect; broad lines denote great and notable events, small and obscure lines denote small and obscure effects.[37]

The deeper the line, the stronger the planetary effect. Pallid lines mean that the effect has passed – though very faint and pallid lines may signify the imminence of death. Straight lines signify directness and rectitude, and where there are two or three, a long and healthy life. Wriggly lines make the individual rapacious and avaricious; broken lines signify death by plague, a short life and a sickly complexion; intersected lines mean great misfortunes, imprisonment and violent death. And so on to the various marks which can occur on the forehead – for instance, a cross-shaped mark indicates death by hanging, shaped like a chain it indicates imprisonment and so on. If a mark occurs on the right side of the forehead it will be favourable, unless it be a bad mark; if it occurs on the left side the effect is negative, and when the mark is a bad one, catastrophically so.

---

tionsschrift, Philipps-Universität, Marburg, 14 April 1982), 220–39; Chapman, A., 'Astrological Medicine' in Webster, C. (ed.) *Health, Medicine and Mortality in the Sixteenth Century* (Cambridge, 1979). A good primary source is Magini, G. A., *De Astrologica ratione ac usu dierum Criticorum . . . ac praeterea de cognoscendis et medendis morbis ex corporum caelestium de cognitione . . .* (Venice, 1607, and Frankfurt, 1608). This work has the distinctions of being lucid and of having a bibliography which summarises iatro-astrological theoretical writing in the fifteenth century.

35 f. 81v.

36 f. 84r.

37 'La largheza della linea dimostra la grandeza delli effetti per ciò che quelle linee che sono contigue, e larghe mostrano anche grande e notabile evento, come per opposito quelle che sono picciole e pure dico oscure, danno piccioli ed oscuri effetti.' (f. 83v).

Some idea of when the bad effects may be expected can be had by measuring the lines. The left-hand third of a given line refers to the first 30 years of life, the central portion to the period from 30 to 60 and the right-hand third denotes the period from 60 to 90 years of age. Within this broad scheme, it is possible to make finer measurements so as to have an approximate idea when a given effect may be expected to make itself felt.[38]

The first section concludes with some advice:

> If we find that someone will die a bad death, but we do not know exactly what kind, we must look at the circumstances and effects. If we find some indication that the individual may be a thief, then we may reasonably assume that his death will be by hanging; if a murderer, he will be beheaded.[39]

The second section examines each of the planetary lines in turn. If the line is straight, firm, unbroken and 'unafflicted' by ramifications, spots, erasures and the like, it signifies a positive influence from the relevant planet. If it is the longest and deepest line, it signifies the dominion of that planet over the individual. Conversely, if the line is weak, broken and obscured by moles and warts, the signification is negative. The planetary influences are defined in the standard astrological way. One example: Mercury's line, when positive, signifies a good intellect and memory, acuity and good judgement, ability in mathematics and astrology, fortune in matters pertaining to the written and spoken word, popularity with foreigners and is indicative of lawyers, teachers and litterati. A poor Mercury line indicates instability and inconsistency, mendacity and garrulousness, restlessness, treacherousness, tendencies to become a thief or a forger and, in women, a propensity for witchcraft and the casting of spells.[40]

Next Magini considers the meetings or intersections (*concorsi*) of each pair of planetary lines. The combined characteristics and destinies correspond, again, to the standard astrological accounts of what is to be expected from the conjunction of each pair of planets. So, for instance, a favourable 'concorso' of the Saturnine and Mercurial lines indicates a thoughtful and melancholic individual with a powerful intellect, a retentive memory and a talent for scholarship. A favourable 'concorso' of the Jovial and Martial lines means a magnanimous and virile person who tends to impatience and anger, who is severe in his judgements and who is a good leader. An unfavourable 'concorso' of the same pair of lines indicates a cruel and unpleasant person who is rapacious, hot-tempered, ungrateful, domineering and so on.

The second section closes with a brief description of the seven ages of man as they fall under the dominion of the seven planets. Once again, this is

38 ff. 86v–88r.

39 'Se troveremo che uno debba morire di mala Morte, e non potrai per aventura sapere di che spezie, considererai le altre circostanze ed effetti; come dire, se vi sarà qualche segno che colui debba essere ladro all'ora potrai affermare assai ragionevolmente che abbia da essere impiccato, se trovi poi che abbia da essere Omicidiale, potrai dire che gli sarà tagliata la testa per giustizia.' (f. 88r)

40 ff. 92r–v.

standard astrological doctrine and its 'locus classicus' is the *Tetrabiblos*.[41] The period of life ruled by a particular planet reveals that planet's influences more intensely than at other times. For instance, the third period of life, which runs from 14 to 22 years of age, is ruled by Venus. If the line of Venus of someone in this age-group is strong and clear, the individual will be prone to libidinousness and he – or she, presumably – 'will spend this period happily occupied with socially acceptable love-making'.[42] Where the Venereal line is weak or malformed the outlook is bleak – overindulgence and forbidden liaisons will lead to bad luck, ill-health and unmentionable problems.[43]

Frequently metoposcopies include a section specifically devoted to the topic of divination from moles and warts on the face or elsewhere on the body. The ps-Spontoni printed versions of Magini's metoposcopy – with the exception of the anomalous, isolated and drastically abridged 1789 edition – do not have this section, though the 1628 Venice manuscript does include it. There is no point in going into this topic here, though it is worth noting that to the characteristically analogical thinking of the medieval and renaissance astrologers – and not just astrologers, to be fair – moles are to the skin what the stars are to the sky. Consequently, once certain areas of the body have been assigned to the dominion of certain planets, the presence there of a mole or wart is analogous to the presence of a planet in a zodiacal sign or an astrological 'domus'. This, and the apparent randomness of such maculations, clearly attracted the theoreticians of divination.[44]

Certain discrepancies emerge when we compare metoposcopy as briefly expounded above with the as-it-were practical metoposcopy which emerges

[41] See *Tetrabiblos*, IV, 10. For the classical antecedents of Ptolemy's ideas on this topic, see Boll, F., *Die Lebensalter: Ein Beitrag zur antiken Ethologie und zur Geschichte der Zahlen* (Leipzig/Berlin, 1913).

[42] 'Se adunque la linea di Venere sarà potente e fortunata, sarà il Nato molto pronto alla libidine, e passerà felicemente questa età usando convenientemente il Coito.' (f. 97v).

[43] f. 97v. Presumably this section on the ages of man was inspired by the obvious reflection that babies do not have lines on their foreheads and old people do, so metoposcopic analysis would have to be restricted to certain age-groups.

[44] A rambling, but elegant treatment of this topic is Lodovico Settala's *De' Nei* (Venice, 1609). This seems to be the only Italian version of his *De naevis* which first appeared in 1606 in Milan, being reprinted there in 1626 and again in Padua in 1628. Thorndike (VIII, p. 453) gives Padua, 1628 and Pavia, 1628 (ibid., VI, p. 506) -- presumably the latter is a slip for Patavii, since there is no other record of an edition in Pavia. The 'locus classicus' for divination by moles etc is a certain Melampus. The text (Greek and Latin) is given by Franz, J. G. F., *Scriptores physiognomoniae veteres ex recensione C. Perush et F. Sylburgh* (Altenburg, 1780) who also provides a guide to the various Latin editions of the work during the Renaissance. De Lavrendière appends to his edition of Cardano's metoposcopy the Greek text and his own translation into Latin (and French), which, he suggests, replace the previous and inadequate translation by Nicolaus Petreius Corcyraeus (Venice, 1552). The identity of Melampus is obscure. Pauly-Wissova only know that he or she is credited with a treatise on mole divination. The name may well be a pseudonym, borrowed from the soothsayer who is mentioned a couple of times in Homer's *Odyssey*. Magini's ideas on moles – which are not very thoroughly worked out – seem to derive from Cardano, and possibly from Nostradamus. The 1605 Venice ms includes a treatise on moles attributed to Nostradamus, and Magini refers to it in the letter to Della Porta (see note 31 above). Whether it is an authentic work by Nostradamus I do not know – the bibliography of his writings is very complex and much corrupted by spurious works; however he does not seem to be credited with a work on this topic so this may be an interesting addition to his oeuvre.

from the illustrations (see Plates).[45] The theory, despite occasionally failing to make what we today regard as elementary, commonsense distinctions, is consistent enough in the light of its premisses. Although character and destiny are sometimes brought uncomfortably close there is very little acount of how to prognosticate specific future events. But in the illustrations – of which we have here a random but representative sample – the coherence is sharply diminished. An individual whose forehead displays lines like those in Number 8 will burn his house down. There are countless similar examples of equally concrete and specific future events which seem only tenuously related to character – murder by thieves, by relatives; falls from heights, from animals; drownings, poisonings, incarcerations, tortures, insanities and venereal diseases.

The second observation which the illustrations almost compel us to make is that it would be extremely difficult – if not impossible – to supply the captions to these specimen heads, were they not already given to us. The information given in the theoretical exposition is simply insufficient for the reader to analyse the illustrations for himself. This, in turn, suggests that the captions and the lines in the illustrations are merely arbitrary. However, if we compare the illustrations from the Florentine manuscript with those of the Venetian 1628 copy, we see that broadly speaking they are the same. So if the captions are arbitrary, at least they are consistently so.

Number 10 reveals another problem. The wretch shown here will be hung in his 34th year. From the wavy line on his head, with a small intersection just after the third wave crest, we infer that there is here some means of dating a future event. Not only is this dating technique not mentioned in the text, it contradicts – or, at least, is different from – the technique which *is* given in the text. Since the intersection occurs in the right-hand third of the forehead, the individual in question (and all others similarly adorned) should be hung sometime between his 60th and 90th birthdays. It is worth noting, moreover, that the Venice manuscript gets this illustration wrong, which suggests that the copyist did not know about the principle involved.

There seem to be a variety of techniques in metoposcopy and Magini, in collecting his illustrations, has presented material which he has not processed in the theoretical exposition. To be fair to him, he does to some extent cover himself with a disclaimer:

> Finally we will examine a great many specific lines on the forehead which have been recorded by various authorities. We will do our best

45 The Florentine ms has over 800 illustrations – done by hand onto printed blank heads – in addition to those contained in the body of the text. Nos 120–720 precede the text, nos 1–228 follow. The numbering system is not entirely clear since overlapping numbers are not duplicates. The Venice mss are also copiously illustrated. The 1628 copy, for instance, has 357 illustrations after the text, plus a further 25 specimen heads purporting to be of real individuals, including Torquato Tasso and Magini himself (ff. 71–74). This cornucopia of illustrations is an important reason for working with the ms instead of the printed version of Magini's metoposcopy, because the latter has only the handful of instructional diagrams embedded in the text.

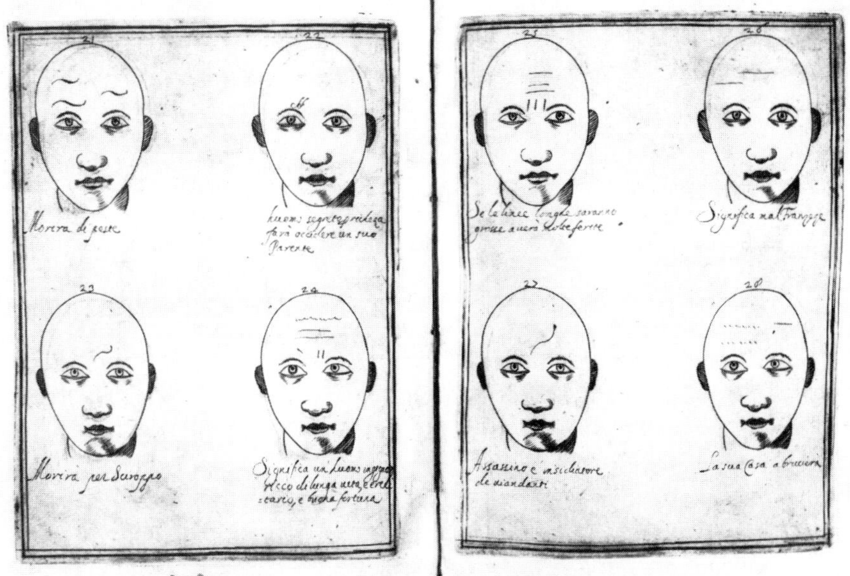

Plate 1. Florence, Biblioteca nazionale, MS serie Targioni-Tozzetti 80 (101) S. xvii, ff. 102v–103r.
1. He will die from the plague. 2. A secretive man who will have a relative murdered for financial gain. 3. He will die by poison. 4. Signifies a very intelligent man, wealthy, long-lived, heir to wealth, lucky. 5. If the long lines are emphasised he will suffer many wounds. 6. Signifies syphilis. 7. A murderer and highwayman. 8. His house will be burnt down.

> to bring some order into the confusion of conflicting ideas, but with the proviso that reliance should only be placed on those signs which conform to the rules we have expounded in the first section. Those signs that do not conform must be regarded as provisional until they are confirmed by experience.[46]

A praiseworthy but somewhat optimistic sentiment.

All this leads us to what is, I think, a reasonable question: how practicable is metoposcopy? If the premisses of astrology are accepted, and assuming the availability of accurate time-pieces, ephemerides and data, astrology is

[46] 'Per ultimo discorreremo molti particolari linee amessi della fronte, che da molti sono state osservate, con quello migliore modo, che in tanta confusione ne abbiamo potuto ottenere; avertendo che molto più s'averà da prestare fede a quei Segni che averanno conformità e concordanza con le regole poste da noi in questo primo libro, ed quei che non concordano tengansi per sospesi finchè dalla esperienza venghino confermate.' (ff. 81r–v).

Plate 2. Venice, Biblioteca nazionale marciana, Cod. marc. ital. II, 4994. ff. not numbered.
1. She will die of a pestilential fever. 2. A secretive man who will have a relative murdered for financial gain. 3. He will die by poison (?).
4. A wealthy and intelligent man, who will enjoy long life, a legacy and good luck. 5. If the lines are long and deep he will have many wounds.
6. He will suffer from syphilis. 7. Traitor and highway murderer.
8. His house will be burnt down.

eminently practicable. This holds, of course, regardless of its truth-value. But is metoposcopy so easily practised? One example of a practical difficulty will make this clearer. A line which ascends from left to right is generally indicative of a positive transition from the unfavourable influence indicated by the starting-point on the left to the favourable influence of the planet ruling the area on the right where the line ends. Conversely, when a line descends from right to left, the interpretation is reversed.[47] Given the premisses, this is theoretically consistent. However, when we are actually

[47] This observation is synthesised from the section on 'concorsi' (ff. 93r–96v). Cf. Plato's *Republic* (614) where the souls of the just travel upwards to the right and those of the unjust go down and to the left. Clearly any study of physiognomics would have to take account of the right/left, good/bad, sacred/profane dichotomy which relates physiology and philosophico-religious thought in such a problematic fashion.

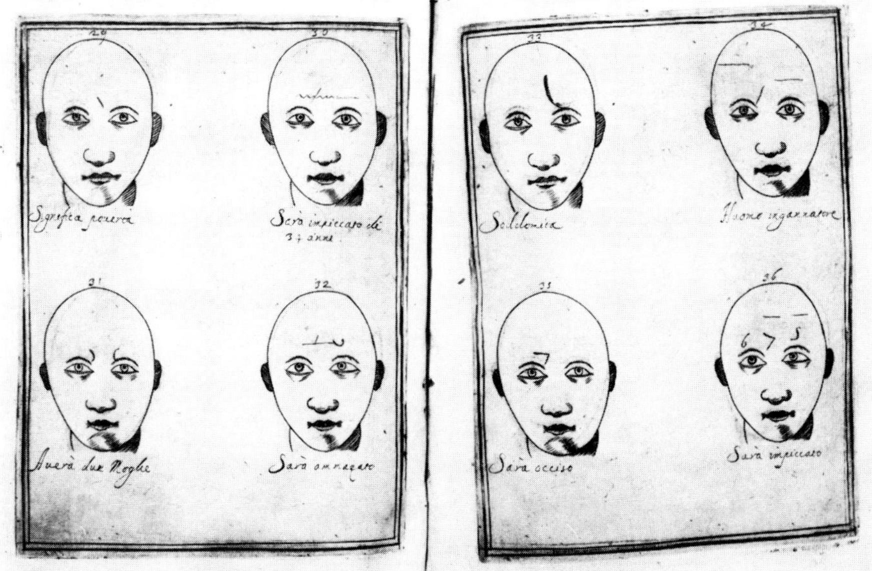

Plate 3. MS Targioni-Tozzetti 80, ff. 103v–104r.
1. Signifies poverty. 2. He will be hanged at the age of 34. 3. He will have two wives. 4. He will be killed. 5. Sodomite. 6. Deceiver. 7. He will be murdered. 8. He will be hanged.

confronted with another person – who has just got out of bed – how do we decide whether a line is ascending or descending? There is nothing in the text, or the forehead, to tell us. Another example of a practical problem is the difficulty of deciding which lines belong to which planets – if there are, say, three lines crossing the forehead all could belong to one, two or three planets.[48]

If we regard metoposcopy as an extreme refinement of the basic astrological notion of melothesia – that each limb, organ and function of the body is ruled by a zodiacal sign and a planet – then it is in a situation similar to that of astrology proper in this period. In the hands of highly competent and scrupulously professional authorities like Magini, astrology had developed to a point where sheer theoretical refinement was becoming an obstacle to its practical application. As Thomas puts it, the astrologer's

[48] The third book of the 1628 Venice ms – which has never been published – briefly attempts to tackle this problem. The metoposcopist should use a pair of compasses to establish the distance between the inner corners of the eyes. This unit should be one fifth of the depth of the forehead from hairline to the 'root' of the nose. Since this does not always work, other units may be tried – such as the distance from the 'root' of the nose to the corner of the eye. When a suitable unit has been found, the five horizontal zones of the forehead have been found too (ff. 9r–10v).

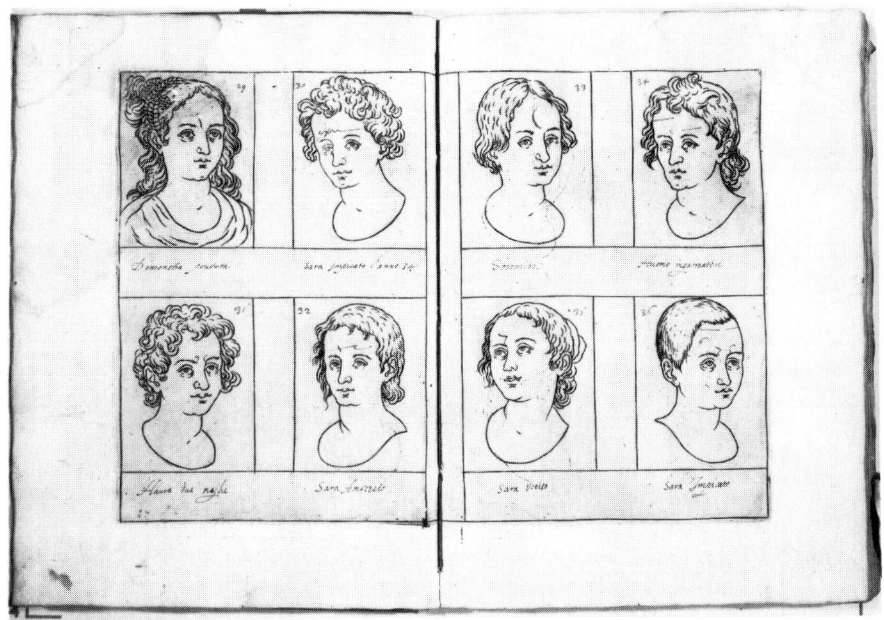

Plate 4. MS Cod. marc. ital. II 4994.
1. Reveals poverty. 2. He will be hanged at the age of 34. 3. He will have two wives. 4. He will be murdered. 5. Sodomite. 6. Deceiver. 7. He will be killed. 8. He will be hanged.

> efforts to sharpen his conceptual tools only meant that he came nearer to reproducing on paper the chaotic diveristy which he saw in the world around him.[49]

The initial physiognomic insight that broad generalisations about people's personalities can be related to certain obvious physical characteristics inspires the hope that ever more minute observation of the body will provide ever more precise information about the individual. But there comes a point where such minute observation is theoretically possible but extremely difficult in practice. Metoposcopy seems perilously close to that point.

We should stress metoposcopy's astrological rationale. Any pronouncement about the significance of marks on the body would be entirely arbitrary unless those marks could be systematically assimilated to some acceptably verifiable conceptual scheme. Astrology, and more generally the notion of a continuity and correspondence between the microcosm and the macrocosm, provides such a scheme. Indeed, it is difficult to think of any other such

[49] Thomas, K., *Religion and the Decline of Magic* (1971), Penguin, 1978, p. 341.

scheme which was available to Renaissance investigators. Although it may strike us as strange, to assimilate bodily markings to an astrological scheme was to provide the best available naturalistic account of bodily markings. Clearly it was much more satisfactory than the rationale of the 'bestiarum humanum' type of physiognomics which relies on anthropomorphic preconceptions. In the brief passage from ps-Aristotle cited above, we see that in order to describe a man with a rounded forehead as dull, we have first to establish that donkeys are dull.

The fact that physiognomics – and more specifically metoposcopy – was assimilated to astrology underlines a further factor: the reluctance of Renaissance scientists to accept that a phenomenon might be a random and meaningless event, not situated in the universal scheme of things.

Finally, the astrologisation of metoposcopy is surprisingly thorough. The intersections (*concorsi*) of the various lines correspond to the astrological aspects of conjunction and, perhaps, opposition. There are various ways of dividing the forehead into the zones ruled by each planet. Magini's method, for instance, has only four lines across the forhead. The area above the right eye is then assigned to the sun and that above the left eye to the moon (and the top of the nose, between the eyes, to Venus). The eyes thus correspond strictly with the sun and the moon, called in astrological terms, the 'luminaries'. Other metoposcopists, such as Filippo Finella, divide the forehead into seven horizontal zones and follow the conventional ptolemaic cosmology, with the line of Saturn at the top and the line of the moon at the bottom, just above the eyes.[50] In fact, Finella takes his reliance on astrology as the intellectual legitimisation of metoposcopy an ingenious step further. He assigns meaning to each line according to the point of view of the astrological house from which one approaches it – a given line can thus mean twelve different things. A long line in the 'situatione Jovis' means, from the point of view of the first house, a long life. The same line from the point of view of the fourth house means 'bona stabilia'. In Finella's case astrology provides the structural framework for the book as well as the rationale for its contents.

We have already noted Magini's failure to distinguish consistently between mood and underlying character, and between character and destiny. There is a similar mingling of ideas at the lexical level. Although a close textual analysis would be rewarding and revealing, a handful of instances will suffice for present purposes. For example, the word *l'uomo* (man, the individual who is the object of the metoposcopist's researches) is often replaced with *il nato* (the native, the word used to refer to the subject of a natal horoscope), as if the discussion was about horoscopes instead of foreheads. As a rule, Magini uses *significare* (to signify, mean) to describe what a line does; thus, say, line x–y 'signifies' that the individual will be such-and-such. Frequently, how-

[50] Finella or Phinella, P., *De metroposcopia; seu methoposcopia naturali* (Antwerp, 1648). This work is rehashed as *De methoposcopia astronomica. De duodecim signis coelestibus* (Antwerp, 1650). The copy of the latter I consulted in Bologna, Bibl. universitaria, was bound with a further work by Finella, *Tractatus de duodecim coelestibus signis in 360 gradibus divisis, cum eorum inclinationibus et naturis* (Antwerp, 1656). Finella's works are usefully discussed by Thorndike, VIII, 457f.

ever, *significare* is replaced with *(di)mostrare* (show), *causare* (cause), *fare* (make) and even *essere* (to be). Sometimes he uses the words *causa* and *effetto* (cause and effect) as we would use them, sometimes as if they were synonymous, and sometimes even as if they were synonymous with *influenza* (celestial influence).

Even if we make allowance for a degree of stylistic variation, this relaxed usage of certain key words reveals, at the least, a less than rigorous theory of divination and its relationship with free will. As is so often the case in astrological writing, we are unsure whether these lines are signs or agents.

We have so far suggested that metoposcopy was not a practical proposition. However, it may well have stayed in Magini's mind as a viable way of analysing people. For instance, in a letter to Galileo, dated January 1612, Magini complains about a certain Gaspare Bindoni who had appropriated a technique that Magini had evolved for the fabrication of very large curved mirrors. He describes Bindoni as

> this man who bears engraved on his face in letters of fire the very idea of brazen arrogance.[51]

Clearly Magini had no doubts about some degree of continuity between interior and exterior features.

We can approach this question of the practicality of metoposcopy from another angle. Suetonius, it has been shown, did not always describe his emperors as they were, but occasionally composed (verbal) physiognomical portraits appropriate to their characters and moral outlook.[52] So, we might expect to find visual artists displaying a lively interest in theories about how personality might be related to external appearance. Wittkower reckons that

> we can take it for granted that (Michelaangelo) and his pupils were as interested in physiognomical experiments as were other Italians of the Cinquecento from Leonardo down to . . . della Porta.[53]

But it is a long step from 'physiognomical experiments' to portraiture and very few seem to have taken it.[54] The nearest that theoreticians of art come to physiognomics seems to be, on the one hand, studies of ideal proportions,

[51] 'Quest'huomo che porta scolpito in faccia, di carattere di fuoco, l'idea della sfacciataggine e dell'arroganza . . .' in *Le opere di Galileo Galilei* (Firenze, 1929–39, ristampa dell'edizione nazionale), XI, p. 259, letter No. 641.

[52] See Couissin, J., 'Suetonius physiognomiste', *Revue des Etudes Latines*, 31, 1953, 234–56 passim.

[53] Wittkower, R., 'Physiognomic experiment by Michelangelo and his pupils', *JWCI*, 1, 1937/38, 183–84.

[54] Bower (1973) discusses the use of physiognomical features in the portrayal of various stereotyped figures, such as bystanders at the Crucifixion who personify greed, ignorance and so on, in a number of Renaissance paintings. Meller, P., 'Physiognomical Theory in Renaissance Heroic Portraits', *Acts of 20th Internat. Congress of Hist. of Art* (Princeton, 1963) II, 53–69, discusses the attribution by artists of 'leonine' features to important subjects, such as the bust of Cosimo I by Benvenuto Cellini in the Museo nazionale, Florence. For a more general discussion

and studies comparing human faces and bodies with animals, on the other. The only example of an art theory treatise to include a chapter on physiognomics before the middle of the sixteenth century is the *De sculptura* (1504) by Pomponio Gaurico, brother of the celebrated astrologer Luca.[55]

There is no space here to discuss the whys and wherefores of this lack of interest – which is, perhaps, surprising in view of the growth of portraiture in the later Renaissance – but one tentative explanation is worth giving, if only because it underlines an important characteristic of physiognomic studies. Studies of ideal proportions and physiognomics tend to share a quality which limits their usefulness, especially to practising artists. They are mostly concerned with types of human body, static and out of context. The complex 'contraposti' and dynamic, unstable poses favoured by artists in the later Renaissance – the kind of dramatically foreshortened figures that we see, for instance, on the vault of the Sistine Chapel – cannot learn much from static, full-frontal studies. The artist, like the aspiring metoposcopist, is concerned with specific individuals in specific circumstances, rather than idealised types.[56]

Francis Bacon, in what may be taken as an appeal for a new and more effective approach to psychology, noticed this static quality in physiognomics. 'In physiognomics', he writes,

> I find a deficience. For Aristotle hath very ingeniously and diligently handled the factures of the body, but not the gestures of the body, which are no less comprehensible by art, and of greater use and advantage. For the lineaments of the body do disclose the disposition and inclination of the mind in general; but the motions of the countenance and parts do not only so, but do further disclose the present humour and state of the mind and will. For as your majesty saith most aptly and elegantly, 'As the tongue speaketh to the ear so the gesture to the eye'. And therefore a number of subtle persons, whose eyes do dwell upon the faces and fashions of men, do well know the advantage of this observation, as being the most part of their ability; neither can it be denied, but that it is a great discovery of dissimulations, and a great direction in business.[57]

---

of some of the problems implicit in the use of physiognomical ideas by artists, see Gombrich, E. H., 'On physiognomic perception' in *Meditations on a Hobby Horse* (London, 1963) and idem., 'The Mask and the Face: the perception of physiognomic likeness in life and in art' in Gombrich, E. H., Hochberg, J. and Black, M., *Art, Perception and Reality* (London, 1972).

[55] This statement comes from Barasch, M., 'Character and Physiognomy: Bocchi on Donatello's St George. A Renaissance text on expression in art' in *J. Hist. Ideas*, 36 (3), 1975, 413–30. Barasch notes that physiognomic literature 'had made no significant impact on art theory and the description of works of art prior to the late sixteenth century'. (p. 426). For Pomponius Gauricus, see the edition of *De sculptura* by Chastel, A. and Klein, R. (Geneva, 1969) which has excellent and informative apparatuses on physiognomics, Gaurico and his brother's contribution.

[56] Proportion studies, it should be noted, tend to yield pride of place to the very high quality anatomical illustrations which begin to appear after Vesalius in the second half of the sixteenth century.

[57] Bacon, *Of the Advancement of Learning* (1605), II, ix, 2 (p. 103 in A. Johnston's edition, Oxford, 1974).

There is no doubt that people in the late Renaissance and early seventeenth century were extremely interested in physiognomics and regarded it as a useful and plausible activity. The enthusiasm – admittedly qualified – shown by so important a figure as Francis Bacon was by no means idiosyncratic. Camden, in what is one of the very rare scholarly discussions of post-classical physiognomics, has shown convincingly and with copious quotation that physiognomy was a mainstream interest in Elizabethan and Jacobean Britain, and we have no reason to suppose that matters were any different elsewhere in Europe. Camden observes that metoposcopy was never quite as popular as physiognomy, chiromancy 'or other pseudo-sciences', and this may be a consequence of metoposcopy's sheer impracticality. And, as our prefatory quotation shows, voices were raised against metoposcopy simply on the grounds of its implausibility.[58]

This point that physiognomics and its various branches were enthusiastically cultivated in the late Renaissance is one good reason for studying – however briefly – a topic as marginal and bizarre as metoposcopy.

Another reason – already mentioned above – is that in its small compass metoposcopy raises many of the questions posed by the larger – and equally neglected – subject of physiognomics: the philosphical problems implied, or obscured, by positing a meaningful relationship between external characteristics and interior, psychological traits, and the problem of evolving a model of human personality which discriminates between underlying and transient characteristics, and so on.[59]

A third reason for looking at metoposcopy is that in the century after 1560 it enjoys a small publishing boom. In addition to the nine editions of Magini's work, the three editions of Hajek, the two editions of Cardano and the two editions of Finella, which have already been mentioned, there are works by several other individuals. They include (in alphabetical order): J. S. Elsholtz, Herman Janszoon Follinus, Samual Fuchs, Cornelio Ghirardelli, Rudolf Goclenius junior, Johannes ab Indagine, Giovanni Ingegneri, Philip Mey (or Meys or Meyens), Christian Moldenarius, Antonio Pellegrini, Richard Saunders (or Sanders), Nicola Spadon, Jean Taisnier, Jean Taxil and

58 Camden, C., 'The mind's construction in the face', *Philological Quarterly*, 20, 1941, 400–12 (p. 409). On post-classical physiognomics, apart from the works already mentioned, notably Denieul-Cormier (1956) and Thorndike (passim) there is little. A brief introduction is given by Cardoner, A., 'La fisiognomia hasta el siglo XIX', *Revista de dialectologia y tradiciones populares*, 27, 1971, 81–95, with useful remarks on the *Secret of Secrets* and Della Porta's physiognomical studies. These are also covered by Scapini, A., *Dalla 'fisiognomica' di Giovan Battista della Porta (sec. XVII) alla morfologia consitutionalistica* (Pisa, 1970). Primary sources are best quarried from Laehr (1900), Schüling (1967) – who goes only as far as 1600 unfortunately – and of course Thorndike.

59 See, for instance, De Angelis, E., 'Rapporti tra mente e corpo dall'età classica al XIX secolo', *Medicina nei secoli*, 19, 1982, 193–213, who comes to the subject from a physiological point of view, being concerned with 'mind' as a function of neural activity, cerebral configuration etc, rather than how it differs from person to person. See also Mora, G., 'Mind-body concepts in the Middle Ages: Part 1. The classical background and its merging with the judaeo-christian tradition in the early Middle Ages', *J.Hist. Behavioural Sciences*, 14, 1978, 344–61.

Henri de Boyvin du Vaurouy.[60] At least twenty authors, and many more than that number of editions and translations of their work, constitute a nucleus large enough not to be ignored.

I shall conclude on a more general note with what is, I think, a further justification for studying metoposcopy.

The historian of astrology tends to feel the reproachful echo of modern academic scepticism about presentday astrology. The best solution to this problem of credibility has been the assimilation of the history of astrology to the history of the sciences in general, and specifically to the history of astronomy. In this way the history of astrology partakes, modestly, of the prestige of the so-called 'hard' sciences. Though difficult to assess, there is, however, a price to be paid for this respectability.

Before the seventeenth century there were not many interdisciplinary boundaries – there were not that many disciplines in any case – and much more intellectual activity could find a common ground in astrology. To some extent the modern institutional reluctance to cross disciplinary boundaries makes this difficult to appreciate. But to give a full account of astrology in the Renaissance and the Middle Ages – and from the point of view of the historian of astrology any sub-division of the period from the mid-twelfth until the early seventeenth centuries creates more problems than it solves – we would have to do what the astrologers of that time did. That is, they attempted, implicitly, to provide a unified theory of knowledge. Astrology in one form or another provided a common ground for the encounter of, say, acoustics and meteorology, architecture and biography, medicine and political propaganda, historiography and jewellery-making, iconology and navigation, town-planning and agricultural science, and so on, to mention just some of the topics on which astrologers felt qualified to pronounce.[61]

[60] Elsholtz, J. S., *Anthropometria sive de muta membrorum corporis humani proportione, & Naevorum harmonia libellus* (Frankfurt a. Oder, 1663); Follinus, H. I., *Physiognomia of te menschenkenner* (Haarlem, 1613); Fuchs, S., *Metoposcopia & ophthalmoscopia* (Strasburg, 1615); Ghirardelli, C., *Cefalogia fisionomica* (Bologna, 1630); Goclenius, R., (jr) *Uranoscopiae, chiroscopiae et ophthalmoscopiae contemplatio* (Frankfurt a. Main, 1608); John of Indagine, see note 21 above; Ingegneri, G., *Fisionomia naturale* (Vicenza, 1615); Mey, P., *Chiromantia medica . . . Nebens einem Tractätlein von der Physiognomia medica* (The Hague, 1667); Moldenarius, C., *Exercitationes physiognomicae* (np, 1616); Pellegrini, A., *Della fisionomia naturale* (Milano, 1612); Saunders, R., *Physiognomie and chiromancie, metoposcopie, the symmetrical proportions and signal moles of the body* (London, 1653); Spadon, N., *Studio di curiosità nel quale si tratta di Fisonomia, Chiromantia, Metoposcopia* (Venice, 1662) (See Thorndike, VIII, 466–7); Taisnier, J., *Opus mathematicum octo libros* (Cologne, 1562); Taxil, J., *L'Astrologie et physiognomie en leur splendeur* (Tournon, 1614) (ref. Thorndike, VI, 169–70); Boyvin de Vaurouy, H., *La phisionomie* (Paris, 1635) (ref. Laehr, 1900). In addition to this certainly incomplete list of printed works about (or containing sections about) metoposcopy there are numerous manuscripts, judging by a brief survey of manuscript catalogues with subject indices. This is only to be expected, especially in areas under the jurisdiction of the Papacy and Inquisition, since the various Indices of Forbidden Books generally have a rubric devoted to extirpating divination. I have found no mention of metoposcopy but chiromancy is invariably listed and that would be sufficient to prevent authorisation to publish, at least until the partial 'thaw' around the accession of Urban VIII in 1623.

[61] Preliminary indications for these various topics may be found in: Boucher, B., 'Giuseppe Salviati, pittore e matematico' in *Arte veneta*, 30, 1976, 219–24 (re Salviati's speculations about an astrological account of acoustics!); for weather-forecasting, see Hellmann, G., 'Versuch einer Geschichte der Wettervorhersage im XVI Jahrhundert' in *Abhandl. d. preussischen Akad. d.*

So, a consequence of the assimilation of the history of astrology to the history of the 'hard' sciences is that we cannot give a full picture of astrology as it was. (There are other problems which we cannot discuss here, such as the distortions imposed on the history of astrology by 'triumphalist' ways of doing the history of the sciences, which seek, and therefore find, models of epistemological progress – astrology remained essentially unchanged from Ptolemy until the seventeenth century.) If astrology was ever a science in any modern sense of the word, it was closer to what we call the social sciences, and the historian of astrology has a valuable contribution to make to the history of these somewhat hazily defined disciplines.

Metoposcopy and, more importantly, physiognomics clearly belong to the history of psychology – or at least, in the first instance, to that part of psychology which is concerned with personality theory (or at it used to be called, ethology, characterology or character typology).[62]

Broadly speaking, the history of psychology finds its starting-point in the later seventeenth century, in the speculations of people like Hume and Descartes. Of course, there is no suggestion that humanity – with the distant exception of a handful of fifth-century Athenians – had spent the previous millennia wantonly ignoring itself. Rather, historians of psychology seem to have detected very little objective psychologising of a kind that would be

---

Wissenschaften (Phys.-math. Klasse) Berlin, 1924, and Zambelli, P., 'Fine del mondo o inizio della propaganda? . . .' in *Scienze credenze occulte livelli di cultura* (Firenze, 1982); Vitruvius, L. B. Alberti, Francesco di Giorgio Martini and Filarete all recommended that architects knew enough astrology to orient buildings accurately and to chose the best time to begin construction; Aubrey's *Brief Lives* and Giovanni Villani's *Cronica* both afford numerous examples of astrologically slanted explanations of personality and behaviour; for astrological medicine see note 34 above; on politics and astrology, see the essay by Zambelli above and Capp, B., *Astrology and the Popular Press: British Almanacs 1500–1800* (London, 1979) especially Ch. 3: 'Almanacs and Politics'; Garin, E., *Lo zodiaco della vita* (Bari, 1976), Ch. 4 discusses the astrological views of history held by Pomponazzi, Palingenius, Agrippa, Bruno and others. Examples of astrologically flavoured history abound in Villani's *Cronica*; Castelli, P. et al (eds) 'Le virtù delle gemme: Il loro significato simbolico e astrologico nella cultura umanistica e nelle credenze popolari del '400', in *L'oreficeria nella Firenze del '400* (Firenze, 1977), sez. VI; apart from the classic study by Warburg, A., 'Italienische Kunst und internationale Astrologie im Palazzo Schifanoja zu Ferrara' (1912) in his *Gesammelte Schriften* (repr. Liechtenstein, 1969, pp. 459–81), see Boczkowska, A., 'The Crab, the Sun, the Moon and Venus: Studies in the iconology of Hieronymus Bosch's triptych "The Garden of Earthly Delights"', *Oud Holland*, 91, 1977, 197–231; Ingram, B. S. (ed.) *Three Sea Journals of Stuart Times* (London, 1936); Castelli, P., '"Caeli enarrant": astrologia e città', *Atti del 2° Convegno internaz. di storia urbanistica, Lucca 1977* (a cura di Martinelli, R. & Nuti, L.) (Vicenza, 1978), pp. 173–93; Webster, J. C., *The Labours of the Months* (Chicago, 1938). For a general discussion of Renaissance astrology, see my thesis (note 27 above).

62 Diamond, S. (ed.) *The Roots of Psychology: A Source Book in the History of Ideas* (New York, 1974) notes that 'theories of temperament have a long, well-knit history, in which the compatibility of various systems is high enough to strongly suggest underlying validity. This continuity continues to persist into the present, though more sophisticated statistical approaches lead us to speak of "factor loadings" and "variables" rather than of "types".' Though slightly dated, an excellent and exhaustive account of personality theory's evolution is given by Allport, G. W., *Personality, a psychological interpretation* (London, 1937), including some 50 different definitions of the word *persona* (pp. 26–50).

recognised by modern psychologists.[63] As far as modern psychologists are concerned that may be fine, but for the historian it is not enough because the historian is primarily concerned with the people and cultures that are the object of his research and only secondarily concerned with their relationship to modern society.

The value of investigating astrology's role as a proto-psychology is two-fold. First, it liberates the history of astrology from the procrustean bed of the history of the harder sciences. Second, it opens a more appropriate and, as far as I can see, neglected research field for the historian of astrology.

Physiognomics has flourished throughout the history of Europe, and indeed still flourishes today, but never was it more rampant than in the seventeenth century – the century in which astrology withered away and in which psychology, amongst other things, first blossomed. Whether or not we agree with the Thane of Cawdor that 'there's no art to read the mind's construction in the face', it is to physiognomics – astrological or otherwise – that we must turn if we want to know what the Renaissance thought about the construction of the mind and how it was concealed, or revealed, by the body.

[63] Jung, C. G., 'The Problem of Types in the History of Classical and Mediaeval Thought' (1921) in *Collected Works* VI, 8–66 (trans. Baynes, H. G., revised Hull, R. F. C.) puts it thus: 'The further we go back in history, the more we see personality disappearing beneath the wrappings of collectivity' (para. 12).

THE REVEALING PROCESS OF TRANSLATION AND CRITICISM IN THE HISTORY OF ASTROLOGY

Jacques E. Halbronn

The structure of astrological thought is revealed in two different ways: through the attacks which are led against it and through the nature of the translations which take place from one culture into another culture. We would like to illustrate this statement with the case of astrological polemics in France, and the process of the transfer of the French astrological 'ensemble', including all its possible facets connected with astrology, to England. (Obviously, the demonstration could use other examples and occur in other fields than astrology.)

Our study will therefore combine two different view-points: the nature of anti-astrological arguments in France (notably those of Calvin, Bodin, Gassendi and Pluche) between the sixteenth and eithteenth centuries – the chronological limits of this study – and the different levels of astrological manifestation, examined through the question of French influence in England in this respect. It seems necessary to distinguish systematically between two levels in the apprehension of astrological literature. This distinction will not be the obvious one between pro- and anti-astrology – which in fact often correspond in their degree and quality of argumentation – but between a small circle, an élite, and a larger and more popular type. In this way we shall arrive at a kind of hierarchy in the approach to and practice of astrology.

In our study on French influence, we shall have to modify a certain approach – which is certainly not restricted to the history of astrology – that typically tends to underestimate or neutralize the French and Norman influences upon English culture, through an exaggerated emphasis upon the Latin and Italian ones.

A Note on French-Italian Astrology in the Sixteenth Century

When one speaks of Italian astrology in the sixteenth century, one should remember to what extent it was interrelated with that of France. Many Italian astrologers spent quite a long time in France and published their works there; so, seen from England, it does often appear that French and Italian astrologers were part of the same process.

For example, let us take some excerpts from John Harvey, of the famous astrological Harvey family, about two Italian astrologers: Francesco Giuntini and Francesco Liberati. The first one is better known under his Latin name of Junctinus. He published a great part of his astrological work in France, as his *Tractatus judicandi revolutiones nativitum*, and his *Speculum Astrologiae* (both in Lyons). In 1577, there appeared in Paris a *Discours sur ce qui menace devoir advenir la comete apparue a Lyon le 12 de ce mois de Novembre 1577 laquelle se voit encore a present par Maistre Francois Junctin*. Therefore it is no surprise to read in John Harvey's *A Discourse . . . concerning Prophecies* (1588) that 'Junctinus [was] framing the celestial scheme to Lyons in France'.

Another and more striking example concerns Liberati. In his *Astrological Addition* (1583), John Harvey wrote that he had in his hands 'a French Almanack or rather discourse of certain Astrological Accidents . . . I can say the lesse for my small skill in French, yet thus farre dare I presume upon the very title or Inscription thereof, that *Docteur Francois Liberati* de Rome is none of the perfect Astrologers either in Italy or in France.' The mixture of French and Italian characteristics in the description of Liberati is noticeable, as is the ascription of a French alamanac to the Italian astrologer.[1] It is not our purpose to analyze this phenomenon in France, but just to underline the fact that France was attractive to many Italian astrologers, who developed very well there under Catherine de Medici (to whom the French astrology Auger Ferrier dedicated his textbook in 1550).[2] The way in which Harvey (for example) would read and quote these works in French is remarkable: 'Et finalement, l'an 1583 sera pluvieux et dangereux pour les biens de la terre et aussi "respectini tumulus et apparatus belli". And therefore, we are hartily to wish and pray with him: Dieu par sa misericorde nous delivre du mal et du danger par les Astres nous sommes menazes et nous envoye la paix et la Sainte Grace which is all that I am presently to note touching the contents of that French discourse.'[3] If we add the works of local French astrologers such as Fine, Mizauld, Ferrier, Dariot and Nostradamus to those of the 'French-Italian' astrologers, we can understand that all this constituted a considerable pole of attraction for English astrology.[4]

Introductory Literature

By this phrase we mean texts intended to give a grounding in astrology to those who wished to master its techniques. It is significant, with regard to its effects, that in Christian Europe this literature was written in both Latin and

[1] Liberati published a text in French in 1577 which is not considered to be a translation: *Discours de la comète . . . par excellent Maître François Liberati de Rome*. This was the same comet which received a commentary by Junctinus.

[2] Eugène Defrance has demonstrated the cohabitation of French and Italian astrologers in French political life around the time of Henri II's death. See 'Les oracles astrologiques de Luc Gauric et de Nostradamus' in *Catherine de Médici, ses astrologues . . .* (1911).

[3] From *An Astrological Addition* (1583).

[4] It should be mentioned that German astrology is also introduced into certain french texts, such as Ferrier's *Judements Astronomiques sur les Nativitez* (1550); the third book is (as he himself declares) is mainly a translation from German authors.

in vernacular languages. The Latin texts were addressed to an élite which took no account of linguistic limits, while the vernacular texts reached a 'lower' cultural level in a particular society. Both types were composed essentially of treatises and other guides or introductions. Alamanacs occasionally provided some explanations, but always quite briefly. On the popular level, there were also descriptions of astrology and its sub-divisions, but they did not give instructions for using the ephemerides.

(a) *The French Origins of English Learned Introductory Literature*

In 1647, William Lilly supplied at the end of his *Christian Astrology*[5] a list of works, mostly in Latin. This list deserves attention.[6] The first observation one could make is that only three of the listed textbooks are in English: *The Judgements of Nativities* (1642), by Auger Ferrier; *The Judgements of the Starres* (1598), by Claude Dariot; and *Introduction to Judiciall Astrology* (n.d.), by Humphrey Baker.[7] Obviously, the first two books are translations from the French; they were published in Lyons, in 1550 and 1557 respectively. As for the third, it too is the work of a French author – Humphrey Baker being only (as he himself states) the translator. The real author was Oronce Finé, who published a first version in 1543, and another and more complete version (which was used by Baker) in 1551, in Paris. (It can be further specified that Dariot's book was translated from the Latin version, while that of Ferrier was not.)[8]

What strikes us is the possibility that Lilly could have borrowed mainly from these three very accessible texts, and less from the others on his list. A

5 Republished in a modern edition by Regulus Press (London, 1985). (*Christian Astrology* is not mentioned by Lynn Thorndike in his work on the history of astrology.) It seems to us that there were two subsequent editions of *Christian Astrology* after the first, rather than the one usually mentioned. The 'second' edition includes, in fact, one book published in 1658, and sometimes erroneously considered to be a different work – *An Introduction to Astrologie. An easie and plain method teaching how to judge upon a Nativity. The second Edition corrected and amended* – and one book published in 1659. The former consisted of the first edition's 'third Book', whole the latter was comprised of its 'first Book' and 'second Book'. (See the National Union Catalogue (pre-1956 imprints), vol. 333, p. 297.) The structure of *Christian Astrology* and the order of its subjects – which also appears with Ferrier – strikes us as redundant, and not very logical. It is therefore easy to understand why this order was reversed in the subsequent editions. After 1659, other textbooks seem to have taken over from Lilly's book, whose reign seems to have been fairly short. It's success in the nineteenth and twentieth centuries, of course, is another story. It could also be mentioned here that in 1653, two texts by Lilly were translated into Dutch: *Engelsche Waerseggher van Mr W. Lelly*, and *Monarchy ofte geen Monarchy in Engelant*. And in 1688, some of John Partridge's polemical texts (against John Gadbury) appeared in French.

6 This list includes books which obviously have nothing to do with the contents of *Christian Astrology*, and corresponds more to an attempt to describe astrological literature in general. It is a catalogue, not a bibliography. To give an example of the problems with his list, Lilly names a text by Hermes Trismegistus on medical astrology, which is in fact merely an editing of John Harvey's translation which had been published in 1583 in the *Astrological Addition*, which does not itself appear in the catalogue.

7 These titles, taken directly from Lilly, are given here in their shortened versions.

8 *Pace* the British Library catalogue entry.

study of the complete title and plan of *Christian Astrology* tends to confirm this idea: the book's first part contains instructions on 'the use of an Ephemeris' – also the object expressed by Finé's work. The book's subtitle states that it is 'a most easie Introduction', an expression belonging to Dariot's treatise. Its second part shows 'how to judge or resolve all manner of Questions', where Dariot proposed to 'give Answere to any Question'. The third part explains how 'to judge upon Nativities', which not only reminds us of Ferrier's title but includes the 'Art of Direction' – precisely the division employed by Ferrier, and the content of his third book.[9]

What further gives us the right to suspect such a design by Lilly, or possibly by his publishers? In 1650, the publisher Lawrence Chapman offered to the public a new book, *Country Astrology* – a 'plain piece of Country Astrology, rude collections of indigested and wholly unpolished experiments'. (No theory, the students want facts.) As a matter of fact, this book is nothing other than Ferrier's text-book under another name (and with an added preface), eight years after the edition of 1642.

We believe that English astrology at this time was faced with a certain perplexity. On the one hand, it appeared necessary to Lilly to claim a degree of autonomy: 'verily the Methode is my owne, its no translation; yet have I conferred my notes with . . .' (a long list follows, including Dariot, Ferrier and Duret).[10] On the other hand, however, the authority of a famous astrologer such as Dariot could be an asset. What makes us sure that Dariot was still popular in mid-seventeenth century England, while he was actually forgotten in France? Six years after Christian Astrology appeared, another text-book came onto the astrological market: *Dariotus Redivivus: or a Briefe Introduction to the Judgement of the Stars*. Paradoxiacally, it may have been Lilly who provoked this success. He included in his book a long passage entitled 'Dariot Abridged', which he included in these terms: 'In regard I have ever affected Dariot his Method of judgement in sicknesses, I have with some abbrieviation annexed it, in a farre more short way and method than heretofore published.'[11]

Another author, a contemporary of Lilly, confirms this francophile tendency of the mid-seventeenth century; Nicholas Culpeper, in his *Astrolgical Judgement of Diseases* (1651), used two books also mentioned in Lilly's list: *De diebus decretoriis secondum Pythagoricam doctrinam*, by Augerius Ferrerius (whom Culpeper changed to Perrerius); and *De ratione et usu dierum critoricum . . . cui accessit Hermes Trismegistus de decubitu infirmorum* (1555), by Thomas Boderius. (The latter title supplied part of Culpeper's own, *viz*. 'decumbiture of the sick'.) Actually, there would have been an earlier English text-book of medical astrology, of French origin, if Dariot's 1598 publisher had realized

9 Actually, the subject-matter of *Christian Astrology* – despite its self-description as 'modestly treated of in three books' – is divided into four books, as is revealed by the inner or textual titles: 'Introduction to Astrology', 'The resolution of all Manners of Questions', 'An Introduction to Nativities' and 'The effects of Directions'.

10 Sig. B2.

11 P. 258.

his ambitions: 'I shall be encouraged thereby to publish and set forth in our language the second part of this Booke of Claudius Dariot entreating more copiously of Physicke and no less needful to be known.' This confirms that the non-translation of a Latin text considerably limited its English audience. Lilly (who called himself 'Merlinus Anglicus' and often used Latin in the titles of his books, like most seventeenth century English astrologers) also mentioned the language problem: 'Some may blame me that I write in the English tongue. Yet I trust that I have offended no man if I write in my own language.'[12] He thereby reached a broader local audience, but with the exception of Robert Fludd – who published in Latin and on the continent – English astrology remained unknown by French astrologers until the end of the seventeenth century. Had Lilly written in Latin, or been translated into Latin, his reputation and impact abroad would have been greater.

A common mistake currently made by Anglo-Saxon historians of astrology is to over-estimate the importance of English astrology in the seventeenth century. That it played a considerable role in English political life is true, but that does not imply that it exerted any influence elsewhere. Such a leading role only took place in the course of the nineteenth century. Until then, English astrology, with its attraction to horary astrology, was (as some astrologers of the period admitted) rather provincial. This false perspective, of course, is basically due to the fact that most historians of astrology stop at the end of the seventeenth century, and refuse to consider the last three centuries in their research.

(b) *The Problem of the Translation of French Texts*

The historian of astrology cannot neglect the problem of language. When astrology penetrates a new cultural space, this requires the construction of a particular vocabulary. Linguistic analysis could therefore provide a very valuable method with which to determine the origin of the astrological influences in a certain area. In the case of English, the terminology is divided into French and Latin expressions; if the names of the zodiacal signs are mostly of Latin origin, the rest of the astrological terms are mostly French. French cannot really be called a properly 'foreign' language in England, considering its role in the formation of modern English. One could therefore suspect that French astrological texts were read by more English readers than Latin ones; and that it was not difficult to find translators for the former texts – men such as Thomas Kelway, whose translation of Ferrier's *Jugemens Astronomiques* (1550) we shall now briefly examine.

Merely reading the contents list makes clear not only how closely the translation has followed the original French, but also the degree of further borrowing from the Arabic: 'of the Giver of Life, called by the Arabs Hyleg' (Ch. VIII, Book I); 'Of the Giver of Years, called by the Arabs Alcoden' (Ch. IX, Book I); 'Of the separator or delimiter, called by the Arabs Algebuthar' (Ch. II, Book III); 'Of the years governed by the planets, called

12 Sig. 000002 [p. 843].

by the Arabs Fridarie' (Ch. VI, Book III). We could also mention the translation of Chapter I, Book I. There are several terms which retain their precise French form: 'Dragon lunaire' remains 'Dragon lunaire' in the English text, 'figure estimative' remains just that, and so does 'opposite signes'. Many other examples are possible. One might almost wonder why a translation was thought necessary . . .[13]

(c) *French Astrology in Seventeenth Century England*

These kinds of influence during the sixteenth century would have been sufficient to leave a profound impression on English astrological terminology. But the influence was to continue, and the umbilical cord remained unsevered in the century that followed. Two names in particular seem to have acquired a reputation as authorities; Noel Duret and (above all) Jean-Baptiste Morin, both French astrologers who had published (or written) the greater part of their work before 1650.

The English champion of Duret (not to be confused with Claude Duret, from the late sixteenth century) was Nicholas Culpeper, whose debts to Ferrier and Bodier we have already mentioned. The complete title of his *Semiotica Uranica* (1651) introduced the name of Duret, together with that of Abraham Ibn Ezra: 'Judgement of diseases . . . one from Aven Ezra by way of Introduction. 2 from Noel Duret by way of direction.' In fact, apart from one reference to Ibn Ezra on the critical degrees, the work essentially followed Duret's plan: 'The Basis of the Story was borrowed from Noel Duret, Cosmographer to the King of France and of the most excellent Cardinal the Duke of Richelieu. (I confess) in some places I have abbreviated him, in others corrected him: let another doe the like by me.' (Duret's principal work was Novae Motuum Ephemeridis Richelianae (1641); its first part was divided into two, 'Astrologia Generalis' and 'Astrologia Specialis', the second of which dealt with medical astrology.)

Turning to Morin, he was better known in England under the Latin form of his name. In his opinion, astrology constituted a 'canon', an integrated whole including all its various techniques – in brief, an outlook radically opposed to that of Kepler, who wished to eliminate certain fundamental parts of the astrological tradition, and whom Morin often criticized on this point. He also considered Ptolemy's *Tetrabiblos* as too vague in its approach to the significations of the planets, in the absence of a proper understanding of the houses. In 1623, he published (in Latin) *Astrologicorum Domorum Cabala*, an English translation of which appeared in 1659: 'The Cabal of the twelve Houses Astrological from Morinus', by George Wharton. It could have been a translation from the French of Morin's *Remarques Astrologiques* (1654, 1657) however, which includes the same text. But in general, many Latin texts by Morin failed to obtain translation into French – especially the huge and

[13] One could make the same point with respect to geographical terminology: partly Latin, partly French; for example, the Latin 'Italia' and the French 'Rome'. The phonetic transcription of Arabic words also often follows the French way.

posthumously published *Astrologia Gallica* (1661). This perhaps contributed to the nature of his influence in England, which was not centered around one specific book, but was more sporadic – though none the less important for that. Some examples from the seventeenth century are as follows. In 1672, Henry Coley, Lilly's friend and successor, gave an excerpt of the *Astrologia Gallica* (Book XIX): 'The astrological axioms and theorems of Morinus', in his *Hemerologiam Astronomicum*. John Gadbury reacted in this way: 'And here give me leave to inform the World that one hath lately pretended to translate Morinus' Theorems and Axiomata: many of which I hope ere long to meet an opportunity soberly to inform the World. But as he is now inhabited in English accompained with idle proverbs, drunken rhymes and other lewd stuff, under the disguise of an alamanack, I cannot but blush to see and I wonder how one ignorant of Morines words should draw to pretend thus to deal by him . . .' In 1679, another major astrologer – William Salmon, in *Horae Mathematica seu Urania, the soul of Astrology* – called the house system devised by Morin 'the catholick way'. In 1697, John Partridge entered the discussion by throwing doubt on the exactitude of Morin's example birth charts.

century later. In 1785, *Astrologia Gallica* inspired one of the first texts of a new generation, namely *The New Astrology* by C. Heydon Jr. (a pseudonym that is probably a tribute to the early seventeenth century astrologer, Christopher Heydon). This work was reissued, with some variants, in 1786 and 1792. As for the 'Cabal', it reappeared at the same time, in John Worsdale's *Genethliacal Astrology* (1796, 1798). The English astrologer described it as 'The definition of the twelve houses of Heaven from the works of the learned and much esteemed Philosopher and Astrologian Morinus . . .'[14]

(d) *The Popular Mode (Alamanacs, Zodiacal Books)*

Morin did not publish almanacs or political prognostications as his contemporary Lilly did. In this respect, he was a kind of anti-Nostradamus. In general, the authors of astrological textbooks in Renaissance France did not find it acceptable to appear in that sort of literature. But another class of astrologers did. Bernard Capp has tried to distinguish clearly between the French and the English almanacs.[15] It is true that the French almanac experienced changes which did not occur in the English one (the reasons for which we cannot study here). But that does not mean that the former did not influence the latter while in a previous stage. If the content of the French almanac became progressively rather bland, the outline still prevailed in the

[14] In 1625, the following text was published, which we consider very typical of how one used to see the relations between the two languages: *Part of Du Bartas, English and French and his owne kind of verse so neare the French Englished as may teach an English-man French or a French-man English, by W. L'Isle.* The text by the French poet Du Bartas was, besides, inspired by cosmological developments.

[15] B. Capp, *Astrology and the Popular Press* (London, 1979).

seventeenth century of a calender emphasizing Moon positions, followed by a second part giving prognostications for the year. In this respect, English and French almanacs follow the same pattern. And the prognostications of Nostradamus, for example, were translated into English – as well as probably read in French – as early as 1559.

In the seventeenth century, new French authors will represent prognosticators and anti-prognosticators in England. In 1652, Jean d'Espagne, a clergyman of the French Church of Westminster – constituted by emigrés – published in both French and English his reactions to the eclipse of that year, under the name of John Despagne: *Considerations Held Forth in a Sermon the 28 of March this year 1652 Upon the Eclipse which happened the day following*. He was preaching against any signification of the eclipse. But ten years later, Pierre Serrurier (Serarius), a Protestant, tried on the contrary to arouse people's spirits. His text was translated into English as *An Awakening to the Wofull World by a voyce in three nations, uttered in a brief dissertation concerning that fatal conjuction of all the planets in one and the same sign, Sagitarius* . . . Later on, of course, Pierre Bayle had his 'Reflections' on the 1680 comet translated too.

But the contribution of France also extends to works which we could call 'Books of the Zodiac', like the *Kalendrier et Compost des Bergiers* (fifteenth century) and the *Livre d'Arcandam*, which gave basic knowledge based on the birth month (thereby with a perpetual basis). The former (as Bosanquet has shown) was translated into English more than once, and a certain *Compost of Ptholomeus* was in fact an adaptation of this French 'Bible' of popular knowledge – opposed, as it were, to the *Tetrabiblos* of Ptolemy. As for the *Book of Arcandam*, it was originally an Arabic text translated into Latin and edited around 1540 by Richard Roussat. Subsequently it was translated into French and then English. One of the English versions included a treatise on physiognomy by the great astrologer Antoine Mizauld – the least known in England of the French astrologer-physicians of the sixteenth century. The edition of 1578 reads: 'The most excellent . . . Booke of . . . Arcandum or Alcandrin to find the fatall destiny . . . of every man . . . Wyth an edition of Physiognomy . . . Now newly tourned out of French . . . by William Warde.' This is an opportune reminder of the many antique or medieval astrological texts that came to England through France.

The first signs of astrologers' interest in mythology also appeared in almanacs, both in their planetary iconography and in certain astrological poems. Against this, however, French and English astrologers throughout the sixteenth and seventeenth century remained relatively ignorant of the planets' mythological connotations – an attitude which was already evident in Ptolemy's treatise, compared with that of Manilius.[16] So, learned treatises for a long time made no attempt to develop a mythologically based symbolism

[16] An exception was the *Dialogues d'Amour* by Léon l'Hebreu, an Italian Jew who combined his account of the planets with the Graeco- Latin pantheon. His work was translated into French in the sixteenth century. Mythology is also to be found in the seventeenth century with Bonneau's and Pagan's astrological works.

of signs and planets, preferring instead a show of empiricism and certain procedures (such as the planetary dignities and debilities) which control the relations between planets and signs according to a relatively sophisticated logic. From the eighteenth century onwards, interest in mythology takes on new proportions in France. Witness *L'Antiquité Expliquée* (1716) of Montfaucon, without however involving matters of astrology. Mythology, particularly its poetic dimension, attracted a new public. But astrology, apart from a few almanacs, scarcely managed to make use of this opportunity. The surrounding culture, however, was increasingly acquainted with mythology, and sooner or later astrology would be obliged to come to terms with it.

As for astronomy, it was a long time before it appealed to more than a relatively small number of people (if we except the 'scientific poetry' of sixteenth century France, which was very much inspired by astronomical themes). In 1596 there was published in London a translation from the French Protestant poet, Guillaume Du Bartas: *First Day of the World Creation* . . . This poetry heralded the prose astronomical literature in the seventeenth century of Fontenelle, and later of the Abbé Pluche, most of it translated into English, and whose target mainly was women readers. One could therefore read astronomy without astrology, and one could practise astrology without referring much to astronomy – that is probably what best characterizes the French eighteenth century. In England too, the same literature appeared, and Halley was making astronomy an exciting predictive activity.

Literature Refuting Astrology

In England, the beginning of the seventeenth century was the theatre of harsh polemics between opponents and defenders of astrology. Sir Christopher Heydon was probably the most eminent astrologer of this time. His *Defense of Iudiciall Astrologie* (1603), published in response to John Chamber, was essentially a refutation of Pico della Mirandola. If France can be considered responsible for a certain impluse in English astrological literature, she also had a role in the opposite camp. In 1606, there appeared an English translation of Jean Bodin's *Commonweale*, which includes an important chapter attacking astrology and denouncing certain astronomical absurdities of Ferrier.

But already in 1560, Calvin had been translated into English with *An Admonicion against Astrology Judiciall and other curiosities that raign now in the world*. According to Calvin, astrology 'hath been rejected by a common consent as pernicious by Mankind. [Yet] at this day it hath gotte the upper hand in such sorte that many whych thynk themselves witty men . . . are as it were bewitched therewith'.[17] However, the historian has to be careful not to attribute merits to the wrong persons. In the same way that Dariot owed a

[17] This passage is from the English edition. Subsequent passages are translated from the French edition.

debt to the canon Richard Roussat – whose work *Les Eléments et Principes d'Astronomie* (1552) he abridged, without saying whose work it was – the Protestant Calvin owed one to the Catholic physician David Finarensis, of Italian origin, who published one year before the former's *Advertissement contre l'Astrologie qu'on appelle Judiciaire* a text which quite clearly influenced Calvin, although it was unacknowledged. This anti-astrological work was one of the first ever written in French by a sixteenth century author – *L'epitome de la Vraye Astrologie et de la Reprouvée* (1547). But it is Calvin who will get the credit in France and – through a Latin version – in England and the rest of Europe.

(a) *The Contest in Exegesis; True and False Astrologies*

If Calvin made numerous allusions to Biblical texts, this was mainly because such texts were appealed to by the astrologers or 'genethliaques'. Confronted with their appeal for support from the prophetic powers of Joseph or Moses or Daniel, Calvin tried to turn their argument back on them: 'So when the genethliaques, [in order] to honour their science, put forward its claim to antiquity, all that they gain is that we know that it is a form of curiosity that God has reproved and which he has strictly forbidden in his Church.' He also had to defuse the astrologers' arguments concerning the appearance of signs in heaven before the final coming, and the star seen by the Magi. For his own part, he does not hesitate to cite the example of Esau and Jacob as twins with opposing destinies, nor to point out a passage in Deuteronomy condemning divination, in order to show that astrology is both erroneous and a forbidden science.

It was not Calvin's purpose to reject astrology as a whole, and this position is an important consideration in his criticism of those who relied on certain Biblical passages that evidently favour astrology. 'Now every man of sound judgement well knows that Moses meant the same as what I have said above, about true astrology. If the stars are signs to show us the season for sowing or planting, for bleeding or giving medicines, for cutting wood, that is not to say that they are signs to show whether we should put on new clothes, or deal in goods on a Monday rather than a Tuesday and so on, things which have no connection with the stars.'

In fact, Calvin was attacking what we call non-astronomical astrology, which is based on the calender rather than true planetary cycles. This criticism ran the risk of rapidly becoming obsolete, when from the mid-sixteenth century onwards astrologers were increasingly to abandon this simplified astrology and explain to their readers how to draw up a genuine map of the sky. (Ferrier and Dariot were especially notable for their success in transmitting methods of doing this, through their seventeenth century English translations.)

In the same way, Calvin rejected the practice of horary astrology (which later disappeared in France, if not in England). 'What is more, they do not content themselves with having tried their hand at a man's nativity, but also lay claim to judge, or rather divine, every enterprise from the present aspect of the heavenly bodies. Thus if anyone has some business a hundred leagues away from where he is, Sir Astrologer will spy out the sky to see when it will be favourable . . .' Thus Calvin employed arguments whose force will be undisputed by seventeenth century astrologers such as Morin. Calvin could not bring himself to condemn astrology completely and radically, as a total error. He could only emphasize its archaisms, its excesses and abuses. He was equally in difficulties when it came to denying significance to eclipses and comets. Here too he was compelled to make distinctions which allowed some doubt to remain: 'However, I do not deny that when God wishes to stretch out His hand to bring about some judgement worthy of memory by the world, he sometimes warns us by means of comets . . .' For Calvin, it was necessary to proceed prudently. The man who stood for a kind of return to the Old Testament could not ignore the bond between God and the heavenly bodies, His faithful interpreters; nor could he overlook the well-established nature of some physical influences. The determining notion was that of true and false astrology, as it was for his contemporary Finarensis, whose published work influenced Calvin. His influence remained in seventeenth century France through another Protestant, David Derodon. The latter's *Discours contre l'Astrologie Judicaire* cited Calvin abundantly. It appeared about 1645 (a truncated edition is the only one to survive), then again in 1663.

(b) *Jean Bodin and the Relation of Astrology to Politics*

There does not seem to exist be a real difference between Catholic anti-astrology and Protestant. After all, did not Calvin use Finarensis's text in his own attack? Almost fifty years afterwards, Jean Bodin came to similar conclusions about astrology in his examination of the astrologers' fiasco in 1524, when a widely predicted flood of Biblical proportions failed to materialize. Bodin found confirmation in this event of the delicacy of 'political' predictions – an opinion derived from his reading of Pico della Mirandola, and in particular Pico's list of mistaken prophecies concerning the return of the Messiah in the fifteenth century.

World (or 'mundane') astrology had emerged from the Middle Ages with an appreciable degree of prestige. It escaped, in part, from the criticisms levelled against individual (or 'judicial') astrology. From the nineth century onwards, Jewish and Arab astrologers – notably Albumasar and Messahala – had erected the technique of the Great Conjunctions, one not to be found in Ptolemy. This involved structuring time according to the aspects between the slower-moving planets – Saturn, Jupiter and Mars – through the signs. This provided a chronological framework on a scale appropriate to the history of religions and nations.

Chapter II of Book IV of Bodin's *Republic* (1576) – concerned, among other things, with the question 'If there is a means of knowing the punishments and ruin of republics in the future' – addressed itself specifically to the events of 1524. It concluded that astrologers are decidedly causes of unrest which interferes with good government.

Incidentally, Bodin attacked Auger Ferrier by noting some astronomical inconsistencies in the latter's *Treatise*, published several times since 1550. A few years later, Ferrier replied in his *Advertissements à M. Jean Bodin sur le quatrième livre de sa République* (1580). Bodin replied in turn, under the assumed name of René Herpin. The quarrel turned bitter, and was only curtailed by Ferrier's death. The parallels between the Bodin/Ferrier exchanges and those of Gassendi/Morin, three-quarters of a century later, are quite striking.

In his writings on astrology,[18] Bodin's overall position in fact attempts a compromise: 'Although Calvin . . . seeing that Melanchthon gave too much praise to astrology, for his part belittled it as much as he could, nonetheless he was compelled to acknowledge the wonderful effects of heavenly bodies, adding only that God is above all this and that he who trusts in God must fear nothing.'

Did this approach influence Kepler? In 1610, the astronomer-astrologer published a work entitled *Tertius interveniens*, in which he pointed out that one must not 'throw the baby out with the bathwater'. The title signifies a man in the middle, arbitrating between Feselius and Röslin who each assumed an extreme position with regard to astrology. It was published in Frankfurt, where editions of Bodin's *Republic* had appeared in (*inter alia*) 1591 and 1609. Kepler could thus easily have had access to the Frenchman's ideas; although, this said, the latter merely sketched the problem. He did not publish any true astrological work, as Kepler and Morin were to do.

Bodin seems to have been heavily influenced by Ibn Ezra, whom he frequently cited, and Maimonides (*e.g.* his *Letter* on astrology, published in Latin in 1555). This could explain why he considered it impious to enquire into religious questions by means of astrology: *e.g.*, the coming of the Messiah, the horoscope of Jesus Christ (as Cardano among others had examined), or even the hour for a prayer to be most efficacious (as Albumasar had proposed). It did not matter whether astrology could actually do these things; the problem was first of all a moral one. There was also the question of intellectual honesty. All the misfortunes of astrology, 'the most beautiful science of the world', arose from its abuse, so that in the end 'astrologer' became synonomous with 'sorcerer'. It was certain that the heavenly bodies exercise influence and determine tendencies, but 'after stating from the horoscope what the temper and natural disposition of the body is, [astrologers] go further and discuss things which do not concern the body, such as marriages, honours, journeys, riches and other such things . . .'

In fact, Bodin continued nontheless to hope that a science of politics could be based on the movements of the heavenly bodies: 'Briefly, if there is any science of celestial things for the changes in Republics, we must examine the meetings of the superior planets for the last 1,570 years, the conjunctions, eclipses and aspects of the interior planets and fixed stars when the previous

[18] This chapter from the *Republic* is not the only writing Bodin dedicates to the subject of astrology. Another and more balanced passage is found in the *Démonologie des Sorciers* (1580), which in 1616 assumed the title *Fléau des Démons et Sorciers*. And there is his *Methodus ad facilem historiam cognitionem* (1566).

great conjunctions took place . . . the natures of the signs and of the planets.'

Thus, the criticisms made by Calvin and Bodin are on the whole moderate; astrology has not yet lost all its credibility. Some of its positions are still secure. But in the seventeenth century, astrology will encounter greater intrangience. Astrology had lost in the meantime part of its credit.

(c) *Gassendi and the Confrontation with Astrology*

The reason for this changed situation was precisely astrology's success. It posed a greater threat than ever, succeeding in attracting dynamic and influential men such as Jean-Baptiste Morin, who abandoned a medical career for that of astrology.

For Gassendi, the question of astrology's limits and methods was an old one. One needs to know it quite well in order to know its flaws, and Gassendi himself, in his youth, had made sufficient study of astrology to discover (in the opinion of his friend Mersenne) its Achilles heel. Moreover, why play on the difference between good and bad astrology? The astrologer Blaise de Pagan had not hesitated to entitle his treatise, openly and without embarrassment, *L'Astrologie Naturelle* (1653) – a term formerly reserved for astrology other than judicial. From the seventeenth century on, the battle was about the uses of astrology, even medical or agricultural. This was apparent as early as 1600 to Francois de l'Alouëte, who went so far as to reject lunar influence in his *Impostures d'impiété des fausses puissances et dominations attribué à la Lune et planèttes* . . . By 1710, l'Abbé Bordelon (in the second edition of his *Entretien de l'Astrologie Judiciaire*) could cite a long list of authors who had rejected the use of astrology in horticulture. As for medical astrology, it also faced fire – notably from Alexandre Tinelis, the Abbé of Castelet, in a *Dissertation concernant des réflections curieuses et nouvelles sur la question si l'astrologie judiciare doit être d'usage dans la pratique de la médecine* (1681). We are now far from Calvin's distinction between true and false astrology – one which is also no longer of interest to astrologers.

Around Gassendi gathered allies, the most redoubtable being the Minim priest Marin Mersenne and François Bernier, a traveller in the Orient. His customary adversary was the astrologer Jean-Baptiste Morin. In 1634, having developed his own line of argument in a passage denouncing astrology, Mersenne alluded for the first time to the as yet unpublished anti-astrological texts of Gassendi: 'I add the following proposition, which I have taken from the Apology which M. Gassendi Théologal of Digne showed me in advocacy of the atoms of Epicurus.'[19] The previous year, both men had taken on the Englishman Robert Fludd, author of astrological and geomantic texts in Latin. This started a fairly voluminous series of polemical exchanges. They reveal that the conflicts over astrology in this period often extended well

[19] For this reason, some astrological treatises may escape identification, in any superficial approach, and thus neglected. For example, the last great treatise on astrology in the seventeenth century appeared under the title *Tableaux des Philosophes* (1693), by Baron Eustache Lenoble.

beyond the boundaries of strictly astrological concerns; and also that learned natural philosophers did not feel they were wasting their time by discussing astrology. Around this time, several learned treatises appeared in support of astrology – most notably Morin's *Astrologia Gallica* (1661) – which also, characteristically, did not confine themselves just to astrology.[20]

Mersenne denounced the way astrologers disagreed among themselves: 'But how can one hope to find any truth in astrology when its chief authors are not in agreement about the houses that form the basis of the science?' In other words, Mersenne presented the debate about domification within the astrological community not as a sign of health, but as proof of confusion. And it is true that astrologers did not present a united front and display unanimity; nor did they attempt to establish a high degree of rationality or coherence in their various pronouncements. Hence Gassendi's related strategy of merely stating '. . . it is not necessary to refute Astrologers' assertions, since they demolish one another . . . These are the principal bases of Judicial Astrology, the emptiness of which is so obvious that in order to refute them it is enough to set them out.' Gassendi apparently thought that astrology had been credited due to people's ignorance of its rules, and that knowing these – far from aiding astrology – would convince people of the correctness of their doubts. That is why such anti-astrological treatises as *Le Tombeau de l'Astrologie Judiciare* (1657) by Jacques de Billy presented actual manuals of astrology, leaving the reader virtually free to judge for himself of the unreasonableness of astrology's principles and categories.

Gassendi published two versions of his critique of astrology: one in *Animadversiones in decimum librum Diogenis Laertii* (1649), the other in a chapter entitled 'De effectibus siderum' in his posthumously published *Opera Omnia* (1658). The latter text formed the basis of an English edition in 1659, under the title of *The Vanity of Judiciary Astrology or divination by the stars*, translated by 'a Person of Quality'. In the same year, Samuel Clarke published an attack on astrology in his *Medulla Theologiae* (Chapter XV), which astrologers also considered to be based on Gassendi's theses. In 1660, there appeared *A brief answer to six syllogistical arguments brought by mr Clark . . . against Astrology and Astrologers*, with the subtitle 'Gassendus's arguments against Astrology . . . retorted and refuted.'

(d) *The Polemic from 1649 to 1654, and Its Later Effects*

Before his death in 1655 (which in no way ended his influence), Gassendi had to face a hard battle against the Beaujolais astrologer Morin, amid rumours about the imminent effects of the eclipse of August 1654. In the course of this he also published other less important texts – what we might call occasional pieces. In one of these ('Lettre de M. Gassendi à M. Morin, 26 Octobre 1649'), he mainly considered the failures of individual rather than world

[20] In 1674, Bernier published the second part of his abridgement of Gassendi's philosophy. He included three chapters, making up the 'refutation of Judicial Astrology'.

astrology. The confrontation between these two men was as remarkable as that between Bodin and Ferrier, in the previous century. In fact, the affair of 1654 concerning the eclipse brought about another conflict between them. In his *Remarques Astrologiques* (1654), Morin criticized a work of Gassendi's that had appeared anonymously: *Sentimens sur l'éclipse du mois d'Aôust prochain* . . . This time, Gassendi had attacked world astrology, following in the footsteps of Bodin, and stressing (for example) that since in any battle there are two opposing camps, one and the same configuration cannot be 'maleficent' for both the vanquished and the victors. But he raised the stakes of the debate by adding a barbed comment, at the end of his pamphelet, concerning 'those who, whether persuaded or not of the truth of what they say, find it amusing to see their crude errors given credit in the world. Men such as the author of the abovementioned article [one Doctor Andréas], who seeing so many people now alarmed on account of his predictions, is no doubt laughing behind the arras, either alone or with his accomplices . . .'

Gassendi was followed by others – such as Bernier[21] – in his strategy of analyzing the reasoning of astrology's defenders, and denouncing their sophistry in drawing extravagent conclusions from accepted facts. This was by now far removed from Calvin's concern to base his rejection of astrology on Scripture – even though Gassendi too was a cleric. His aim was to forearm his readers against astrological partisans. And as has been pointed out, he exploited the quarrels among astrologers: for example, by quoting Cardano to the effect that countless 'scoundrels and charlatans have so much spoiled and corrupted this Art, by the impostures and foolishness they have introduced into it, that not the least vestige of it remains.' Here was an astrologer acknowledging how little faith should be accorded to his own tradition! The astrological camp appeared divided into intolerant castes, thus providing their critics with invaluable ammunition. For his attacks on astrology, Gassendi was commended by such men as René de Ceriziers, in 1654.[22] Overall we see a tradition, indeed a solidarity, arising among opponents of astrology, with its particular heroes among the ranks of its members.

In 1697, Pierre Bayle – in his *Dictionnaire Historique et Critique*, which was twice translated into English in the eighteenth century – devoted one article to Morin and another to the German astrologer Johannes Stoefler, who had predicted the 1524 'Deluge'. Between that controversy and the stir about the eclipse of 1654, more than 130 years had elapsed without seeing any great change in the hold of astrology over the mass of people. In fact, Bayle in 1697 was pessimistic about uprooting astrology. Even Bodin is not considered a sincere critic; he is described as 'a credulous man and infatuated with astrology'. Moreover, Bayle had himself witnessed the rumours about the comet of 1680 (which he addressed in his *Réflexions sur la comète* . . ., translated into English in 1708). At the end of the seventeenth century, French astrology does not really show consistent signs of decline. It is not

[21] *Examen du Jugement de l'Argolin sur l'éclipse du mois d'aoust de l'an 1654* (Paris).
[22] *Entretiens de l'astrologie judiaire* (1710).

easy to assess this situation; for one thing, the division between texts for and against astrology was more subtle than is generally believed, and some texts hostile to astrology can also be perceived as defending it. (One thinks of Thomas Aquinas.) A case in point is Jacques Rohault's *Traité de Physique* (later published in England in a Latin edition [1697] and in English [1723]). Although Rohault's chapter on astrology severely limited its field of operations, it was interpreted by Dennis de Coetlogon, writing in English in 1745, as supporting astrology.[23] However, the tradition of anti-astrology was by now well-established. In 1737, Bougerel – a biographer of Gassendi – was criticized by the Abbé de La Varde for failing to stress Gassendi's essential role in this affair: 'Permit me, Sir, to ask you whether it would have been beyond the scope of your book to show Gassendi as the leader of those who fought against Astrology with such ardour and success: he who above all in 1649 had gained so complete a victory over Morin?' Even at the end of the eighteenth century, Gassendi was cited in esoteric circles as one who had tried to deal astrology a fatal blow.

(e) *The Popular Mode (Comedies and Satires)*

Attacks on astrology often had a forbidding character which failed to undermine its appeal for large sectors of the population. In such cases laughter is often a more effective weapon, and the strategem devised at the beginning of the eighteenth century by Jonathan Swift is well known. He drew up an almanac under the name of Isaac Bickerstaff, which also appeared in translation on the Continent, in which he pretended that the astrologer John Partridge was dead. When the latter believed he had put an end to these stories in the next edition of his almanac, Swift averred that this proved nothing: many authors of this kind of literature continued to appear on title-pages long after their deaths. The almanac-makers thus found themselves victims of their own 'immortality'. But Swift had been preceded in this genre by (among others) François Rabelais. The latter's *Pantagruèline Prognostication* was a sort of prognostication 'for all years', which reveals all the truisms and banalities typical of this kind of astrological discourse. It didn't appear in English until the late seventeenth century, serving as an attack against Lilly. Astrology was also parodied in plays, such as Thomas Corneille's 'le Feint Astrologue' (also translated into English).

Swift was a contemporary of the Abbé Laurent Bordelon, who was the author of *Les Imaginations Extraordinaires de Monsieur Oufle*. This concerned a rather bourgeois gentleman confronted with occult initiation, and astrology occupied a central part of the work, called 'criti-cosmic reflections' – which shows clearly the marriage of rigorous refutation with riposte. The central character was intoxicated with works on astrology and magic (of which, moreover, we are given a substantial list). This text appeared in English in

23 *Universal History of Arts and Sciences* (London).

1710, just when the effects of Swift's sarcasms were being felt. Curiously enough, Bordelon was himself taken in by *Bickerstaff* and denounced it in the course of another text.[24]

Astrology and History

From the eighteenth century on, astrology meets the discipline of history. There were at least two reasons: on the one hand, there were those who wished to understand astrology's origins and the causes of its empire (now that it seemed to be breaking up); on the other hand, the rising tide of archaeology increasingly encountered documents and monuments of an astrological character, such as the Graeco-Egyptian planisphere of Bianchini, presented to the Académie Royale des Sciences in 1708. Such was the situation in the world of learning. In the popular mode, there was a tendency to locate the sources of astrology in a more or less mythical past. Credence was given to astrology's claim to give an account of history, both human and cultural, which explains the continued appeal of Nostradamus's prophecies.

Was modern history of astrology born in eighteenth century France? In any case, the decline of astrology contributed to the development of its own historical study. In 1749, a time when French astrology was really discredited, the Royal Academie des Inscriptions et Belles Lettres chose as the theme for its prize in 1751 the problem of astrology's origins. The breakup of astrological domination had paradoxically good effects; now that mankind, in this 'Siècle des Lumières', had shaken off its terrible addiction, one was ready to admit how important astrology had been in the past.[24]

One of the important figures here was the Abbé Pluche, author of the *Histoire du Ciel* – or, to give it its full title, 'History of the Heavens, in which is discovered the origin of idolatry and the errors of philosophy regarding the formation and influences of celestial bodies'.[25] Pluche's primary goal was to undermine the foundations of astrology by reawakening the world of gods and heroes that had been pushed aside. 'The history of the birth of this supposed science,' he wrote, 'is its refutation, for all Astrology is no more than a false interpretation of certain signs that have been misunderstood.'[26] In 1739, he published not only his *Histoire du Ciel* but more work which was to cause a stir among historians of astronomy, if not of astrology. Volume IV of the *Journal des Sciences et des Beaux Arts* was devoted to the question of the invention of the zodiac, and many other studies followed this attempt.

For Pluche, the zodiac was emphatically not devised as a structure for divinatory use. There was therefore no sense in anyone saying that they had

[24] It is not our purpose here to examine the causes of such a decline. But a striking point which might be mentioned is the apparently parellel process in France and in England. Some French historians of astrology think that Colbert's decision not to admit astrology into the Royal Académie des Sciences might have had effects beyond the borders of France.

[25] Interestingly, in the title of the 1743 edition, the term 'influences' has been dropped.

[26] From *Histoire du Ciel*, Chapter VI.

been born under such-and-such a sign. Referring to the sign on the Ascendent (Eastern horizon) – better known as the 'horoscope' – he asks how one can sensibly say that one had been born, say, 'under the sign of the Ram'? Actually, Pluche's criticisms were a little anachronistic; such reference to the zodiac was already controversial in the seventeenth century. In fact, he seems to have known astrology mainly through the attacks of Sextus Empiricus and the *Saturnalia* of Macrobius, which he cites repeatedly. It is easy to understand how the Abbé got so out of phase. In 1739, such astrology belonged to the 'dead' past, and he was not dealing with its most recent manifestations, those of the late seventeenth century. He bolstered his attack on astrologers' use of the zodiac by emphasizing the effects of the precession of the equinoxes – a point which will be constantly repeated, notably by Voltaire, as a demonstration of why astrology must eventually fail, given this flaw in the system.

One of the Abbé Pluche's major preoccupations was to distinguish closely between astronomy and astrology, and in this he was followed by all the historians of the Revolutionary period, from Bailly to La Lande and Delambre. For astronomers to some extent felt affected by the disfavour attached to astrology. As Pluche put it, 'Astronomy does not degenerate into dreams and superstitions except when religion is perverted.'[27] Thus we can understand why the historians who followed him could not avoid taking a stand with regard to judicial astrology. Even with the 'disappearance' of astrologers, they and astronomers continued sometimes to be equated in the public mind. It should not be overlooked that astronomers were in charge of assigning names to newly-discovered celestial phenomena – including, from 1781 onwards, the new heavenly bodies in the solar system. With Pluche, astronomy became the prime opponent of astrology, which it had not been up until that time. Henceforth it is in this area that for the next two centuries, anti-astrological discourse will be fostered. Historians of astronomy committed themselves to removing the stigma of astrology, by purifying their discourse of everything that might be a reminder of the link between the two activities. Thus La Lande, in his *Astronomie* (1764), supplied a list of the astronomers of past centuries, but almost never pointed out that several of the names also belong to the history of astrology. In fact, the history of astronomy arrayed itself against not only astrology, but the history of astrology. In his 'Discours sur l'Origine de l'Astrologie' – and note the development beyond Derodon and his 'discours contre l'Astrologie judiciaire' – Jean-Sylvain Bailly made the following revealing remarks:

> The reader cannot expect us to supply details of the rules by which, throughout so many centuries, scoundrels cheated those who were eager for fabulous stories. But this science was long identified with the one whose history we are writing; it sustained astronomy in the centuries of barbarism, when sciences held no appeal; the desire to know the future, the belief that one could predict it, caused new

27 From *Spectacle de la Nature*, Volume IV, Part II.

> observations to be made and old ones to be preserved. We shall attempt to discover the origin of an error to which our weakness seems to cling; it is the longest sickness that has ever afflicted human reason . . . It is said that astrology is the daughter of ignorance and the mother of astronomy. This is how ideas become confused. Astronomy certainly came first; she is the wise mother of a foolish daughter; it was necessary to know the heavenly bodies before ascribing to them power over us, it was necessary to have an idea of their motions and their cycles before relating these to the destiny of men, and the chain of events in life . . . We shall not be blamed for having shed light on the origin of this pretended science, which deserves the low estate to which it has fallen . . . In a century when sciences and reason are both equally cultivated, Astrology is despised and finds no defenders.

In fact, Pluche – who was followed by Bailly, and others in the next century such as the astronomer Camille Flammarion[28] – inaugerated the era of post-mortem criticism. For a century and a half, astrology was considered only in the past tense (that is, roughly from 1730 to 1880). There apparently remained only vague denatured traces of a discredited learning. Before Pluche, the opponents of astrology were intent above all on destroying it, on leading their readers away from a very present temptation. For him, there was no such danger. Astrology had become empty, and all that remained to be decided was the verdict of history on disciplines which might, in the absence of the chief suspect, be considered as 'accomplices', as 'receivers'. In other words, the case against astrology threatened, even compromised, the brand image of astronomy; thus it had to be handled very carefully.[29]

The history of astrology was not always to be confined to a limited circle of specialists, however, as was shown by the extraordinary interest in Eygpt in the nineteenth century (especially after Bonaparte's oriental campaign). The zodiac of the Temple of Dendera became a matter of widespread discussions, and in 1825 London welcomed an exhibition in Leicester Square entitled: '. . . Epitome of the celebrated sculptured Zodiac of Dendera, so famous in Eygptian Antiquity, on which it is conjectured the present system of astronomy was founded.'[30] The popular impact of this zodiac can be compared with that of the Rosetta Stone.

It should also be mentioned that astrology also encountered history in another way. The French Revolution revived old astrological texts: some from the early fourteenth century by Cardinal Pierre d'Ailly, many from the sixteenth century by Turrel, Roussat and of course Nostradamus. Several of these works had clearly mentioned the end of the eighteenth century as a

[28] Flammarion was also the author of a history of the heavens, and an essay on 'The Greatness and Decay of Astrology'.

[29] Note however that in 1781, Charles-François Dupuis put forward the thesis in Volume IV of La Lande's *Astronomie* that astrology was a universal and eternal religion, a cause rather than effect of mythological aberrations. He expanded this idea in his *L'Origine de Toutes les Réligions* (1795).

[30] See the British Library catalogue entry; the reference text has disappeared.

turning point. One could in some ways compare the astrological climate of France around 1800 with that of England in the mid-seventeenth century. Astrologers were active on both sides in war.[31]

Conclusion

Our aim was not an exhaustive description of French anti-astrology or the influence of French astrology upon English, but to give a certain orientation to research. Obviously, it requires a rich corpus to be able to achieve such an aim. Otherwise the possibility of seeing clear trends would be insufficient. Also, as Richard Lemay had pointed out in his study on Albumasar, astrology never stands alone. It is connected with other cultural aspects which have preceded or will follow the astrological wave from one cultural area to another.

Finally, we would like to propose the use of the expression 'astrological culture' to describe the 'ensemble' of facts that develop and evolve together, including attacks and expositions of all sorts. Anti-astrology is part of astrological culture: similarly the French influence in England was that of a whole astrological culture, and not limited to just certain aspects.

[31] In fact, Nostradamus became most significant in England at the end of the seventeenth century. In the sixteenth century only his 'Prognostications' were known there (as Fulke's *Antiprognosticon* (1559) shows).

SELECTED BIBLIOGRAPHY

Pierre Bayle, *Pensées Diverses sur la Comète* (Paris, 1911) (Intro. by A. Prat).

Jean Bodin, 2 volumes, *Actes du Colloque Interdisciplinaire d'Angers* (Paris, 1985).

Eustace Bosanquet, *English Printed Alamanacs and Prognostications. A Bibliographical History to the Year 1600* (London, 1917).

Jean Calvin, *L'Advertissement contre l'Astrologie Judiciaire* (Geneva, 1985) (Intro. by O. Millet).

Bernard Capp, *Astrology and the Popular Press. English Almanacs 1500–1800* (London, 1979).

J. Collin du Plancy, *La Fin des Temps Confirmée par des Prophéties Authentiques Nouvellement Recueillies* (Paris, 1871).

A. Dassier, *Eloge historique et Critique d'Auger Ferrier, Médecin Toulousain (1513–1588)* (Toulouse, 1847).

Jacques Halbronn, *Introduction Bibliographique à l'Histoire de l'Astrologie Francaise* (thesis, Université de Paris XII, 1987).

——, *Le Monde Juif et l'Astrologie* (Milan, 1985).

——, Introduction to *Morinus's Astrological Remarks* (Paris, 1975).

——, 'The Astrologer Marquis (Nicolas Bourdin), in *Astrological Review*, Winter 1973–74 (N.Y.).

J. Halbronn and Serge Hutin, *Histoire de l'Astrologie* (Paris, 1986).

Elizabeth Labrousse, *L'Entrée de Saturne au Lion. L'Eclipse du 16 Août 1654* (Hague, 1974).

——, 'Quelques Sources Réformées des Pensées Diverses' in *Mélanges de Littérature Offerts à René Pintard* (Starsbourg, 1975).

Sidney Lee, *French Influence in Renaissance England. An Account of the Literary Relations of England and France in the Sixteenth Century* (Oxford, 1910).

Richard Lemay, *Abumashar and the Rediscovery of Aristotelian Philosophy* (Beirut, 1962).

Frank Manuel, *Shapes of Philosophical History* (N.Y., 1970).

Johnston Parr, *Tamburlaine's Malady and Other Essays on Astrology ...* (Alabama, 1953).

Albert Schmidt, *La Poésie Scientifique au XVI Siècle.*

NEWTON'S COMETS AND THE TRANSFORMATION OF ASTROLOGY

Simon Schaffer

> Genius without the improvement, at least, of experience, is what comets once were thought to be: a blazing meteor, irregular in his course and dangerous in his approach, of no use to any system and able to destroy any. (Henry St John, Viscount Bolingbroke, 1740)

Between 1681 and 1705, Isaac Newton and Edmond Halley transformed the place of comets in astronomy. The prediction that the comet of 1682 would return in roughly 75 years was perhaps the least important of these transformations. Yet historians have often treated Halley's prediction as one of the most crucial achievements of this programme, and, furthermore, they have viewed the prediction as a deadly blow to any form of astrological view of comets' significance. Keith Thomas has argued that 'the Newtonian system could not accommodate the concept of celestial influences. The gigantic structure of astrological explanation accordingly collapsed'. Similarly, others have insisted that with Halley's Comet 'the ideas of astrology could at last be completely rejected. The unpredictability of comets had always made them appear ominous'. In a revealing turn of phrase, the return of the comet in 1758 is described as 'a posthumous confirmation of his prophecy'.[1] Prophecy and celestial influence were, in fact, the very essence of the transformation marked by the programme of Newton and Halley. In this paper, I shall show how the work of the 1680s and 1690s redefined the meaning of comets for astronomers and for natural philosophers. Halley and Newton did not reduce the moral and theological function of comets. They made them an integral part of a natural philosophy whose task was to locate the restorative, transformative and prophetic effects of astronomical signs.

Before 1681 Newton's interest in cometography was extremely intermittent. As Guerlac has suggested, the comet of 1664–5 may well have aroused his concern for astronomy in general. John Conduitt recorded that Newton 'sate up so often in the year 1664 to observe a comet that appeared then that he found himself much disordered and learned from thence to go to

[1] K. Thomas, *Religion and the decline of magic* (Harmondsworth: Penguin, 1972), p. 418; S. Toulmin and J. Goodfield, *The Fabric of the Heavens* (Harmondsworth: Penguin, 1963), p. 262. For Newton as the end of astrological law, see also L. Thorndike, 'The true place of astrology in the history of science', *Isis*, 1955, *46*:273–278, p. 273; T. G. Cowling, *Isaac Newton and astrology* (Leeds: Leeds University Press, 1977).

bed betimes.'[2] He followed the comet between 17 December 1664 and 23 January 1665, recorded his sightings in his Commonplace Book, and introduced himself to existing astronomical scholarship, particularly that of Gassendi, Ward, Snell, Vincent Wing and Thomas Streete. The comet itself was very bright and also retrograde: it drew much attention from the European astronomical community. In January 1665, the Jesuits organized a conference on comets at the Collège de Clermont in Paris, in which the consensus view of comets as transient but celestial phenomena was propounded. Pierre Petit dedicated a dissertation on comets to the King; John Beale collected notes on comets and their effects on earthly events for John Evelyn; Samuel Pepys attended a lecture at Gresham College 'where . . . Mr Hooke read a second very curious lecture about the late Comett'. Hooke argued that comets such as this might return, 'which is a very new opinion', Pepys recorded. Later published as *Cometa* (1678), this view was to play a major part in Newton's own ideas on the problem.[3]

This comet provided the main opportunity for testing the fashionable cometary theories of the time. Two issues were of importance: firstly, were comets permanent or transient objects; secondly, what was the significance of comets for events on Earth? Comets were valuable commodities for theologians, radicals and almanac makers, astronomers and natural philosophers. To describe their function was to describe the proper practice which should treat them. Astronomers were compelled to decide whether comets were to be treated like planets or as ephemeral. Kepler had distinguished between permanent objects in closed orbits and transient bodies, such as comets, which would move linearly. J. A. Ruffner has shown that after Kepler's texts on comets (1619–20) the link between permanence and closed orbits was very influential. Newton's astronomical sources used this classification. Seth Ward asserted that comets were not ephemeral, and so moved in circles; in *Cometographia* (1668). Hevelius argued that 'a perennial object gyrates in an orbit, but objects which are going to be destroyed, on the contrary, are thrown in straight lines'.[4] Newton's main sources before 1681 were Hevelius and his critics of the 1660s, together with the 1678 editions of

[2] H. Guerlac, 'Newton's first comet', in *Newton on the continent* (Ithaca: Cornell University Press, 1981), pp. 29–40; John Conduitt in Kings College Cambridge Keynes MSS 130.10 f. 4v, cited in R. S. Westfall, *Never at rest: a biography of Isaac Newton* (Cambridge: Cambridge University Press, 1980), p. 104.

[3] J. E. McGuire and M. Tamny, *Certain philosophical questions: Newton's Trinity notebook* (Cambridge: Cambridge University Press, 1983), pp. 296–304; Guerlac, 'Newton's first comet', 35–38; R. Latham and W. Matthews, *Diary of Samuel Pepys* (London: Bell, 1970–1983), 11 Vols, Vol. VI, p. 48; T. Birch, *History of the Royal Society* (London, 1756–1757), 4 Vols, Vol. II, pp. 69–72; R. Hooke, *Cometa*, in R. T. Gunther, *Early science in Oxford* (Oxford: privately printed, 1930), Vol. VIII, pp. 209 ff; *Journal des scavans*, 1665, *1*:51–60; L. Thorndike, *A history of magic and experimental science* (New York: Columbia University Press, 1923–1958), 8 Vols, Vol. VIII, p. 325; C. Webster, *From Paracelsus to Newton* (Cambridge: Cambridge University Press, 1982), p. 40.

[4] J. A. Ruffner, 'The curved and the straight: cometary theory from Kepler to Hevelius', *Journal for the history of astronomy*, 1971, *2*:178–194; C. D. Hellmann, 'Kepler and comets', *Vistas in astronomy*, 1975, *18*:789–796; Seth Ward, *De cometis* (Oxford, 1653), p. 31. J. Hevelius, *Cometographia* (Danzig, 1668), p. 562 (both cited in Ruffner).

Hooke's *Cometa* and Wallis's publication of Jeremiah Horrox. Just as importantly, during 1664 Christopher Wren developed techniques for computing the linear projection of cometary paths and communicated this both to Wallis and to Hooke, who published Wren's method in *Cometa*. It was this model on which Newton drew in his own calculations on the comet of 1680–1 as late as autumn 1685. By 1680, in fact, Newton had already convinced himself that the standard astronomical dichotomy implied that comets were transient and moved in straight lines in space.[5]

The events of 1679 and after transformed Newton's view and his conception of the proper manner in which comets should be analyzed. This transformation involved a series of re-definitions of the practice of which cometography should be a part. Firstly, Hooke and others argued, against orthodoxy, that comets were bodies which had once been stable members of the solar system, but were now transient projectiles wasting away into the interplanetary aether. Since transient, their orbits would be open, but since possessed of a similar magnetic constitution to the planets, this process 'may continue yet for many ages before it be quite dissolved into the Aether'. Hooke used the tradition of the magnetical philosophy developed by William Gilbert as the basis of his analysis of the forces working on comets' orbits. In *Cometa* he used the observations of the 1664 comet to show that many astronomical models could save the phenomena: since the comet was subject to magnetic action and since its physical constitution affected its true path, comets must become a concern for natural philosophers rather than astronomical model-builders alone.[6] In November 1679, Hooke drew Newton's attention to some of these views, and to the importance of a dynamic analysis of astronomical phenomena, whether these forces were magnetic or otherwise. In 1681, too, John Flamsteed was also to use a magnetic force model for cometography in his own exchanges with Newton. In this context, Newton himself came to accept that cometography must be approached within natural philosophy.[7]

[5] McGuire and Tamny, *Certain philosophical questions*, pp. 410–418; Cambridge University Library, MSS Add 4004 ff. 103–4 (Waste Book); Jeremiah Horrox, *Opera posthuma*, ed. John Wallis (London, 1678), pp. 310–21; S. P. Rigaud, *Correspondence of scientific men of the seventeenth century* (Oxford, 1841), 2 Vols, Vol. II, p. 605; J. A. Bennett, *The mathematical science of Christopher Wren* (Cambridge: Cambridge University Press, 1982), pp. 65–69; D. T. Whitesider, *The mathematical papers of Isaac Newton* (Cambridge: Cambridge University Press, 1967–1980), 8 Vols, Vol. V, pp. 524–31 and Vol. VI, pp. 81–85.

[6] Hooke, *Cometa* (1678), in Gunther, *Early science in Oxford*, Vol. VIII, pp. 247 and 229; Hooke to Wren, 4 May 1665, in C. Wren, Jnr, *Parentalia* (London, 1750), p. 220; J. A. Bennett, 'Hooke and Wren and the system of the world', *British journal for the history of science*, 1975, *8*:32–61. Compare Huygens to Kinner, 16/26 November 1654, in C. Huygens, *Oeuvres complètes* (The Hague: Nijoff, 1888–1950), 22 Vols, Vol. I, p. 307.

[7] Hooke to Newton, 24 November 1679, 9 December 1679, 6 January 1680, 17 January 1680, in H. W. Turnbull, J. F. Scott, A. R. Hall and L. Tilling, eds, *Correspondence of Isaac Newton* (Cambridge: Cambridge University Press, 1959–1977), 7 Vols, Vol. II, pp. 297–8, 304–6, 309, 313; Flamsteed to Crompton for Newton, 15 December 1680, 3 January, 12 February and 7 March 1681, *ibid.*, pp. 315, 319–20, 336, 351. See A. Koyré, 'An unpublished letter of Robert Hooke to Isaac Newton', *Isis*, 1952, *43*:312–337 and R. S. Westfall, *Force in Newton's physics* (London: Macdonald, 1971), pp. 424–431.

A second and equally important aspect of this change was the astrological use of comets. A sharp distinction was to be drawn between comets in astromony and natural philosophy, and their use by almanac makers or quacks. In May 1677 Flamsteed told Richard Towneley that the comet of that year might return in 12 years' time – if so, 'it will wholly overthrow ye conjectures [and] fearfull predictions of ye Astrologers.' In 1683 he published a paper on the conjunctions of Saturn and Jupiter and the comet of 1680–1. He argued that the term 'great conjunction' should be redefined to deprive it of astrological significance; he pointed out the deficiency of existing tables used to predict such events and touted those in preparation at Greenwich Observatory; once again he attacked 'our astrologers' fearfull predictions of direful events'.[8] The coincidence of the political crisis of 1679–81 and the great comet of winter 1680–1 made the proper interpretation of such objects an urgent problem. In November 1679 John Aubrey recommended astrological texts on the great conjunctions to Edmond Halley, but Halley observed that the recent prosecution of the Catholic astrologer John Gadbury during the Popish Plot made this 'a very ill time for it'. One of the Plot's protagonists, Israel Tong, published his *Northern Star* on the comet which linked its coming with the civil struggle. In January 1681 John Tillotson wrote that 'the Comet hath appear'd here very plaine for several nights . . . What it portends God knows: the Marquess of Dorchester & my Ld Coventry dyed soon after'. John Edwards, a Presbyterian preacher, published *Cometomantia* in which he contested judicial astrology of comets but argued instead for a natural philosophical interpretation of their significant effects, including the fall of monarchies. Across Europe, the comet and the crisis prompted detailed analysis of these boundary disputes. In Rotterdam Pierre Bayle launched his spectacular attack on judicial astrology; in London, William Lilly accurately saw the comet as a presage of his own death.[9]

The Catholic threat and the comets of the early 1680s were both highly significant concerns for Newton. The imminence of Papism prompted much of Newton's prolific work on church history and prophecy in the 1680s; the comets of 1680–1 and 1682 were to become the prized specimens of his new cometography. From this moment, Newton began working on the comet

[8] Flamsteed to Towneley, 11 May 1677, Royal Society MSS.LIX.c.10; John Flamsteed, 'An exact account of the three late conjunctions of Saturn and Jupiter', *Philosophical transactions*, 1683, *13*:244–258, p. 244.

[9] Halley to Aubrey, 16 November 1679, in E. F. MacPike, *Correspondence and papers of Edmond Halley* (London: Taylor and Francis, 1937), pp. 47–48; M. Hunter, *John Aubrey and the realm of learning* (London: Duckworth, 1975), p. 140 n. 3; C. Webster, *Paracelsus to Newton*, p. 41; I. Tong, *The Northern Star* (London, 1680); J. Edwards, *Cometomantia* (London, 1684); J. P. Kenyon, *The Popish Plot* (Harmondsworth: Penguin, 1974), p. 278; J. Howard Robinson, *The great comet of 1680: a study in the history of rationalism* (Northfield, Minn., 1916); D. Parker, *Familiar to all: William Lilly and astrology in the seventeenth century* (London: Jonathan Cape, 1975), p. 258; John Tillotson to Nelson, 5 January 1681, British Library MSS Add 4236 f225; Thorndike, *History of magic*, Vol. VIII, pp. 338–340. For previous political uses of comets, compare J. R. Christianson, 'Tycho Brahe's German treatise on the comet of 1577: a study in science and politics', *Isis*, 1979, *70*:110–140 and D. C. Allen, *The star-crossed Renaissance* (Durham, N.C.: Duke University Press, 1941), pp. 178–181.

problem in detail for the first time. An example of the combination of these interests was Newton's exchange with Thomas Burnet after December 1680. This began as Burnet was preparing his *Sacred Theory of the Earth* for the press. Burnet's book combined attention to Scriptural history and prophecy with more immediate political and natural philosophical concerns. Newton sent him critical comments on the use of Mosaic prophecy and a series of chemical analogues of the Earth's formation culled from his writings of the 1670s on optics and alchemy. On 13 January 1681, Burnet also informed Newton that 'wee are all soe busy in gazeing upon yee Comet & wt doe you say at Cambridge can be ye cause of such a prodigious coma as it had'.[10] Two features of the comet drew astronomers' attention. Firstly, one comet had been visible in November 1680 and had then disappeared behind the Sun. A second object had appeared in December and was visible until February or March 1681. Given the cometographers' assumption that comets' paths were linear, this suggested that these were really two distinct bodies. Edmond Halley was in France during the winter, and wrote to Robert Hooke that while in Paris 'the general talk of the virtuosi . . . is about the Comet'; he himself had 'tryed but without success to represent the Observations by an equable Motion in a right line'. Secondly, the spectacular tail of the comet prompted further speculation on its physical composition and the possibility of constructing a natural philosophical account of comets' motions.[11]

At this stage Newton subscribed to the orthodox view. Comets were transient bodies, quite distinct in status from the planets. They therefore moved in straight lines, and the comet of November was a different object from that which re-appeared the following month. But he was presented with a series of challenges to this view. Hooke had argued that a cometary theory demanded a physical account of its composition and motion. He gave comets a status in the heavens much more like that of planets. Edmond Halley, in France, had convinced himself that no linear trajectory could save the phenomena, and in May 1681 he told Hooke about Cassini's idea that this comet orbited the Earth with a period of two and a half years. Cassini identified the comet of 1680 with those of 1577 and 1665. Halley told Hooke that he was very sceptical of this model, but that ''tis very remarkable that 3 Cometts should soe exactly trace out the same path in the Heavens and the same degrees of velocity'. He also told Flamsteed that the comet might lose its gravity and yet be attracted away from a linear path by the pull of the Sun. Halley's failure to produce a convincing linear path, and his introduction to the concept of cometary returns was to have a profound effect on his loyalty to Newton's ultimate model of comets and their meaning.[12]

Flamsteed also had well-developed preconceptions about the true character of comets. His views were presented in the lectures on astronomy he gave at

[10] Burnet to Newton, 13 January 1681 and Newton to Burnet, January 1681, *Correspondence of Newton*, Vol. II, 329–34; M. C. Jacob and W. A. Lockwood, 'Political millenarianism and Burnet's *Sacred Theory*', *Science studies*, 1972, *2*:265–279.
[11] Halley to Hooke, 5/15 January 1681 and 19/29 May 1681, in MacPike, *Correspondence and papers of Halley*, pp. 48–50.
[12] Halley, *ibid.*, pp. 49–50.

Gresham College when deputizing for Walter Pope between 1681 and 1684, and were transmitted to Newton by James Crompton, a fellow of Jesus College Cambridge, where they prompted some important suggestions in cometography. In 1677 Flamsteed had already announced that he held that comets 'make their returns as in stated times & move about ye fixed stars at a vast distance'. This would be a powerful argument against judicial astrology, Flamsteed claimed; once again, it would make comets more like planets, and would demand an analysis of the forces which made comets deviate from linear paths. In 1680–1, Flamsteed told both Towneley and Halley that his expectation, 'when I had onely heard of it', that the comet of November would reappear from the Sun's vicinity, had been happily confirmed. Now he revised his views of the action which produced this path and the comet's tail. Firstly, Flamsteed combined a Cartesian idea that comets were the products of ruined planets from other vortices with the magnetic force discussed by Wren and Hooke. In his lecture on the comet which was given at Gresham College on 11 May 1681, Flamsteed collated observations of the comet, including those reported by Halley in France and by Newton, and argued for a cometary path bending very sharply when it neared the Sun. He then referred this sharp inclination of the comet to the action of the Sun's magnetism, for 'as the Earth is a vast magnet attracteing all things within its little vortex [,] so is the Sun to all that move in his greater, attracteing the planets to him & repelling them'. Flamsteed suggested that the comet might be attracted towards the Sun in its approach and then repelled afterwards as it moved in the aether. He sent this view to Cambridge, and as late as May 1683 he lectured at Gresham on the opinion that 'Comets . . . being vast bodys of matter wee can not easyly conceave how they should otherwayes be generated but admitting them to be the planets of some vortex whose centrall sun is thus extinct'. Then, as Flamsteed told Halley in February 1681, after nearing the Sun, the comet would be repelled 'as ye North pole of ye loadstone attracts ye one end of ye Magnetick needle but repells ye other'. Secondly, however, Newton pointed out for Flamsteed that even if these two bodies were the same, it was nevertheless difficult to see how very hot bodies such as the Sun and the comet at perihelion could preserve their magnetism. Flamsteed was compelled to suppose that 'the attraction of the Sun may be of a very different nature' and that the comet might carry a humid atmosphere which would stream behind it under the influence of solar heat like 'ye smoke from a Chimny'.[13] Newton continued to deny that the comet of November and that of December were the same object. Yet he did comment on this force, and Flamsteed's notion of a humid tail acting like chimney-smoke was to be a valuable resource later in the 1680s.

In March and April 1681, Newton was presented with arguments which challenged the orthodox model of linear and ephemeral cometary paths, and

[13] Flamsteed to Towneley, 15 December 1680, Royal Society MSS LIX.c.10: F.1.50; Flamsteed to Halley, 17 February 1681, *Correspondence of Newton*, Vol. II, pp. 337–9; Flamsteed to Crompton for Newton, 15 December 1680, 3 January, 12 February and 7 March 1681, *ibid.*, pp. 315–17, 319–20, 336, 351; E. G. Forbes, *The Gresham Lectures of John Flamsteed* (London: Mansell, 1975), pp. 20–35, 105–16, 360.

which drew on the magnetic philosophy to analyze such motions. Hot bodies lost magnetism and any humid atmosphere; so Flamsteed's force must be analogous to but different from magnetism, and the tail must be burnt away as the comet passed the Sun. In April, Newton wrote a suppressed draft in which he put forward an argument which Flamsteed might use to explain the deviation from a linear trajectory. He suggested an attractive force in the Sun, which would work on the comet *throughout* its path. At perihelion the centri*fugal* force of the comet would overpower this attraction and it would begin to recede from the Sun. This made comets just like planets, since in December 1679 and in December 1680 Newton had already suggested that a continual imbalance between a centrifugal force and some sort of gravitational attraction might account for both planetary and lunar motions. D. T. Whiteside has shown that these Borellian models reveal that Newton had not yet consistently formulated any notion of a combination of a centri*petal* attractive force with linear inertia to account for celestial motions. But the exchanges with Flamsteed on comets now prompted important changes in these views. Newton came to see that comets were to be granted the same status as planets; that their orbits must deviate considerably from linear paths under the influence of some attractive force; and that their tails would perform very significant functions in the heavens. When the status of comets changed this way, so did Newton's definition of the proper practice of cometography.[14]

Between spring 1681 and autumn 1684 Newton decided that comets should be treated in the same manner as planets. Both types of objects moved in closed elliptical orbits round the Sun. Two crucial moves were made in this period. Firstly, Newton now asserted that comets and planets experienced no sensible resistance from an interplanetary aether. So Kepler's area law could legitimately be applied to their motions. In a manuscript on comets composed before 1683, Newton still suggested that the matter of the heavens circulated following the comets' paths. His first insistence that Kepler's area law held in such motions was written in the tract *De motu* in autumn 1684. In an expansion of this tract composed between November 1684 and January 1685 Newton added that 'all those sounder astronomers think that comets descend below the orbit of Saturn . . . those therefore are indifferently carried through all parts of our heaven with an immense velocity and yet they do not lose their tails nor the vapour surrounding their heads, which the resistance of the aether would impede and tear away'.[15]

[14] Newton draft for Flamsteed, 28 February 1681, *Correspondence of Newton*, Vol. II, p. 341; Newton draft for Flamsteed, April 1681, *ibid.*, pp. 358–62; Newton to Hooke, 13 December 1679, *ibid.*, p. 307; Newton to Burnet, 24 December 1680, *ibid.*, cited 329–34; D. T. Whiteside, 'Before the *Principia*: the maturing of Newton's thoughts on dynamical astronomy', *Journal for the history of astronomy*, 1970, *1*:5–19, pp. 13–14; Whiteside, *Mathematical papers*, Vol. VI, pp. 9–14, esp. p. 11 n. 32. Compare Newton to Flamsteed, 16 April 1681, *Correspondence of Newton*, Vol. II, p. 364: 'to make ye Comets of November & December but one is to make that one *paradoxical*'.

[15] Newton, 'De motu corporum in gyrum', in A. R. Hall and M. B. Hall, *Unpublished scientific papers of Isaac Newton* (Cambridge: Cambridge University Press, 1962), p. 277 and Whiteside, *Mathematical papers*, Vol. VI, p. 49; 'De motu sphaericorum corporum in fluidis', in Hall and

From now on, Newton would use this argument as a weapon against any Cartesian notion of an interplanetary aether. Secondly, Newton now insisted that all comets move in closed paths. They had become permanent members of the solar system. In the comet manuscript of the early 1680s, he noted that if a comet returned it would move in some sort of oval. Again, by autumn 1684 he was convinced that comets' paths were elliptical and that 'it enables the [comet]ary orbits to be defined, and thence their times of revolution ... from the magnitude, eccentricity, [peri]helia, inclinations to the plane of the ecliptic and nodes of the orbits compared with one another we may know whether the same comet returns time and again'.[16] Newton's task was clarified: he had asserted that comets experienced no resistance, and he assumed they moved in ellipses. It was now necessary to develop a method by which the orbits' parameters could be calculated and their periods determined.

This task occupied Newton between early 1685 and summer 1686. In October 1685 he was still trying Wren's linear method as an approximation, using data sent by Flamsteed for the comet of 1680–1. Flamsteed commented ironically on Newton's belated concession that the comet of that year was one and the same body: 'this is what he before contended against with some virulency, but he had no mind to remember it'. But Newton was dissatisfied with the details of this approach, and told Halley in June 1686 that in 'Autumn last I spent two months in calculations to no purpose for want of a good method'. At the same time, in late summer and autumn 1685, Newton drafted his *System of the world* as a proposed second part of this magisterial treatise. Here Newton amplified his preconception that comets moved in very eccentric closed orbits. He outlined the three standard hypotheses of current cometography, that 'they are generated and perish as often as they appear and vanish', or that 'they come from the region of the fixed stars', or, finally, that they orbit the Sun 'in very eccentric orbits'. Newton commented that 'so far as I could hitherto observe the third case obtains', and that comets move in ellipses very close to parabolas.[17]

Hall, *Unpublished papers*, pp. 285–6 and Whiteside, *Mathematical papers*, Vol. VI, pp. 79–80. Compare J. Herivel, *Background to Newton's Principia* (Oxford: Oxford University Press, 1965), pp. 257–74, 294–303. On knowledge and importance of Kepler's laws for Newton, see J. L. Russell, 'Kepler's Laws of Planetary Motion', *British journal for the history of science*, 1964, *2*:1–24; C. A. Wilson, 'From Kepler's Laws so-called to universal gravitation: empirical factors', *Archive for history of exact sciences*, 1970, *6*:89–170, esp. 151–157; V. Thoren, 'Kepler's second law in England', *British journal for the history sciences*, 1974, 7:243–258.

[16] Newton, 'De motu corporum in gyrum', in Hall and Hall, *Unpublished scientific papers*, pp. 283–285 and Whiteside, *Mathematical papers*, Vol. VI, pp. 57, 59, 61 and 58–60 n. 79.

[17] Newton's computations of approximations to a curved path by rectilinear approaches are printd in Whiteside, *Mathematical papers*, Vol. V, pp. 524–31 and Vol. VI, pp. 81–85. Flamsteed's comments are printed in *Correspondence of Newton*, Vol. II, p. 421 n. 4 ('he would not grant it before see his letter of 1681') and in A. Chapman (ed.), *Preface to John Flamsteed's Historia Coelestis Britannica (1725)* (London: National Maritime Museum, 1982), p. 160. See Newton to Halley, 20 June 1686, in *Correspondence of Newton*, Vol. II, p. 437 and 'System of the world', in A. Motte, rev. F. Cajori, *Sir Isaac Newton's Mathematical Principles of Natural Philosophy and his System of the World* (Berkeley: University of California, 1962), p. 615 and Whiteside, *Mathematical papers*, Vol. VI, p. 483.

This conclusion was established before Newton had any firm evidence of its veracity. In September 1685 he told Flamsteed that he had 'not yet computed ye orbit of a comet'. His work hitherto had relied on 'rough ways of computing'. He was compelled to appeal to an assumption that comets' tails point away from the Sun. His argument that comets' orbits were scarcely ever inclined more than 40° to the ecliptic was soon falsified. But in a section of the *System of the world*, Newton showed how a parabola could be used as a good approximation and how in the case of the 1680–1 comet any deviation from this parabola would be due to 'the true orbit in which it was carried [being] an ellipse'. Ultimately, using an incredibly accurate graphical method, Newton was able to place an improved approach in the lemmas and proposition 41 of *Principia*, Book 3. Thus the assumption of autumn 1684 and 1685 had, by 1687, already acquired the status of a phenomenally guaranteed truth. It was now to be put to effective use.[18]

In the *System of the World* and in the third book of the *Principia* published in 1687 the assumption of the elliptic character of comets' orbits was placed at the centre of his method. This assumption generated two further cardinal principles of his cometography. Firstly, Newton argued that parabolas could and must be used as approximations to the true orbit: 'the orbits will be so near parabolas that parabolas may be used for them without sensible error'. This view was re-emphasized in the review of the work which Edmond Halley published in the *Philosophical Transactions*.[19] Secondly, once the approximate parabola had been calculated the period of return could be ascertained by the historic method of comparison with previous cometary transits. The statement in *De motu* to this effect was set in the *System of the World* and again at the end of proposition 41 of *Principia*, Book 3. Halley also pointed out in April 1687 that this 'method of determining the Orb of a Comet deserves to be practised upon more of them, as far as may ascertain whether any of those that have passed in former times may have returned again'. This approach would allow Newton and Halley to estimate a period from the comparison of two comets with similar elements, and thus correct the original parabolic approximation due to its differences with the true ellipse.[20]

The argument from the assumption of closed, elliptical orbits needed considerable emphasis. Newton and Halley both agreed that it was well-nigh impossible to distinguish parabola and ellipse for one transit of a comet. This fact allowed the use of a parabolic approximation. In 1728 Henry Pemberton argued that 'the comparing together different appearances of the same comet is the only way to discover certainly the true form of the orbit; for it is impossible to determine with exactness the figure of an orbit so exceedingly

18 Newton to Flamsteed, 19 September 1685, *Correspondence of Newton*, Vol. II, p. 419; Newton, 'System of the World', in Motte and Cajori, *Newton's Principles*, pp. 615, 619, 626; Whiteside, *Mathematical papers*, Vol. VI, pp. 483 n. 7, 485 n. 10, 498–504.

19 Newton, *Principia*, Book 3, Prop. 40, Cor. 2, in Motte and Cajori, *Newton's Principles*, p. 498; Halley's review in I. B. Cohen and R. E. Schofield, eds, *Isaac Newton's papers and letters on natural philosophy* (Cambridge: Cambridge University Press, 1958), p. 410.

20 Newton, 'System of the world', in Motte and Cajori, *Newton's Principles*, p. 620 and *Principia*, in *ibid.*, p. 532; Halley to Newton, 5 April 1687, in *Correspondence of Newton*, Vol. II, p. 474.

eccentric from single observations taken in one part of it'. Voltaire made the same claim ten years later.[21] So two problems preoccupied cometographers. They sought the presumptions which forced the assertion that comets moved in ellipses, since these were indistinguishable from parabolas. Secondly, they traced functions which comets must play in the divine cosmos Newton had mapped. Because their status had dramatically changed, there must be a range of such functions. 'Now is revealed what is the bending path of horrifying comets', Halley wrote in the Lucretian ode which prefaced the *Principia*. Once dispossessed from popular divination, comets must now play play a different role. But in his ode Halley also wrote that 'we no longer marvel at the appearances of the bearded star'. W. R. Albury has pointed out the link between this view and Epicureanism. For Newton, 'the philosophy of Epicurus and Lucretius is true and old, but was wrongly interpreted by the ancients as atheism'. Ancient philosophy supported his view of comets, as we shall see. Furthermore, Richard Bentley, editor of the 1713 edition of the *Principia*, carefully excised this line from Halley's poem. It was restored by Pemberton in 1726. This well illustrates the deep concern of interpreters in rightly formulating the way in which comets should affect men on Earth. Their benevolent or marvellous role must be intimately related to their closed orbits and spectacular tails.[22]

The task of substantiating comets' permanence was performed by Newton and Halley from spring 1694, after Newton's 'black year' of 1693. Other problems included the break between Halley and Flamsteed, exacerbated when plans for the publication of the Greenwich Tables ran into trouble in early 1692, and when Halley failed to win the chair in astronomy at Oxford.[23] The successful candidate for the Savilean professorship, David Gregory, evidently encouraged the plans for a fresh edition of the *Principia*. He reported in May 1694 that Newton was planning further examples of the mainly graphical methods for computing parabolic approximations to closed orbits. In July, Gregory noted that Newton 'says that this discussion about comets is the most difficult of the whole book'. Indeed, in 1705, when assessing candidates for the Plumian chair at Cambridge, Newton stated that the 'construction of a Comets Orb from 3 Obseervations' was to be 'the Tryall of the Persons that pretend to this Profession'.[24] The programme of

[21] Henry Pemberton, *View of Sir Isaac Newton's Philosophy* (London, 1728), p. 193; Voltaire, *Elements of Sir Isaac Newton's Philosophy* (London, 1738), p. 329: Voltaire's English editor insisted that the parabolas themselves were 'matters of fact'.

[22] Halley's poem is discussed in W. R. Albury, 'Halley's *Ode* on the *Principia* of Newton and the Epicurean revival in England', *Journal of the history of ideas*, 1978, *39*:24–43; the changes are in MacPike, *Correspondence and Letters of Halley*, pp. 204–6; Newton's comment on Lucretius is in *Correspondence of Newton*, Vol. III, p. 338.

[23] For this period, see Westfall, *Never at rest*, pp. 533–550 and F. Manuel, *Portrait of Isaac Newton* (London: Muller, 1980), 213–225; for Halley and the Savilean chair, see S. Schaffer, 'Halley's atheism and the end of the world', *Notes and records of the Royal Society*, 1977, *32*:17–40.

[24] Gregory memorandum, July 1694, *Correspondence of Newton*, Vol. III, p. 385; Newton's comment on the Plumian chair is discussed in *ibid.*, Vol. IV, p. 473 and W. Hiscock, *David Gregory, Isaac Newton and their circle* (Oxford: Clarendon, 1937), p. 33. Gregory's proposed edition is discussed in I. B. Cohen, *Introduction to Newton's Principia* (Cambridge: Cambridge University Press, 1971), pp. 188–199.

revision got going in earnest in October 1694, when Newton was in London and Greenwich negociating with Flamsteed over lunar observations and other material. Newton dined with Halley, a meeting which, as we shall see, had very important consequences. Halley then visited Newton in August 1695 in Cambridge and discussed 'a designe of determining the Orbs of some Comets for me'. Flamsteed maintained his inveterate hostility to Halley: 'I know him and you doe not therefore am resolved to have no further concern with him'. Newton had to transmit Flamsteed's invaluable data to Halley for his work.[25]

Between 7 September and late October 1695, Halley worked energetically to show that the comets of 1680–1 and of 1682 moved in ellipses, and to calculate their periods. In early September, Halley said he had an exact arithmetic method which allowed him then to search the record for similar comets, and which did not rely on the '*operationes partim Graphicas*' such as those presented in the 1687 *Principia*. On 28 September Halley told Newton that 'I am more and more confirmed that we have seen that Comett [of 1682] now three times since ye yeare 1531'. By 7 October Halley suggested that the 1682 comet was the same as that of 1607, but, significantly, invited Newton 'to consider how far a Comets motion may be disturbed by the centers of Saturn and Jupiter . . . and what difference they may cause in the time of the Revolution of a Comett in its so very Elliptick Orb'. This work was extraordinarily difficult, since, as Halley pointed out, observers such as Hevelius or Cassini often added '8 or 9 minutes to the places observed'. This gave Flamsteed's data their value. Furthermore, the remaining inequalities could indeed be attributed to perturbations of the comets' paths by the larger outer planets. The commitment to comets as permanent bodies sustained this tortuous analysis.[26] By 17 October 1695 Halley had considered the angle at the Sun formed by the parabolic orb of the comet of 1680–1, and had 'satisfied' Newton that 'the Orb of the Comet of 1680–1 is Elliptical'. Halley was delighted that Newton agreed with his results for these two critical comets, and announced his intention on 21 October of working on 'the Orbs of all the Cometts that have been hitherto observed', an essential prerequisite if the closure of these paths was to be confirmed.[27]

By the winter of 1695–6, therefore, Halley and Newton had established at least two closed and periodic cometary paths. On 3 June 1696 Halley told the Royal Society that the comet of 1682 and that of 1607 were the same, 'having

[25] Newton to Flamsteed, 24 October 1694, *Correspondence of Newton*, Vol. IV, p. 34; Flamsteed to Newton, 25 October 1694, *ibid.*, pp. 36–37; Flamsteed to Newton, 24 February 1691, *ibid.*, Vol. III, p. 203; Newton to Flamsteed, 14 September 1695, *ibid.*, Vol. IV, p. 169.

[26] Halley to Newton, 7 September 1695, *Correspondence of Newton*, Vol. IV, p. 165; Halley to Newton, 28 September 1685, *ibid.*, pp. 171–2; Halley to Newton, ?7 October 1685, *ibid.*, pp. 173–5.

[27] Halley to Newton, 15 October 1695 and Newton to Halley, 17 October 1695, *Correspondence of Newton*, Vol. IV, pp. 176–8 and 180–1; Halley to Newton, 21 October 1695, *ibid.*, pp. 182–3. Compare Flamsteed to Newton, 23 July 1695 and 17 September 1695, *ibid.*, pp. 153 and 170 for Flamsteed's comment on the need for 'the Doctrine of gravity'.

a period of about 75 years'. The following month, he explained how a parabolic approximation could save the motion of the comet of 1618 to the accuracy of the coarse data available.[28] These announcements prompted a sustained period of work from Halley, interrupted only by his appointment at Chester Mint under Newton's patronage and his voyages of 1698–1700 on the magnetic survey. The need was for a table of all the reliable comet reports with which to compare current parabolic elements. In June 1698 Halley told David Gregory that he had worked on at least 14 or 20 comets in the solar system; he also suggested a dynamic account of their formation and progress. Gregory continued to report on Newton's planned revisions on comets.[29] In March 1703 Newton said that 'the Comet 1680–1 its Orb would be exacter or agree better with Observations if in stead of Parabolick it were made or assumed Elliptick', just as Halley had suggested in 1695. In March 1705, Halley completed his *Synopsis of the astronomy of comets*, the most widely distributed of his works, saving perhaps his geo-magnetic chart. The *Synopsis* contained a table of 24 comets since 1337, a second table which derived the perihelion distance and inclination of a comet's parabolic approximation from its mean motion, and the first publication of the prediction of a return in 1758. The parabolic table was derivable because of the similarity of all parabolas, and, using these resources, any cometographer could know whether a comet 'has appear'd before, and consequently . . . determine its period and the axis of its orbit, and . . . foretell its return'. Cometography now had the resources to become a generalized research programme. It was in this form that it reached France as an onslaught on the hypothesis of vortices and, thanks to Delisle and Lalande, as a definite cometary prediction. In July 1706 Newton told Gregory he would print Halley's tables in the *Principia*, and Gregory published them in his *Elements of astronomy* in 1715.[30] The 1713 edition of the *Principia* included Halley's prediction for the 1682 comet, and also the suggestion that the 1680–1 comet would return in 'more than 500 years'. After further work in 1724–5, Newton printed Halley's identification of the comet of 1680–1 with those of 44 B.C., 531 and 1106. Finally, an expanded treatment of the 1682 comet, and a re-statement of the periodicity of the 1680–1 comet were included in

[28] Royal Society Journal Book, in MacPike, *Correspondence and Papers of Halley*, p. 238.

[29] Royal Society Gregory MSS 247 f. 62 (29 June 1698); Gregory memorandum, ?July 1698, in *Correspondence of Newton*, Vol. IV, pp. 276–7.

[30] Gregory memorandum, March 1703, *Correspondence of Newton*, Vol. IV, pp. 402–3; Edmond Halley, *A synopsis of the astronomy of comets* (London, 1705) (I have used the edition in *Miscellanea curiosa* (2 Vols, London, 1706), Vol. II); for the formation of Halley's Comet as a definite prediction, see J. Lalande, *Tables astronomiques de M. Halley*, 2nd ed. (Paris, 1759); J. B. Delambre, *Histoire de l'astronomie au 18e siècle* (Paris, 1827), esp. pp. 130ff and J. N. Delisle, *Lettres sur les Tables Astronomiques de M. Halley* (Paris, 1749). For the use of Halley's cometography against the French, see Brook Taylor to John Keill, 26 April 1719, *Correspondence of Newton*, Vol. VII, p. 37 and des Maizeaux to Conti, 11 September 1720, *ibid.*, p. 100. See Hiscock, *Gregory, Newton and their circle*, p. 37; David Gregory, *Elements of astronomy* (London, 1715), Vol. II, pp. 881–905, discussed in *Correspondence of Newton*, Vol. VII, pp. 294–5. Compare P. Broughton, 'First predicted return of Comet Halley', *Journal for history of astronomy*, 1985, *16*:123–133.

the version of the *Synopsis* appended to Halley's posthumous tables in 1749.[31] Historians have assumed that these predictions sum up the research programme mounted by Newton and Halley in their cometography.

Yet the real significance of this research programme cannot be understood within such an isolated context. The work of Newton and Halley in the 1690s was intimately linked with their re-definition of comets' functions in the universe. A connected set of projects described these functions and re-inforced the claim that comets were crucial members of the solar system, moving in closed orbits round the Sun. These projects included the analysis of the stability of the solar system; the scriptural history of the Earth, including the Deluge and the end of the world; the change in mass of the planets and the Sun; and, finally, the maintenance of vital activity throughout the cosmos. Firstly, in December 1692 Newton corresponded with Richard Bentley on the Cartesian model of cometary origin. Newton rejected the position which had been partly espoused by Hooke and Flamsteed that comets might be the ruined stars from other vortices. The argument against this notion depended explicitly on Newton's axiom that comets orbited the Sun in closed orbits. The message of cometary motion was equally clear: 'this must have been the effect of Counsel', and so 'the Growth of new Systems out of old ones' was impossible 'without the mediation of a divine Power'.[32] Throughout the 1690s, therefore, comets were part of a divinely planned system. Newton repeated his arguments sent to Bentley in a conversation with Gregory in May 1694, when beginning his planned revisions on comets for the *Principia*. He said 'that the great eccentricity in Comets in directions both different from and contrary to the planets indicates a divine hand: and implies that the Comets are destined for a use other than that of the planets'. In his work of 1694 and after Newton made this use ever more clear.[33]

As I have indicated, one of these clarifications involved systemic stability. In 1687 Newton had shown that 'comets are a sort of planet', so just as planets nearer the Sun were smaller, so comets 'which in their perihelion approach nearer to the Sun are generally less magnitude'. During his work of the 1690s, Newton identified the purpose behind this correlation. He inserted an addition in the 1713 edition, declaring now that comets which came nearer the Sun were smaller just so 'that they may not agitate the Sun too much by their attractions'. In January 1703, Newton repeated these ideas for Gregory's

31 A. Koyré and I. B. Cohen, *Isaac Newton's Philosophiae Naturalis Principia Mathematica: the third edition with variant readings* (Cambridge: Cambridge University Press, 1972), p. 733 (note at 506:19) for 1680–1 return in second edition; p. 721 (note at 500:11) for 1680–1 return in third edition; p. 756 for 1682 return. See Edmond Halley, ed. John Bevis, *Edmundi Hallei Astronomi dum viveret Regii Tabulae Astronomicae* (London, 1749).

32 Newton to Bentley, 10 December 1692, and 25 February 1693, in *Correspondence of Newton*, Vol. III, pp. 234 and 253–5. For the 'Platonic' cosmogony discussed here, and incorporated in Newton's proposed 'classical scholia' for the *Principia*, see A. Koyré, *Newtonian studies* (Cambridge, Mass.: Harvard University Press, 1965), p. 207 and I. B. Cohen, 'Galileo, Newton and the divine order of the solar system', in E. McMullin, ed., *Galileo: man of science* (New York: Basic Books, 1967), pp. 207–31, p. 225.

33 Gregory memorandum, May 1694, *Correspondence of Newton*, Vol. III, p. 336.

benefit. Summarizing the results of the 1690s, Newton argued that 'the Comets seem not to be Fixt Starrs extinguished, as M des Cartes imagined, but constant bodys moving about the Sun as regularly as the Planets'.[34] However, the completion of the work with Halley showed the great significance of comets and prompted ever more detailed contemplation of their function and origin. As Kubrin has pointed out, Gregory adopted Newton's 1694 suggestion that comets might disturb the satellites of Jupiter and Saturn and turn them into planets, and published it in his *Astronomiae physicae et geometricae synopsis* (1702). This book also contained Newton's essay on ancient philosophy, which, as we shall see, played a significant part in the 1690s programme. Soon after this publication, in March 1703, Gregory also recorded that 'the Comet whose Orbit Mr Newton determins may sometime impinge on the Earth', and he was referred to Origen's view of the destruction of worlds in support. Origen was a source on whom Newton drew plentifully in his studies of ancient natural philosophy, and the notorious memorandum by Conduitt made in 1725 also drew on these ideas.[35]

A second and equally important role for these bodies emerged in the work of Edmond Halley. As we have seen, in 1681 Burnet produced his *Sacred theory*, which launched a prolonged series of political and theological works on sacred physics which analyzed scriptural history in natural philosophical terms and in the context of the crisis of the 1680s. Here Halley found fruitful resources for his own new views on cometography.[36] In February 1687, Robert Hooke read the Royal Society a paper which attributed the Deluge to a change in the figure of the Earth. Wallis told Halley that his colleagues at Oxford 'seemed not forward to turn ye world upside down' on Hooke's invitation. Instead, Halley proposed that the Deluge might have been 'performed naturally by the casual shock of some transient body, as a comet or the like'.[37] The publication of the *Principia* and Halley's rejection at Oxford because of suspicions of his heterodóxy concentrated his attention on this problem. His work on the comets recommenced after talking with Newton in October 1694. In the following December, Halley read a paper on the

34 Koyré and Cohen, *Newton's Principia with variant readings*, p. 747 (note at 517:31–35); Royal Society Gregory MSS 247 f. 79r.

35 D. Kubrin, 'Newton and the cyclical cosmos', *Journal of the history of ideas*, 1967, *28*:325–46, p. 341; Gregory memorandum, May 1694, *Correspondence of Newton*, Vol. III, p. 336; David Gregory, *Astronomiae physicae et geometriae elementa* (Oxford, 1702), p. 481; Gregory memorandum, March 1703, *Correspondence of Newton*, Vol. IV, pp. 402–3; Kings College Cambridge, Keynes MSS 130.11.

36 For scriptural history, see Jacob and Lockwood, 'Political millenarianism'; R. Porter, 'Creation and credence: the career of earth theories in Britain', in B. Barnes and S. Shapin, eds, *Natural order* (London: Sage, 1979), pp. 97–124.

37 For Hooke's lectures see R. T. Gunther, *Early science in Oxford* (Oxford: privately printed, 1930), Vol. VII, pp. 701–2 (January–February 1687) and 710–711 (February 1688); R. Waller, *Posthumous works of Robert Hooke* (London, 1705), pp. 343–4, 350–4, 403–16. See John Wallis to Halley, 9 April 1687, in MacPike, *Correspondence and papers of Halley*, pp. 80–82; Edmond Halley, 'An account of some observations lately made at Nuremburg', *Philosophical transactions*, 1687, *16*:403–6; British Library MSS Add 4478b ff. 142–150 (on Halley's paper).

Deluge and the comet. He rejected the accounts both of Burnet and of Hooke, since they relied on 'a preternatural *digitus Dei*'. Comets were well fitted to perform this divine office, 'the Almighty generally making use of Natural Means to bring about his Will'. He spelt out the effects of such a collision and pointed out that 'the Earth seems as if it were new made out of an Old World'. The following week, after a conversation with 'a Person whose Judgment I have great Reason to respect', Halley conceded that such a comet might not have caused the Flood, but it could well have reduced some former Earth to the chaos from which this creation emerged. Halley linked the periodic returns of such comets with a divine plan, since 'in due Periods of Time, such a Catastrophe may not be unnecessary for the well-being of the future World', by restoring the Earth's fertility.[38]

William Whiston's better known cometary theory was published in his *New Theory of the Earth* in 1696. His very similar model of the Deluge prompted a complex priority dispute with Halley, who was compelled to dissociate himself from Whiston's notoriously suspect views. In January 1707 Halley reminded the Society of his thoughts on the Deluge and the comet '2 years before Mr Whiston's book was published', and in May 1724 he published these thoughts in the *Philosophical transactions*. Whiston now used Halley's calculations on the long-period comet of 1680–1 to argue that this was the body which had produced the Deluge, that a comet had indeed produced the pre-Adamite chaos, and that comets 'seem fit to cause vast Mutations in the Planets' and were 'capable of being the Instruments of Divine Vengeance' and 'of purging the outward Regions of them in order to a Renovation'.[39] Whiston's *Astronomical principles* was enormously influential on subsequent English theologians and natural philosophers. Throughout the eighteenth century, the arguments Whiston and Halley presented on the destructive and restorative function of comets under God's guidance acted as sources for a variety of astronomical systems.[40] While critical of the details of Newton's chronology and church history, Whiston's lectures on 'astronomy & sacred architecture' given in the London coffee houses disseminated these conceptions to a wide audience. Furthermore, since Newtonian cometography had revealed the true purpose and future of the cosmos, their

[38] Edmond Halley, 'Some considerations about the cause of the universal Deluge', and 'Some farther thoughts on the same subject', *Philosophical transactions*, 1724, *33*:118–125; Newton to Flamsteed, 24 October 1694, in *Correspondence of Newton*, Vol. IV, p. 34.

[39] Halley's comment on Whiston, 8 January 1707, Bodleian Library MSS Rigaud 37 f. 89; William Whiston, *Vindication of the New Theory of the Earth* (London, 1698), Preface; Whiston, *Astronomical principles of religion, natural and revealed* (London, 1717), pp. 23, 139–156ff. Compare Webster, *Paracelsus to Newton*, pp. 40–41 (though we may now have enough evidence to move beyond Webster's concern that 'whether Newton believed in more adventurous speculations concerning comets, depends on our evaluation of the worth of the enigmatic Conduitt memorandum').

[40] For Whiston and Halley as cosmological sources in the eighteenth century, see M. A. Hoskin, 'The English background to the cosmology of Wright and Herschel', in W. Yourgrau and A. Breck, eds, *Cosmology, history and theology* (New York: Plenum Press, 1977), pp. 219–232 and S. Schaffer, 'Fire and evolutionary cosmology in Wright and Kant', *Journal for the history of astronomy*, 1978, *9*:180–200.

revelation in the *Principia* must mark 'an eminent prelude and preparation to those happy *times of the restitution of all things, which God has spoken of by the mouth of all his holy prophets since the world began*'.[41]

At exactly the same time as his work on the Deluge, Halley was also considering the change in mass of the planets. We have seen that Halley suggested in December 1694 that a comet could restore the fertility of the Earth at periodic times. This paper was closely connected to those he had written between 1691 and 1693 on the motion of the Earth, its mass and its age. In late October 1694, as we have seen, Newton had talked with Halley about these issues. On 24 October Halley discussed the motion of the Earth and the Moon with Flamsteed, and one week later he told the Royal Society that Newton had said that 'there was reason to conclude that the bulk of the Earth did grow and increase . . . by the perpetuall Acession of new particles'. Halley supposed that 'this Encrease of the Moles of the Earth would occasion an Acceleration of the Moons motion, she being at this time attracted by a stronger Vis Centripeta than in Remote Ages'.[42] Halley's argument for the secular acceleration of the Moon was published in 1695 in a paper on archaeological work at Palmyra, and his claim was reproduced in the 1713 edition of the *Principia* and in Pemberton's *View* in 1728. However, the significance of this claim was its link with cometography, for the *Principia* argued that it was from comets' tails that this increase in terrestrial mass must come, and thus produce the lunar acceleration. Despite the scepticism of Flamsteed, expressed forcefully in a letter to Abraham Sharp in February 1710, this too became a vital function for the system of comets.[43]

The full model of cometary significance and their divine role, therefore, required an analysis of the composition and function of their tails. Newton and Halley gave comets' tails two specific roles in the solar system. Firstly, in 1687 Newton had used his lengthy work on comets' tails first drafted for the *System of the World* to argue that 'comets seem to be required, that, from their exhalations and vapours condensed, the wastes of the planetary fluids spent upon vegetation and putrefaction, and converted into dry earth, may be continually supplied and made up'. Since earthly life was a process which

[41] William Whiston, *Remarks on Sir Isaac Newton's Observations upon the prophecies of Daniel* (London, 1734), pp. 297–302; *idem.*, *Memoirs of the life and writings of William Whiston*, 2nd ed. (London, 1753), Vol. II, p. 34; 'Notes on Whiston's Astronomy Lectures by Thomas Morell', British Library MSS Burney 522 f. 2; 'Lectures on astronomy and architecture', Royal Astronomical Society MSS Add 88 p. 39; for Whiston's views, see E. Duffy, ' "Whiston's affair": the trials of a Primitive Christian 1709–1714', *Journal of ecclesiastical history*, 1976, *27*:129–150 and J. E. Force, *William Whiston: honest Newtonian* (Cambridge: Cambridge University Press, 1985), ch. 2.

[42] Royal Society, Journal Book, 31 October 1694, cited in Kubrin, 'Newton and the cyclical cosmos', p. 337. For the work Halley pursued on the motion of the Earth and the eternity of the world, see S. Schaffer, 'Halley's atheism and the end of the world'.

[43] Cohen and Koyré, *Newton's Principia with variant readings*, p. 758 (note to 526.30); Henry Pemberton, *View of Sir Isaac Newton's Philosophy*, p. 246; Edmond Halley, 'Some account of the ancient state of the City of Palmyra', *Philosophical transactions*, 1695, *19*:174–5; Flamsteed to Sharp, 11 February 1710, in F. Baily, *An account of the Reverend John Flamsteed* (1st ed. 1835–1837) (London: Dawsons, 1966), p. 225.

exhausted 'the fluids, if they are not supplied from without, must be in continual decrease, and quite fail at least'.[44] This function for comets was preserved in all three editions of the *Principia*, and it was explicitly the exchanges with Flamsteed and Halley and the difficulties with cometary returns and the formation of tails in interplanetary space which acted as resources for this model. The formation of the tails in a space devoid of other resistance continued to pose a problem, but it could not be abandoned because of the significance Newton gave to the processes of exhalation and condensation in periodic comets. In May 1694, as he began his revisions to this section, Newton told Gregory that 'celestial matter' behind the comet might become hotter, and therefore lighter than the other 'matter' through which the comet moved: 'hence the comparison with smoke rising in a chimney'. But as late as May 1725, Henry Pemberton was still having trouble with Newton's model here. 'Since the heavens are void of any matter that can give a sensible resistance to the progressive motion of this vapour', Pemberton asked, 'what is that *aura aetherea* which by its motion in ascent can carry this vapour along with it?' There was no mechanical answer to this puzzle.[45]

The clue to Newton's view here lies in the argument that the true composition of the comets' tails was *not* purely mechanical. Newton's published texts on the transformation of comets' tails into restorative and humid matter on Earth drew on his much earlier work of the 1670s, which had formulated an alchemical cosmology in which transformation was a direct result of the activity of nature as 'a perpetuall circulatory worker'. In a manuscript entitled 'Of natures obvious laws & processes in vegetation', composed early in the 1670s, Newton argued that 'this Earth resembles a great animall or rather inanimate vegetable, draws in aethereall breath for its dayly refreshment & vitall ferment & transpires again the grosse exhaltations'. Such vegetation was 'the sole effect of a latent spt & . . . this spt is ye same in all things'. In active matter there was 'an exceeding subtile & inimaginably small portion of matter diffused through the mass wch if it were separated there would remain but a dead & unactive earth'.[46] All these ideas were reproduced in Newton's brief remarks in his 'Hypothesis' sent to Oldenburg in December 1675, and touched upon in his references to 'a certain secret principle in nature' and to 'ye more tender exhalations & spirits yt flote' in air in his letter to Boyle in 1679.[47] They were also placed in the

44 Motte and Cajori, *Newton's Principles*, p. 529.

45 Gregory memorandum, May 1694, in *Correspondence of Newton*, Vol. III, p. 316; Pemberton queries, May 1725, in *ibid.*, Vol. VII, pp. 323–5.

46 Newton, 'Of nature's obvious laws and processes in vegeatation', Burndy MS 16, ff. 1, 3, 6; discussed in P. M. Rattansi, 'Newton's alchemical studies', in A. Debus, ed., *Science, medicine and society in the Renaissance* (New York: Science History, 1972), 2 Vols, Vol. II, 167–182; R. S. Westfall, 'Alchemy in Newton's career', in M. Righini Bonelli and W. Shea, eds, *Reason, experiment and mysticism in the Scientific Revolution* (New York: Science History, 1975), pp. 219–221; B. J. T. Dobbs, 'Newton's alchemy and his theory of matter', *Isis*, 1982, 73:511–528.

47 Newton to Oldenburg, 7 December 1675, in *Correspondence of Newton*, Vol. I, pp. 363–364; Newton to Boyle, 28 February 1679, *ibid.*, Vol. II, pp. 292 and 294.

context of cometography in this section of his cosmology after 1687, and amplified in the final queries to the *Optice* in 1706. J. E. McGuire has drawn attention to Newton's revisions to the Definitions in the *Principia* in which he listed just those substances into which comets' tails were held to transmute, and where Newton commented that 'vapours and exhalations on account of their rarity lose almost all perceptible resistance and in the common acceptance often lose even the name of bodies and are called spirits'.[48]

Thus comets served a fundamentally divine and spiritual office – the restoration of vegetative life. At the end of the second edition of his *Principia*, Newton charted the path by which comets' tails could change into 'terrestrial substances'. In June 1705 he explained to David Gregory that this passage was not to be referred to 'the real fluid of water so restored', but to 'that subtle spirit that does turn solids into fluids. A very small Aura or particle of this may be able to do the business'. This was to be connected with Newton's comment in the *Principia* that 'it is chiefly from the comets that spirit comes, . . . so much required to sustain the life of all things with us'. The action of comets' tails could not be referred to pure mechanism. While Pemberton remained sceptical whether any mechanism could account for Newton's view of their formation, he did agree that 'the most subtle and active parts of our air, upon which the life of things depends, is derived to us, and supplied by comets'. Astrological fears of comets were groundless, but their purposes were manifold and divinely planned – they had become the prime transmitters of activity in the cosmos.[49]

Pemberton also pointed out the second function which Newton granted comets' tails. This was the restoration of the stars. During his correspondence with Flamsteed in 1681 and with Halley in 1695, Newton had been compelled to consider the heat experienced by the comet of 1680 and its extreme proximity to the Sun, since this had prompted his denial of its magnetic character. By 1698 he had concluded that it was periodic and that therefore it would experience the resistance of the Sun's atmosphere over several revolutions. In 1713 Newton published the conjecture that its period was over 500 years, and added a comment that it would be 'attracted somewhat nearer to the sun in every revolution' and 'will at last fall down upon the body of the Sun'. Such events would recruit the activity of wasting stars, and would also produce the sudden appearances of novae. In March 1725, Newton was visited by Conduitt in Kensington and told him in detail of these ideas: Newton suggested that the comet might fall into the Sun in 5

[48] J. E. McGuire, 'Body and void and Newton's *De mundi systemate*', *Archive for history of exact sciences*, 1967, *3*:206–248, p. 220, citing Cambridge University Library MSS Add 3965.13 f. 422r; *idem*, 'Transmutation and immutability: Newton's doctrines of physical qualities', *Ambix*, 1967, *14*:69–95; *idem*, 'Force, active principles and Newton's invisible realm', *Ambix*, 1968, *15*: 154–208. See the excellent study of the relation between cometography and matter theory in Sara Schechner Genuth, 'Comets, teleology and the relationship of chemistry to cosmology in Newton's thought', *Annali dell'Istituto e Museo di Storia della Scienza di Firenze*, 1985, *10*:31–65.
[49] Koyré and Cohen, *Newton's Principia*, p. 758; Hiscock, *Gregory, Newton and their circle*, p. 26; McGuire, 'Transmutation and immutability', p. 87; Pemberton, *View*, p. 245; Genuth, 'Comets, teleology', n. 57, which challenges the import of Gregory's memorandum.

or 6 transits, that this would extinguish life on Earth, and that the novae reported by Hipparchus, Tycho and Kepler might be such events. These phenomena were then discussed in great detail in a lengthier addition to the third edition of the *Principia* in the following year. But Conduitt pointed out that Newton had not referred to the more disastrous effects of a comet impinging on the Sun, to which Newton replied 'that concerned us more, and laughing added he had said enough for people to know his meaning'.[50] Through the texts of Gregory, Whiston, Halley and Pemberton, eighteenth century natural philosophers and theologians knew much of these views, and used the stipulations to treat comets as the most divinely significant objects in the heavens. In 1742 Maupertuis commented on the views of Gregory and Newton that 'one of the greatest astronomers of the century has spoken of comets in a manner which re-establishes them in all the reputation of terror where once they were'. In 1749 Buffon depicted God's hand directly guiding a comet in its collision with the Sun which he supposed had formed the planets. In 1754 James Ferguson suggested that comets would be an appropriate residence for astronomers, and that they were fitted for 'recruiting the expended fuel of the Sun; supplying the exhausted Moisture of the Planets; causing Deluges and Conflagrations for the Correction and Punishment of Vice'. Ultimately, in his *Cosmological Letters* of 1761, J. H. Lambert was able to portray astronomers as those specifically concerned with the system of comets, and thus as 'authorized prophets'.[51]

Yet Newton had not said, nor would ever announce publicly, that much of his own work involved the correlation between these divine functions and meanings and ancient prophecy, philosophy and theology developed in the tradition of the *prisca*. In autumn 1685 Newton outlined a series of claims that geocentrism and the doctrines of solid spheres and a resisting interplanetary fluid were later corruptions of pristine truths about the system of the world. In the Vestal temples these ancients had constructed an analogue of the true, heliocentric and vacuist system of the heavens. He placed this argument at the head of his draft *System of the world*, and then suppressed its publication. As P. M. Rattansi, J. E. McGuire and P. Casini have pointed out, such claims appear in the 'classical scholia' discussed with Gregory in 1694, just as the work on comets recommenced, and were revealed in drafts for queries in the *Optice* (1706), in annotations of the General Scholium from 1713 and in a

[50] Newton for Flamsteed, ?April 1681, in *Correspondence of Newton*, Vol. II, pp. 358–62; ms. on comet of 1680–1, *ibid.*, p. 167; Koyré and Cohen, *Newton's Principia*, p. 757; Motte and Cajori, *Newton's Principles*, pp. 530 and 541; Conduitt memorandum, Keynes MSS 130.11 cited *in extenso* in D. Castillejo, *The expanding force in Newton's cosmos* (Madrid: Ediciones de arte y bibliofilia, 1981), pp. 95–97; Pemberton, *View*, p. 246.

[51] P. de Maupertuis, 'Lettre sur la comete', in *Oeuvres* (Lyons, 1768), Vol. III, pp. 209–56, on p. 240; Buffon, *Histoire naturelle* (Paris, 1749), tome premier, p. 127; James Ferguson, *An idea of the material universe* (London, 1754), pp. 26–27; J. H. Lambert, *Cosmological letters* (1761), ed. S. L. Jaki (Edinburgh: Scottish Academic, 1976), p. 63. For comets as divine signs and agents, see R. Long, *Astronomy* (Cambridge, 1764), Vol. II, p. 563; J. Cowley, *Discourse on comets* (London, 1757), pp. 29–36; J. Hill, *Urania* (London, 1754), s.v. 'Comets'; R. Yate, 'A new theory of comets', *Gentleman's magazine*, 1743/4, *13*:193–5.

draft preface for the final edition of the *Principia*: 'The Chaldaeans once believed that the planets revolve around the Sun in almost concentric orbits and the comets in very eccentric orbits. And the Pythagoreans introduced this philosophy into Greece . . . This philosophy was discontinued (it was not propagated to us and gave way to the vulgar opinion of solid spheres). I did not discover this, but endeavoured to restore it to light by the power of demonstration'. In 1685, in the *System of the world*, Newton emphasised cometography in ancient science. The Egyptians and their followers argued for a heliocentric and vacuist cosmos: 'but, above all, the phenomena of comets can by no means tolerate the idea of solid orbs. The Chaldeans, the most learned astronomers of their time, looked upon comets (which of ancient times had been numbered among the celestial bodies) as a particular sort of planets, which, describing eccentric orbits, presented themselves to view only by turns, once in a revolution, when they descended into the lower parts of their orbits'.[52]

By 1685, therefore, Newton had already established that comets moved in closed orbits and that this was good Chaldean astromony. He was using a very common strategy of interpretation of ancient texts, arguing that within these texts his achievements of the 1680s lay concealed beneath such ambiguous utterances. He found sources in Plutarch, Macrobius, Origen, Diodorus Siculus, Diogenes Laertius and others, and methodological inspiration in mythographers such as Natalis Conti. Both Halley and Flamsteed, for example, discussed such sources, and used these methods to engage in an analysis of ancient history of astronomy, and traced the significance of the Chaldeans' cometography. In 1705 Halley wrote that the Egyptians and Chaldeans had predicted cometary returns, but probably only as 'the result of meer Astrological Calculation, than of any Astronomical Theories of the Celestial Motions'. In his lecture at Gresham College in May 1681, Flamsteed cited Seneca's report that Apollonius was 'a good natural philosopher' who had 'studied amongst the Chaldeans', and that Apollonius got his notion of interplanetary comets and their returns from them: Flamsteed commented on 'his desire to heighten his owne repute by encreasing theirs'. In his astronomical tables, Flamsteed speculated that 'the Chaldeans received their astronomy from the Subjected Jews, who having a knowledge of Dialling cannot be thought ignorant of the courses of the Sun and Moon'.[53] But

[52] P. M. Rattansi and J. E. McGuire, 'Newton and the "Pipes of Pan"', *Notes and records of the Royal Society*, 1966, *21*:108–43; P. Casini, 'Newton: the classical scholia', *History of science*, 1984, *22*:1–58, which prints the classical scholia on pp. 25–38; Motte-Cajori, *Newton's Principles*, pp. 549–50; Cambridge University Library MSS Add 3970 ff. 292v & 619r (for queries of *Optice*); Koyré and Cohen, *Newton's Principia*, pp. 761–2 (notes at 528.13–22 and 529.1–14) for General Scholium; Cambridge University Library MSS Add 3968.9 f. 109 in Whiteside, *Mathematical papers*, Vol. VIII, pp. 458–9 n. 49 (draft preface, after 1716); Koyré and Cohen, *Newton's Principia*, pp. 803–7, for scholia insert in *Principia*, incorporated in Casini, 'Newton: the classical scholia', p. 23, n. 49.

[53] Newton's sources include Plutarch, *Opera quae extant omnia* (Frankfurt, 1599); Macrobius, *Opera* (Leyden, 1628); Origen, *Contra Celsum* (Cambridge, 1658); N. Conti, *Mythologiae* (Cologne, 1612); editions in J. Harrison, *Library of Isaac Newton* (Cambridge: Cambridge University Press, 1978), nos 1331, 1013, 1209, 439 and discussed in Casini, 'Newton: the

Newton's concern with this ancient knowledge was individual and profound. He used the *prisca* as a weapon in his attack on both Cartesians and Leibnizians who challenged the doctrines of universal gravity and empty space. Most importantly, we have already shown that the decisive change in Newton's comets occurred between 1681, when he still claimed they probably moved rectilinearly and that their motions would be disturbed by some interplanetary aether, and late summer 1684, when in the tract *De motu* he emptied the heavens and asserted, before any but the most rudimentary calculations, that comets' orbits were elliptical. We have also seen that once these axioms were established, Newton remained committed to the programme, steadily enriching cometary significance and their divine role. The appearance of this Chaldean cometography in autumn 1685, and again, just when the work on comets recommenced, in July 1694, is not without significance.[54]

In the very years when Newton was breaking definitively with orthodox cometography, he was also engaged in the composition of his most fundamental study of ancient theology and natural philosophy, the *Philosophical origins of gentile theology*. Newton began working on this treatise in 1683–1684, just before his composition of *De motu*, in late summer 1684. He re-worked it in 1694 and after, when he recommenced work on comets. The *Philosophical origins* contains the detailed exposition of all the doctrines which later appear in the classical scholia and the drafts for the *Principia* and the *Opticks*. Extracts were to be used in the massive history of the church and later published in a very garbled form in *Chronology of ancient kingdoms amended* (1728). The planned treatise of the mid-1680s argued that a pristine natural philosophical religion had existed, in which the true world system was well known. This was expressed in symbolic form in the Pythagorean

---

classical scholia', p. 7 and pp. 38ff. For Diogenes Laertius, compare McGuire and Tamny, *Certain philosophical questions*, pp. 20–21. See Edmond Halley, *Synopsis*, p. 1; Forbes, *Gresham Lectures of Flamsteed*, p. 105 (citing G. B. Riccioli, *Almagestum novum* (Bologna, 1651), Vol. II, p. 3); Chapman, *Preface to Flamsteed's Historia coelestis*, p. 32. Compare the claim of Ramus that 'the ancient astrology of the Babylonians, Egyptians and Greeks, even before Eudoxus', was without hypotheses: P. Ramus, *Scholae mathematicae* (Basel, 1569), pp. 49–50. I owe this reference to Anne Blair.

54 For the *prisca* as a weapon against Leibniz, see the draft for Conti, 26 February 1716, in A. Koyré and I. B. Cohen, 'Newton and the Leibniz-Clarke correspondence', *Archives internationales de l'histoire des sciences*, 1962, *15*:63–126, p. 73; for the revival of these concepts in the classical scholia and the discussions with Bentley, in 1692–3 and Gregory in 1694, see *Correspondence of Newton*, Vol. III, pp. 333, 338, 384. For such tradition in the seventeenth century, see D. P. Walker, *The ancient theology* (London: Duckworth, 1972), pp. 175–93 (on Herbert's *De religione gentilium*) and pp. 254–263 (on Ramsay's *Cyrus*); C. B. Schmitt, 'Perennial philosophy from Agostino Steuco to Leibniz', *Journal of the history of ideas*, 1966, *27*:505–532; D. B. Sailor, 'Moses and atomism', *Journal of the history of ideas*, 1964, *25*:3–16. For the link between perennial philosophy, a Euhemerist interpretation of myth, and alchemy, see H. J. Sheppard, 'The mythological tradition and seventeenth century alchemy', in A. Debus (ed.), *Science, medicine and society in the Renaissance* (New York: Science History, 1972), 2 Vols, Vol. I, pp. 47–59. For remarks on the relation between the *prisca* and natural philosophy, see J. E. McGuire, 'Neoplatonism and active principles: Newton and the *Corpus hermeticum*', in R. S. Westman and J. E. McGuire, *Hermeticism and the scientific revolution* (Los Angeles: Clark Library, 1977).

harmony of the spheres and the Vestal temple ceremonies. Newton's euhemerist interpretation of gentile religions showed their common root and how they were corrupted. The first chapter explained that 'gentile theology was philosophical and especially regarded the astronomical and physical science of the system of the world, and that the 12 major gods of the nations are the seven planets and the four elements and the quintessence'. Corruption followed when such symbols were misinterpreted, generating beliefs in stars as Gods and the transmigration of souls. Thus in chapter XI Newton outlined what 'the true religion of the Noachids was before it began to be corrupted by the cult of false gods, and that the Christian religion was not more true nor any less corrupted'.[55]

Correct interpretation of these symbols and ceremonies allowed access to this true cosmology. The temples' central fire symbolized a commitment to heliocentricity and the void: Newton explained that ''twas one designe of ye first institution of ye true religion to propose to mankind by ye frame of ye ancient Temples, the study of the frame of the World as the true Temple of ye great God they worshipped'. Newton dealt similarly with Pythagorean harmonics and the empirically generated inverse square law: 'by the harmony of the spheres is to be understood the harmonic motion of the solar soul by which the planets are driven in their spheres by the Sun'. But the gentiles corruptly interpreted these views. Heliocentrism and vacuism gave way to solid spheres and aethereal resistance: 'taking literally the views of the philosophers that the celestial spheres emitted harmonic sounds when moving against each other, they believed that the heavens were not fluid, but that the planets and the fixed stars inhered in solid orbs, and this opinion was first introduced by Eudoxus just before the time of Aristotle'. The false astronomy was a direct consequence of idolatry and false interpretation. After Eudoxus, corrupt philosophers identified each orbital soul with a specific deity, so by abandoning true celestial mechanisms the gentiles allowed idolatry to flourish. Most revealingly, Newton wrote in the mid-1680s, 'because of the solidity of the spheres they located the comets beneath the sphere of the Moon and supposed them to be meteors'.

[55] For the *Philosophical origins of gentile theology*, see Westfall, 'Newton's theological manuscripts', in Z. Bechler, ed., *Contemporary Newtonian Research* (Dordrecht: Reidel, 1982), pp. 129–143 and *idem*, 'Isaac Newton's *Theologiae gentilis origines philosophicae*', in W. Warren Wagar, ed., *The secular mind: transformations of faith in modern Europe* (New York: Holmes and Meier, 1982), pp. 15–34. I have used Jewish National Library, MSS Yahuda 16 and 17.3 as sources. Compare F. Manuel, *Isaac Newton: historian* (Cambridge: Cambridge University Press, 1963), pp. 103–121 and *idem*, *The religion of Isaac Newton* (Oxford: Clarendon, 1974), pp. 53–79. The chapter headings are in Yahuda MS 16 f. 1r and f. 43v. The date *ante quem non* seems to be given by Humphrey Newton's arrival as amanuensis between late 1683 and early 1684: see Westfall, *Never at rest*, p. 343, n. 31 and Whiteside, *Mathematical papers*, Vol. VI, p. xii, n. 5. Its initiation certainly predates the *System of the world* (1685), the *Principia* (completed 1687) and the classical scholia (1693–1694), and thus certainly shows Newton working *at the same time* as his establishment of celestial mechanics. This therefore calls into question the remark of Casini, 'Newton: the classical scholia', p. 15, that 'the record of the Ancients and the reinterpretation of their "fictions" ' were 'superadded to a completed work'.

Restoring cometography would destroy this corrupt religion of polytheism and astral transmigration.[56]

In *Philosophical origins of gentile theology* Newton linked idolatry and false cometography. Idolatry was the central tool in Newton's analysis of how philosophers err, why dispute occurs, and how church and state become corrupt. Cabbalists, Gnostics and neo-Platonists were singled out as sharing a common idolatry (in their doctrines of transmigration and animation) and a common error which concealed the true system of the world (having 'received the Aristotelick system of the heavens' or, like Origen, holding that 'there were many successive worlds' and that aether and matter could transmute).[57] False cometography suffered from the same problem: it connected corrupt natural knowledge with worship of planetary souls as real divinities identified with temporal kings and heroes. Newton posed two issues in the interpretation of idolatry. Firstly, idolatry was defined as the illegitimate attribution of God's power over his works to inferior surrogates, and so false cometography was idolatrous. Newton's clearest definition of idolatry was presented at exactly the same time as the *Philosophical origins* in two sermons of the mid-1680s on 2 Kings xvii, which described the Israelite worship of Baal and the multiplication of gentile religions. In these sermons Newton insisted that idolatry did not misuse God's attributes but his power. In terms identical with those of the 1713 General Scholium, he wrote that 'ye wisest of beings requires of us to be celebrated not so much for his essence as for his actions'. His actions were visible in nature, so true natural knowledge destroyed idolatry. Failure to make this distinction still led Christians into error: 'his works . . . may & have been & still are too frequently ascribed to his creatures, the deified Heroes with their Idols among Gentiles', (and here Newton turned to the modern abuses) '& ye saints wth their reliques & images among too many Christians'.[58]

Secondly, therefore, the reformation of astronomy had direct moral, theological and political functions. Newton's campaign, begun in the 1670s, against Athanasians, and his interpretation of the prophecies, showed that idolatry and false power went in step with corruption. False powers were attributed to natural objects because corrupt men seized false power. In his

[56] On temples and the symbol of the world, Yahuda MS 17.3 ff. 10r and 11; Yahuda MS 41 f. 8r ('The original of religions', a draft for 'Philosophical origins'); Keynes MS 146 ('The original of monarchies' a later text of 1693: see Manuel, *Newton Historian*, p. 220); on Pythagorean harmony, Yahuda MS 17.3 f. 1v, 2v; on its literal misinterpretation, Yahuda MS 17.3 f. 12r. The source for the Vestal temples is Plutarch, cited in *System of the world* (Casini, 'Newton: the classical scholia', p. 20, n. 1) and in Yahuda 17.3 f. 1v. The source for the Pythagorean harmony is Macrobius, *In somnium Scipionis*, cited in the classical scholia (Casini, 'Newton: the classical scholia', pp. 32–36, 43) and in Yahuda MS 17.3 f. 1r, and also the pseudo-Plutarchian *De placitis philosophorum*, cited in Yahuda MS 17.3 f. 2v and in classical scholia (Casini, 'Newton: the classical scholia', p. 25 and p. 37).

[57] Yahuda MS 15.7 ff. 127 and 189bv. This is a history of the church mostly composed after 1715: see Castillejo, *Expanding force*, chapter 3, and Manuel, *Religion of Newton*, pp. 68–74. For Origen's ascription of the corruption of worlds to Democritus, see Newton's reference in the classical scholia, in Casini, 'Newton: the classical scholia', p. 27.

[58] The sermons are in Yahuda MS 21 ff. 1–2; see Westfall, *Never at rest*, p. 354; Manuel, *Religion of Newton*, pp. 21–22.

Paradoxical questions on Athanasius Newton wrote of 'monstrous Legends, fals miracles, veneration of reliques . . . & such other heathen superstitions', and later linked these with gentile theology: 'the old heathens first commemorated their dead men then admired them, afterwards adored them as Gods then praised them . . . so as to make them Gods celestial'.[59] These studies of church history were exactly contemporary with the Popish Plot and the crisis of the 1680s. 'Never was Pagan Idolatry so bad as the Roman, as even Jesuits sometimes confess', Newton had already written in the 1670s. By the mid-1680s, with the composition of the *Philosophical origins*, the link between gentile idolatry, corrupt natural philosophy, the Athanasian exploitation of deceptive signs with which to win allegiance, and the Papist threat, had been clearly established.[60]

In the mid-1680s, Newton argued that the natural philosophers of the ancients had been their Priests, such as the Chaldeans in Babylon. It was from Egyptian priests that Greeks learnt science. In the *Philosophical origins*, he explained that when 'the stars were declared to move in their courses in the heavens by the force of their souls and seemed to all men to be heavenly deities', then 'gentile Astrology and Theology were introduced by cunning Priests to promote the study of stars and the growth of the priesthood and at length spread through the world'.[61] Priestcraft was based on the corruption of true natural philosophy. Monarchs and diviners based their power on this corruption: 'so ready was ye ambition . . . of Princes to introduce their predecessors into ye divine worship of ye People to secure to themselves the greater veneration from their subjects as descended from ye Gods, erected such a worship & such a priesthood as might awe the blinded & seduced people into such an obedience as they desired'. Divine right and Papist monarchy stood condemned: 'here then we have the true original of ye corruption of ye religion of Noah and ye true cause of its spreading so early & so generally. For this policy of ye kings of Egypt soon took with ye kings of other nations'. Astrology was merely a variant of this idolatrous politicking: 'astrologers, augurs, auruspicers &c are such as pretend to ye art

[59] 'Paradoxical questions concerning the morals and actions of Athanasius and his followers', Keynes MS 10, partly published in H. McLachlan, *Theological writings of Isaac Newton* (Liverpool: University of Liverpool, 1950), pp. 61–110; church history of late 1670s and early 1680s, Yahuda MS 18 f. 3r on 'old heathens'. Compare Westfall, *Never at rest*, p. 345.

[60] Yahuda MS 14 f. 9v, manuscript of early 1670s on Arian propositions, against Roman idolatry; compare Yahuda MS 1.4 ff. 67–68, an early interpretation of Revelation, in which idolaters are identified; attacks on Papism in the 1680s in this context, Yahuda MS 9.2 f. 99 and Yahuda MS 10.2 ff. 1–15.

[61] Yahuda MS 17.3 f. 9r on cunning priests in the *Philosophical origins*. For priests as skilled natural philosophers – a commonplace in this period – compare Yahuda MS 41 f. 8r ('thence it was that the Priests anciently were above other men well skilled in the knowledge of the true frame of Nature') with Edward Stillingfleet, *Origines sacrae* (London, 1662), pp. 103–4 and John Beaumont, *Considerations on a book, entitled the Theory of the Earth* (London, 1693), p. 86 ('the Antediluvian Patriarchs as well as the Postdiluvian, were in their respective times, the most absolute Masters of the aforesaid Science [astronomy] of any Men on the Earth, and . . . from them, it has been convey'd down in its Pureness to us'). See the excellent discussion, including the use of Philo of Alexandria by More and by Boyle, in H. Fisch, 'The scientist as priest: a note on Robert Boyle's natural theology', *Isis*, 1953, *44*:252–65.

of divining . . . without being able to do what they pretend to . . . and to believe that man or woman can really divine . . . is of the same nature with believing that the Idols of the Gentiles were not vanities but had spirits really seated in them'. Cunning men in church and state gained power by the appropriation of divine knowledge to themselves and by the false attribution of divine powers to material intermediaries.[62]

Thus in the period of 1685, when such threats seemed very real, Newton composed a treatise on ancient philosophy in which he charged that false worship of what had really been symbols of proper natural philosophy had destroyed the doctrine of comets and planets which he would now restore. In detail, that Chaldean doctrine specified permanent and unimpeded motions of comets and planets in free space focussed on the Sun. This was precisely the doctrine which Newton managed painfully to establish in his work on astronomy and celestial mechanics between 1681 and 1686. In this sense, therefore, Newtonian cometography transformed astrology in at least two ways. Firstly, it changed the practice which should deal with comets from popular divination to theologically oriented natural philosophy, giving comets a profound but scarcely less dramatic function and prophetic meaning. They would cause the Deluge, terminate and restore life on Earth and rejuvenate the Sun and the stars. Secondly, as we have argued, Newton challenged a specific form of corrupt idolatry which attributed the wrong spiritual power to the heavens and to rulers and priests on Earth, and thus generated false philosophy and false cometography. As superior priests of Nature, philosophers had now restored truth. In just that sense, he argued, there was after all something to be said for the moderns against the ancients: ''tis ye nature of man to admire least what he is most acquainted with & this makes us always think o^{r} own times the worst. Men are not sainted till their vices be forgotten'.[63]

62 Bodleian Library, New College MSS 361.3 f. 32 (a set of notes for 'Philosophical origins'), for which see Manuel, *Newton Historian*, pp. 114–116, for princely corruption; Yahuda MS 15.7 f. 133v for astrologers and diviners.

63 Yahuda MS 18 f. 3r: see Manuel, *Religion of Newton*, p. 8.

SAVING ASTROLOGY IN RESTORATION ENGLAND: 'WHIG' AND 'TORY' REFORMS

Patrick Curry

This paper concerns astrology in Restoration England, and in particular efforts to reform and thereby save it.[1] The period in question was a critical and difficult one for astrologers; the preceding two decades, we know with hindsight, were the heyday of their art. There was a Society of Astrologers in London, patronized by a powerful Parliamentarian figure, which met several times a year for a feast and learned sermon.[2] Astrologers' advice was sought along the entire political spectrum, from Charles II to leading Levellers and Ranters.[3] During the Civil War, both sides had employed astrological propagandists, the most famous of whom was William Lilly.[4] Parliament appointed the astrologer John Booker as licenser of astrological books in 1642 (replacing the bishops). Almanacs flourished; by 1660, that of Lilly was selling a steady 30,000 copies a year, and total sales have been estimated at 40,000 a year, or roughly an almanac for every one family in three.[5]

With the conservative settlement of the Restoration, however, astrologers came under considerable pressure – both direct and indirect – resulting from their prominent role in the Interregnum. The resulting associations with sectarian radicalism, civil disorder and sedition, and 'atheism', made astrology highly suspect among the recently re-empowered aristocracy and upper class and their adherents. The fearsome Royalist Roger L'Estrange was unleashed as 'surveyor of the imprimery', *i.e.* censor. In this capacity, he emasculated alamancs by Lilly (whose sales fell to 8,000 in 1664) and others, and incarcerated the astrologer John Heydon and the publisher of radical and

[1] For a fuller account of the subject-matter of this paper, see my Ph.D. thesis 'The Decline of Astrology in Early Modern England, 1642–1800' (University of London, 1986), which I hope eventually to publish. When both primary and secondary sources exist, I have usually chosen for this paper to cite the latter, in the interests of easier accessability for the reader; for fuller documentation, again, see my thesis. Also, two semantic points: in general, 'science' should be understood as natural philosophy, and 'astrology' as judicial astrology.

[2] E. A. Josten, *Elias Ashmole (1617–1692): His Autobiography and Historical Notes, his Correspondence and Other Contemporary Sources* ... (Oxford, 1966), see 'Society of Astrologers'; and Ashmolean MS. 423, f. 168.

[3] K. Thomas, *Religion and the Decline of Magic* (Harmondsworth, 1971), 317–72.

[4] See the afterwords in the recently re-printed edition of Lilly's *Christian Astrology* (London, 1985).

[5] B. Capp, *Astrology and the Popular Press: English Almanacs 1500–1800* (London, 1979), 23, 44.

astrological literature, Giles Calvert.[6] Among other written attacks, the official statement of position of the new Royal Society described astrology as 'a disgrace to the Reason, and honor of mankind', and one which 'withdraws our obedience, from the true Image of God the rightfull Sovereign . . . this melancholy, this frightful, this Astrological humor . . .'[7]

Such opprobrium and its effects were soon felt by astrologers. In 1672 the Reading astrologer-physician Joseph Blagrave complained that his business was falling off; even those who still consulted him came 'for the most part privately, fearing either loss of reputation or reproaches from their Neighbours . . .'[8] The astrologer-divine John Butler wrote to Elias Ashmole from Oxford, in 1680, 'But do you hear the news from Alma Mater? All Astrology must be banished . . . Then what shall become . . . of us all Astrologers?'[9] And seventeen years later, John Partridge sadly remarked that 'Astrology now is like a dead Carkass.'[10] Many astrologers, perhaps most, now saw the only hope for astrology in its reform. Understandably, they approached the problem in internalist terms, that would permit them to undertake the crucial enterprise: to locate and display the true essence of astrology. This would then enable the unreliable and extravagent elements – which, in the astrologers' view, were imperiling the reputation of the whole – to be jettisoned. As the astrologer Richard Saunders put it in 1677, 'except the pure Quintessence be extracted from those faeculent dregs, this Science . . . is likely to perish . . .'[11]

Of course, there were other changes around 1660 which also undermined astrology, and which will be mentioned later. But this paper emphasises social and political factors – both because they are its main concern, and because these factors have (in my view) been underemphasised in standard accounts.[12] In all, then, astrology entered the Restoration inauspiciously. The urgency felt by the reformers in that context is not difficult to understand; and without impugning their genuine love of astrology, which in most cases is difficult to doubt, their zeal was not disinterested.

When one looks at these reformers, they fall fairly clearly into two distinct and conflicting camps, which are easily differentiated in terms of both their social interests and their approaches to astrological reform. (It would be extraordinary, of course, if there were no exceptions; these will be discussed later.) Within each programme, its adherents worked together to further the common aim, and in conscious – often bitter – opposition to the other. For reasons that will become apparent, I have termed these two the scientific or 'Tory' reform, and the Ptolemaic or 'Whig' reform. In examining them, it is

[6] On L'Estrange, *Dictionary of National Biography* (Oxford, 1921–22) [hereafter, *DNB*], Vol. 11, 997–1007; on Heydon, *DNB* Vol. 9, 768–69; on Calvert, *Biographical Dictionary of British Radicals* (Brighton, 1985) [hereafter, *BDBR*], Vol. 1, 119–20.

[7] T. Sprat, *History of the Royal Society* (London, 1667), 364–65.

[8] J. Blagrave, *Astrological Practice of Physick* (Reading, 1672), 'Epistle' (n.p.).

[9] J. Butler, . . . *Astrology a Sacred Science* . . . (London, 1680), 'Epistle' (n.p.); in Ash. MS. 303.

[10] J. Partridge, *Defectio Geniturarum* (London, 1697), sig. Bv.

[11] R. Saunders, *The Astrological Judgement and Practice of Physick* (London, 1677), 172.

[12] Notably Thomas and Capp.

as important to include their social, political and religious committments as their attitudes to astrology; to do otherwise results in a seriously impoverished understanding of the latter.

Turning first to the reformers whose hopes lay with science – that is, natural philosophy – there were principally three. All of them drew their inspiration from the hope that Bacon himself had expressed for a purified and corrected 'astrologia sana', a sane astrology.[13]

Joshua Childrey (1625–70) was educated at Royalist Oxford, partly during the Civil War. A convert to 'my Lord Bacon's philosophy' since 1646, upon the Restoration he became (in turn) a beneficed clergyman, an arch-deacon under Seth Ward at Salisbury Cathedral, and finally a rector in Dorset.[14] He was a competent astronomer who debated tidal theory with John Wallis in the pages of *Philosophical Transactions*.[15] Perhaps because of this, he combined Baconianism with an equally strong committment to the Copernican concept of the solar system. His programme for astrology was in fact an amalgam of these two positions. It was first announced in *Indago Astrologica* (1652); this was followed by a completely heliocentric ephemeris, *Syzygiasticon Instauratum* (1653) – the first of its kind. Put at its simplest, Childrey advocated an astrology based on the planets' heliocentric positions and aspects, 'quod naturam' (as they are) rather than 'quod nos' (as they appear to us). For 'though Asronomy be corrected, yet Astrology (which judges mostly by the Aspects) remains yet uncorrected . . .'[16] The other arm of correction was to be the scrupulously empirical gathering and evaluation of data, modelled on Baconian 'histories'. 'For Astrology,' as Childrey put it, 'wants its History as much as any other part of Philosophie; It being the only Via Regia, to its Perfection; and all other wayes but by-wayes.'[17] Childrey accompanied these prescriptions with a tough-minded rejection of anything in astrology which apparently lacked a physical basis or correlate: zodiacal signs, houses, elements, rularships, and so on. But the response to Childrey's efforts must have been disappointing (even allowing for the fact that they were always more programmatic than substantive). Most astrologers were sceptical of taking aspects as if (as one put it) we lived on the Sun.[18] And while Henry Oldenburg, the Secretary of the Royal Society, continued to be a faithful correspondent, Childrey received no real degree of encouragment from that quarter either – not even for his astrometeorological suggestions, traditionally the area where natural philosophers were willing to give astrology most credence.

13 The main discussion of this group to date has been that of M. E. Bowden, in 'The Scientific Revolution in Astrology: The English Reformers, 1558–1686' (unpublished Ph.D. thesis, Yale University, 1974). Bacon's main discussion is in *De Augmentis Scientiarum*, Book III, chapter 4.

14 *DNB* 4, 250–51; Capp, 301.

15 *Philosophical Transactions of the Royal Society of London* [hereafter *PT*], 10 October 1670, 2061–74.

16 *Indago Astrologica*, 9.

17 *Syzygiasticon Instauratum*, 2.

18 'A Brief Account of the Copernican Astrology', appended to J. Gadbury, *A Diary . . . for . . . 1695*.

Nobody could say that the work of *John Goad* (1616–89) lacked substance. It was probably the most determined attempt ever to test the influence of planets on the weather, and simultaneously to try to refine and reform astrology through the analysis of cosmic and meteorological correlations. Goad was also educated at Oxford, where in 1646 he 'performed divine service under fire of parliamentary cannon'. In 1661 he became the headmaster of Merchant Taylors School, a position he held until 1681, when – falling victim to the Popish Plot furor – he was found to be 'popishly and erroneously affected', and dismissed. Three years before his death, he openly declared himself to be a Catholic. By this time his two closest friends were Elias Ashmole and John Gadbury (both firm Royalists, one a nominal Anglican and the other a crypto-Catholic).[19]

Goad's *magnum opus* was *Astro-Meteorologica* (1686). It constituted of thirty years of weather observations, and their attempted correlation with planetary positions. Like Childrey, he tried to rely only on aspects; and equally, he was adament about extending our knowledge through Bacon's approach, arguing against accounts of the weather as simply uncertain or 'casual' that 'Casualty is inconsistent with Science, so inconsistent that it is not to be pleaded by any lovers of Learning.'[20] Goad struggled manfully in his nascent attempts at statistical evaluation, but his attempts to find definite and reliable correlations came painfully unstuck. The problem was at least two-fold: the recalcitrance of the data, resulting from the geographical and temporal variability of weather conditions, plus a lack of precise measurements; and the difficulty of any pioneer – in this case, one trying to subdue the new terrain of probablistic knowledge, waiting to be cleared between certain, mathematical knowledge on one hand and completely random phenomena on the other. Other such pioneers at this time were John Graunt – whose ideas may have influenced Goad – and William Petty, both of whom received more recognition at the time, and since, than Goad. This was due to their more obviously useful (potentially speaking) and less controversial choice of subject-matter, although that is no reason for historians to continue imitating the Royal Society and its royal patron.

Astro-meteorologica found a mixed reception. It was briefly discussed by the Royal Society, where it was criticized by Jonas Moore (speaking for John Flamsteed), but defended by Thomas Henshaw. Flamsteed's attitude itself seems to have been ambivalent, since a letter survives in which he commends Goad, 'whose conjectures come much nearer truth than any I have hitherto met with'. The book was slammed by William Molyneux, but seems to have impressed Robert Plot. Otherwise, it received only a slightly baffled (but by no means completely negative) review in a scholarly journal, *Acta Eruditoruim*. When a Latinized and condensed version of Goad's conclusions appeared posthumously in 1690, it attracted no surviving comment.[21]

[19] *DNB* 8, 18–19.

[20] *Astro-Meteorologia*, 15.

[21] T. Birch, *The History of the Royal Society of London* . . . (London, 1656–57), Vol. 3, 454–55; Royal Society MS. 243 (F1), letter 35 (4 July 1678); R. T. Gunther, *Early Science in Oxford* (Oxford, 1923–45), Vol. 12, 305; *PT* 15, 930–31; *Acta Eruditorum* (Leipzig, January 1688), 22–24; Goad, *Astro-Meteorologia Sana*.

The third remaining reformer of this kind was *John Gadbury* (1628–1704). Gadbury is perhaps particularly instructive, since his life demonstrates two parallel kinds of changes, even transformations: in his political and religious views, emphasizing the general drift to conservatism in the very late 1650s and 1660s, and in his attitude to astrology, subject in this period to similar kinds of pressures. In 1648–52, Gadbury was a sometime Leveller and member of the 'Family of Love'. By the Restoration, however, the first of his considerable production of books and almanacs showed markedly Royalist tendencies. This brought him into bitter conflict with the Parliamentarian astrology Lilly, and later the Whig, John Partridge. He was twice arrested (and acquitted) on suspicion of involvement in Catholic plots. He died in 1704, a nationally notorious Jacobite and crypto-Catholic, besides astrologer.[22]

Gadbury was a deeply committed advocate of a reformed Baconian astrologer. In conscious distinction to Lilly and others, he called his own predictions 'conjectures', not prophecies, and insisted that 'the influences of the stars are purely Natural . . .'[23] He decried the more apocryphal elements in astrology, and praised the value observations and experiments. (His understanding of the latter, although not identical with the current modern definition, stands up well in relation to his contemporaries.) In this experimental spirit, Gadbury called for cooperative research on astrology, kept weather records, and published a collection of nativities comparing that of each person with their chief 'accidents'.[24] Despite his industry and high hopes, however, Gadbury came close to admitting defeat in this poignant passage from his last almanac:

> I am very much ready to part with any Errors, upon an assured conviction that they are such; yet I shall not, cannot, wholly Renounce, or bid Good Night to Astrology. . . . I have been a daily observer of Aireal Variety for almost 35 Years, as the Noble Lord Bacon directs as necessary: And though I have met with several Similitudes of Verity . . . yet, I must freely own to have met with other Arguments too hard for me to bring under a Regimental Order of Experience.[25]

With the death of Gadbury, we can regard the scientific or Baconian attempt to reform astrology as moribund. Outside the circle of its three leading members, any sympathetic Fellows of the Royal Society – such as Ashmole, Aubrey, Beale, and to a lesser extent Boyle – had themselves died by 1700, and in this respect they went unreplaced. The nearest person to a successor was Gadbury's colleague *George Parker* (1654–1743). As we might by now expect, Parker was a Tory and High Church advocate, who became Partridge's chief rival after Gadbury's death. He edited and published Gadbury's last twenty-year ephemeris. Significantly, however, he made no

22 *DNB* 7, 785–86; Capp, 308.

23 *Astrological Predictions . . . for . . . 1679*, 3–4.

24 *A Diary . . . for . . . 1665; Collectio Geniturarum* (London, 1662).

25 *A Diary . . . for . . . 1703*, A1–A3.

attempt to carry on the reform programme, but concentrated on producing almanacs aimed at the learned end, so to speak, of the popular market. But as a late member of that group nonetheless, he confirms its social and intellectual character, in various ways: his almanac for 1690 carried a commendation, for its astronomical accuracy, by Edmond Halley; and in 1704, Parker took over the editing of Eland's *Tutor to Astrology*, earlier editions of which had been published and sold by Joseph Moxon, FRS. (Moxon was responsible for reviving the defunct Society of Astrologers in 1682–83.)[26]

The social character of these reformers – Royalist, later Tory, and High Anglican or Catholic, to a man – has been noted, and so has the Baconian, statistical approach to reform that they favoured. But is there any (historically speaking) necessary connection between these two facets, or is it mere coincidence? The latter view would stretch credibility indeed. The conservative cast and interests of the Royal Society in this period – roughly, 1679–88 – are unmistakable, and the scientific reformers directly shared those values. With respect to consistency, more than reductionism, it is not surprising that they also shared a committment to Baconian ideology and (in areas where probabilistic induction was required, as distinct from experimental demonstration) the passion for quantification and correlation. In the sensitive post-Interregnum climate, such an approach was precisely intended to provide (as one historian put it) 'a form of knowledge free from the distorting effects of controversy and conflict' – and for astrology no less than more central concerns. The production of such knowledge was a key task for Restoration natural philosophy, and among its prime movers were the Royalist and moderate ex-Parliamentarian members of the Royal Society, who were simultaneously leading members of the restored Anglican Church. This was the new milieu within which the reformers were trying to obtain a niche, and hence a new lease of life, for astrology. They were also particularly affected – for methodological reasons – by the new techniques of probabilistic reasoning whose commonest application was political arithmetic: 'a distinctly Restoration doctrine, concerned . . . with supplementing the power of sovereign authority, and facilitating its exercise . . .'[27]

The 'scientific' reformers were not the only ones, however. There was another smaller but (if anything) more vociferous group, slightly later but overlapping with the first. It was led by the chief rival and personal enemy of both Gadbury and Parker (though Childrey and Goad also came in for criticism), *John Partridge* (1644–1715). His convictions were passionately Whig (Country Whig) and Protestant (Dissenter), and his popular almanac

[26] On Parker, *DNB* 15, 233–34; and Capp, 249, 322–23; Parker's *Mercurius Anglicanus* (1690); on Moxon, see Josten, I, 247 and IV, 1705.

[27] P. Buck, 'Seventeenth Century Political Arithmetic: Civil Strife and Vital Statistics', *Isis*, 68 (1977), 67–84, 67, 80; M. Hunter, *Science and Society in Restoration England* (Cambridge, 1981), esp. chapter 5; J. R. Jacob, 'Restoration Ideologies and the Royal Society', *History of Science* 18 (1980), 25–38; S. Schaffer and S. Shapin, *Leviathan and the Air Pump: Hobbes, Boyle and the Experimental Life* (Princeton, 1986), esp. chapter 7.

provided an excellent platform for their expression – usually in colourful language, with no holds barred when it came to opponents. He was forbidden to publish at the beginning of the Exclusion Crisis; and when James II acceded, he wisely departed the country for Holland, to return joyfully with William and Mary. By this time, Partridge was probably the most famous, or notorious, astrologer in England. But he was also and equally committed to a particular approach to astrology, as we shall see. He therefore carried on a vitriolic battle in print with the Tory, Anglican/Catholic and scientifically reformist astrologers, whose astrology and politics he detested equally. In 1707, however, his fortunes dipped when he had the dubious honour of being satirized by Jonathan Swift. Swift's pseudononymous *Bickerstaff* almanac predicted the imminent death of Partridge – including the date, time and means – by apparently astrological means. The day following that of the predicted demise (29 March 1707), Swift followed this up with an annonymous pamphelet, entitled 'The Accomplishment of the First of Mr Bickerstaff's Predictions . . .' which graphically described the death-bed scene, but criticized Bickerstaff for having been almost four hours astray in his reckoning. As a result, Partridge became a laughing-stock in educated and literary circles, both British and Continental. At the same time, he was obliged to cease publishing his almanac for two years, due to a disagreement with the Company of Stationers. It continued to sell well when resumed, however, and when he died he left two thousand pounds.[28]

Turning to his astrology, Partridge was the acknowledged and indefatigable leader of a radically different reform programme, less well known than the one discussed earlier. This reform was to be carried out by shedding all the accretions and innovations in astrological doctrine since Ptolemy; whether medieval, Arabic, or 'scientific', these were viewed *en tout* as corruptions, which had brought astrology into disrepute. The answer – in direct opposition to the first group's modernistic and future-oriented outlook – was to return to an ancient purity in the pre-lapsarian past: in this case, an Aristotelian and specifically Ptolemaic purity. It is also important to appreciate that from this point of view, astrology was a completely rational activity; but rational in the old Aristotelian sense. It was equally opposed to the 'irrational', magical astrology of, say Lilly, and the scientifically rational astrology of Gadbury. The latter was viewed, in fact, as a upstart corruption of true rationality. Partridge's two books (in 1693 and 1697) thus variously attacked Kepler, that 'witty man, and an Enemy to Astrology', Gadbury, 'an ignorant Reformer of Astrology', Parker, and heliocentric astrology in general. His substantive work consisted of individual nativities, attempting to demonstrate the validity of 'the true primitive Astrology'.[29] The text for this reform was, obviously, Ptolemy's *Tetrabiblos* – duly Englished and

[28] *DNB* 15, 428–30; Capp, 323. A good description of the Bickerstaff episode is in J. G. Muddiman, *The King's Journalist* (London, 1923), 248–55.

[29] *Defectio Geniturarum* (London, 1697), 3, A2, 93–94, 62v; *Opus Reformatum* (London, 1693), A3.

published in 1701 by Partridge's principal colleague, John Whalley. And the 'patron saint' of the movement was the Italian monk and astrologer, Placido Titi.[30]

John Whalley (1653–1724) was an astrologer, almanac-writer and (least surprise of all) Whig partisan who became active in Dublin in 1682. His almanacs appeared in the 1680s and '90s. He was obliged to leave Dublin during its occupation by James II, in 1689. Whalley's annotated translation of *Ptolemy's Quadripartite* was the first in English. He took the opportunity therein to warn against Gadbury's and Coley's 'voluminous spurious stuff', adding that 'Young astrologers from hence ought to take care what they read.' Later in life, Whalley apparently dropped astrology, and turned to publishing a popular newspaper-cum-scandal sheet in Dublin.[31]

The final (if honourary) member of the Ptolemaic reform programme was the curious figure of *Placido Titi* (or Placidus de Titis) (1603–68). He was a Benedictine monk who published two books (in 1650 and 1657) advancing the thesis that astrological influences are 'natural, manifest and measurable'. The similarity of language to that of the scientific reforms is entirely misleading, however; his intent was purely Aristotelian. He boldly wrote, 'I desire no other guides but Ptolemy and reason', and it is not surprising that his ecclesiastical superior felt obliged to insert a reminder of God's status, over that of the planets, as first cause. Placido also left his name on a new method of house division, based on time sectors rather than space, which he claimed to have derived from Ptolemy's prescriptions. Adopted by Partridge and Whalley, this method was immediately controversial among English astrologers, as an emblem of Placido's larger claim – advanced by his English advocates – to have re-discovered 'the true and natural Astrology'.[32]

As with the scientific reformers, the question must be asked of their rivals: what is the connection between their social character, including politics, and their astrology? In particular, why should Whigs, of all people, have been so apparently backward-looking, and – during the Whig, Latitudinarian and Newtonian ascendancy from 1689 to 1702 – so anti-science? This puzzle is actually fairly easy to clear up. Regarding the first half, it was a central part of Whig ideology that there existed an Ancient Constitution, which justified the rulership of a king by the consent of the people (without actually rendering the relationship contractual). Articulated principally by Sir Matthew Hale, this idea was bitterly disputed by Tory scholars. The problem was, naturally, particularly acute around the time of the Glorious Revolution, and for some time afterwards. Furthermore, the experience of Whigs as the 'Country' party led them to be obsessed with the issue of

30 L. Thorndike's apt phrase, although mistakenly applied to Placido in relation to all late seventeenth century astrologers; *A History of Magic and Experimental Science* (N.Y., 1923–58), chapter XXXII, p. 302.

31 *Ptolemy's Quadripartite* (London, 1701), A3r, A6r; A. Webb, *A Compendium of Irish Biography* (Dublin, 1878), 560.

32 *Primum Mobile*, a recently reprinted translation (London, 1983) of *Tabulae Primi Mobilis* (Padua, 1657), 47, 14; also *Physiomathematica* (Milan, 1650).

'Court' corruption.[33] It seems plausible and consistent, then, that for the Whig reformers Ptolemy's *Tetrabiblos* was seen as astrology's Ancient Constitution. Such an attitude would also have resonated with Partridge's and Whalley's uncompromising Protestantism, Biblical and highly anti-clerical. In their view, Ptolemy was a warrant to condemn and sweep away all the monarchical-Popish-tyrannical corruptions of astrology, brought about by astrologers who had strayed from the fundamental text. Indeed, this programme on the part of Partridge and Whalley – who were clearly to the 'left' (as we would say) of the Latitudinarian and Junto Whigs – was so far backward-looking as to be radical, in the original sense of the word: a return to a former pristine state. (Here, I am reminded of another advocacy of a return to Aristotelain purity, perhaps impelled by radical motives: that of Henry Stubbe in the early 1670s.)[34]

At any rate, these reformers had more in common with the vocal but powerless Stubbs than with the hero of the Latitudinarian Whigs, Newton; which brings me to the second half of the above-mentioned puzzle, their hostility to natural philosophy. In this connection, one must again recall Partridge's and Whalley's positions, as spokesmen for values which were clearly radical-Country Whig and almost certainly Dissenter. (This was neatly expressed in Partridge's first book, when he recalled with pleasure one of Cromwell's victories which had included the capture of nine parsons.)[35] Given such an overall marginal position, both socially and intellectually, it is hardly surprising that they found even Latitudinarian Newtonian natural philosophy, let alone the still more conservative Royalist strand, unattractive.

Having finished this tour of the two reform programmes, it may reasonably be asked: who have I left out, and who doesn't fit into this schema? The answer is, of astrological reformers: surprisingly few. Certainly one figure, more supporter than active reformer, approaches anomalous status. That is Sir Edward Dering (half-brother to the somewhat better-known Lord Commissioner of the Treasury), a London merchant and staunch Royalist, who cheered Charles II in exile with astrological predictions of the latter's eventual triumph. One of Gadbury's books was dedicated to '(my ever Honour'd Friend) Sir Edward Dering'. But one of Partridge's books also carried a laudatory preface by Dering; and his alamanac for 1697 claimed Dering as a patron.[36] Dering's ecumenism was underlined by his support of the briefly resurrected London Society of Astrologers; he acted as a Steward at the last Astrologers' Feast, in 1683.[37]

I should also mention *The Marrow of Astrology* (1683), wherein John Bishop thanks Robert Boyle for 'many and signal Favours', but goes on to extoll the

33 J. G. A. Pocock, *The Ancient Constitution and the Feudal Law: A Study in English Historical Thought in the Seventeenth Century* (Cambridge, 1957).
34 J. R. Jacob, *Henry Stubbe, Radical Protestantism and the Early Enlightenment* (Cambridge, 1983).
35 Partridge (1693), 38.
36 Josten, I, 241–42; V, 1678, 1712; Partridge (1697); *Merlinus Liberatus* (1697); Gadbury, *Cardines Coeli* (London, 1684).
37 Josten, I, 250.

virtues of Placidean astrology.[38] But Bishop does not seem to have been very active (this is his only surviving publication), and there is no record of Boyle's side of it. Finally, another possible exception is the recently-discovered Rye astrologer, Samuel Jeake (1652–1699); at the time of writing, it is too soon to say.[39]

In the schema's favour, it is worth noting that the astrologer-astronomers associated with the early Royal Society – uniformly staunch heliocentrists, and sympathetic to reformist (*e.g.* Keplerian) astrology – were equally uniformly Royalists and conservative Anglicans. I am referring to Vincent Wing (1619–68) and Thomas Streete (1621–1689), although a similar point could be made of Richard Townley.

The last decade of the seventeenth century and first of the eighteenth saw a series of bitter, often scurrilous, exchanges between the leading reformers of each camp: Partridge and Whalley on the one hand, and Gadbury and Parker on the other. These attacks took in both personal and astrological levels – often simultaneously, as when Partridge accused Parker of whipping his wife 'the heliocentric way'. In reply to the charge of 'Adulterous Innovations' (in Whalley's words), Parker assailed the assumption that no progress had been made since classical times. He asked why Ptolemy should be thought infallible in astrology, 'and yet so deficient in astronomy . . .'[40] A late contribution to the debate was Richard Gibson's *Flagellum Placidianum, or A Whip for Placidianism* (1711). Gibson taunted the Ptolemaic astrologers for practising 'Monkish Astrology' (referring to Placido's role), and trying 'to scourge us (the true Roman method) for our Heresie . . .'[41] But after the Bickerstaff episode, Partridge seems to have lost some of his taste for polemic; while Whalley, in Dublin, went on to experiment with non-astrological newsheets. Not long after its rival scientific reform, the Ptolemaic reform too – despite their deep differences – had expired as completely.[42]

Otherwise, post-Restoration astrologers simply carried on their (slowly shrinking) practices, and some still produced an annual almanac: for example, the astrologer-physicians Richard Saunders (16113–75) and William Salmon (1664–1713), and Henry Coley (1633–1707). Others continued to engage in polemical prophecy, such as the Protestant John Holwell's prediction of the imminent collapse (in 1682) of Catholic Europe, and the Catholic John Merrifield's scathing reply.[43] Although he was William Lilly's adopted son, and heir to his practice and alamanac, Coley – by contrast – steered clear of such internecine bickering. He seems genuinely to have been, in Parker's

38 Written with Richard Kirby; Part II, 92.

39 See Michael Hunter's forthcoming study of Jeake, to be published by Oxford University Press.

40 Partridge, *Flagitiosis Mercurius Flagellatus* (London, 1697), 28; Whalley, 93; Parker, *Mercurius Anglicanus* (London, 1697), C8r.

41 Pp. 2, 8.

42 Allowing for revivals in the twentieth and eighteenth centuries, respectively.

43 Holwell, *Catastrophe Mundi* (London, 1682); Merrifield, *Catastasis Mundi* (London, 1684).

words, 'a person of quiet and peaceable disposition.'[44] Even so, he could not escape the fate overtaking astrology at this time. John Aubrey records a revealing incident in a letter to Anthony à Wood, dated 21 October 1693: 'I forgot to tell you if I just called upon Mr Coley as I was going out of Town; and he is very angry with you, because you term Astrologers *Conjurers*.' It was precisely the collapse of both astrological reforms that left astrologers in this defenseless position, with no avenues of appeal.[45]

I have left till now the most interesting problem of all: the reasons for the failure of both programmes to win the approval of the authorities. That failure in turn relates to astrology's decline in status, if not necessarily popularity – quite a different matter. (I need hardly add that it explains nothing to refer anachronistically or positivistically to astrology being, 'after all', false. We now hold the Earth to orbit the Sun, but for centuries the reverse hypothesis was equally common knowledge; the 'truth' of the matter explains neither that fact, nor its reversal some time after Copernicus. Equally, to posit the 'progress' of science is an empty gesture – besides being both anachronistic and circular, it cannot account for the specificity of such historical changes.)

Concerning the crucial context for both efforts, I have already argued that after the Restoration, the climate for astrology took a serious turn for the worse. Among the propertied and enfranchised classes, memories remained fresh of sectarian 'enthusiasm' and its apparent consequences – including the roles of prominent radical and/or Parliamentarian astrologers like Booker, Lilly and Culpeper. Subsequent events, like the Exclusion Crisis, Popish Plot, and Glorious Revolution of 1688–89, re-stimulated such fears, and ensured that the search for stability continued in the following decades. At the same time, astrologers like Gadbury and Partridge (in their polemical roles) ensured that astrologers – while perhaps less influential – remained vocally evident in the turmoil. Given all this, there is every reason to conclude that astrology, in the final decades of the century, had lost none of its older associations with threats (real or perceived) from Catholic, pantheistic, and/or deist heresies. Such an interpretation is borne out, for example, in a book by the influential Anglican divine and scholar, Henry More. The substance of his *Tetractys Anti-Astrologica* had been written in 1661; but he decided, significantly, to re-publish it in 1681. In addition to trotting out the Augustinian explanation of correct astrological predictions by the intervention of 'Aiery Goblins,' More was at pains to convey that astrology had 'whole Religions come on and go off according to the Configurations of Heaven, and that Christ himself and his Religion is subjected to the same laws . . .' This clearly implied a

> God, who keeps his Throne in the Eighth Sphere, and intermeddles not with humane affairs in any particular way, but aloof off hands down,

44 *An Ephemeris . . . for . . . 1699*, 93v. Coley sponsored an isolated reform tract by John Kendall, . . . *The Measure of Time in Directions* . . . (London, 1684). There are other such texts that I haven't mentioned here.

45 Bodleian Wood MS. F.51, f. 6v.

> by the help and mediation of the Celestial Intelligences and the power of the Stars, some general casts of Providence upon the generations of the Earth.

In his own words, this was plainly 'Aristotelian Atheism', and exactly the same concern that Boyle had expressed fifteen years earlier.[46]

It is hardly surprising that the same kinds of fears were shared by contemporaneous natural philosophers. Their ambivalence has been noted; but even where outright hostility to astrology was expressed, it proves impossible cleanly to separate the criticisms (as 'Whiggish' modern historians would like) into boxes marked 'epistemological', 'political', and so on.[47] For example, John Flamsteed objected to 'the practices of Astrologers', notably their *ad hoc* justification of predictions, and their endless mutual disagreements. But such criticisms derived their force from the fact that the former precisely parelleled the illuminationism and antinomianism of the sects, while the latter reflected their endless fission. That is why Flamsteed could object in the same breath to 'pernicious predictions of ye Weather *and* State affaires', and their esteem among 'the superstitious Vulgar'.[48] Similarly, the mathematician-astronomer David Gregory, in an unpublished manuscript written in 1686, objected to astrology as uncertain and fraudulent. Seamlessly, he added that

> it is hardly believable what great damage those predictions invented by Astrologers and ascribed to the stars have caused to our best kings, Charles I and II . . . Therefore we prohibit Astrology to take a place in our Astronomy, since it is supported by no solid fundament, but stands on the utterly ridiculous opinions of certain people, opinions that are so framed as to promote the attempts of men tending to form factions.[49]

In this context, the onus was clearly on the scientific reformers, especially, to demonstrate that astrology was *not* just a purely plastic subject, which could be endlessly manipulated by 'irresponsible' interpreters for extreme political or religious ends. That is, they had to show, to general educated satisfaction, that it was possible to produce stable (and if possible, useful) empirical regularities. And that was what – despite heroic efforts, the nominal support of Bacon, the ambivalence of many natural philosophers, and the relative sophistication of their methodology – they failed to do. (That failure was, of course, constituted just by the agreement among those with the necessary authority that the reformers had indeed failed. This in turn confirmed the suspicions and fears about astrology just described.) Certainly

[46] *Tetractys Anti-Astrologica* . . . (London, 1681), 134, 4, 1; *cf.* R. Boyle, letter to Stubbe (9 March 1666), in T. Birch, *The Works of the Honourable Robert Boyle* (London, 1772), I, lxxix.

[47] *Eg.* R. F. Jones, *Ancients and Moderns* . . . (Gloucester, MA, 1961), and B. Vickers (ed.), *Occult and Scientific Mentalities in the Renaissance* (Cambridge, 1984).

[48] The identity of Flamsteed as the author of this anti-astrological polemic was discovered by Michael Hunter; see his paper in this volume. My quotes are from pp. 289–90; my emphasis. (It should be noted that Dr Hunter places a different interpretation on the passage quoted.)

[49] Christ Church College, Oxford MS. 133, ff. 47–50; translated from the Latin.

if anyone could see in the stars anything they wanted, without any apparent control by 'nature', then astrology was worse than useless in the search for knowledge that could be publically demonstrated or verified, and that would therefore underwrite a consensus. Once it was firmly in place, this perception of astrology set the final seal on its marginality – which remains more-or-less intact today. But I would like to stress that it was only at this time, and in relation to the failed scientific reform of astrology, that it *became* 'occult' in the modern sense of the word. Up till now, the issue had been (to many if not most minds) undecided. (At the risk of indelicacy, may I point out the corollary: that modern astrology, in its essentials, is partly the creation of early modern science.)

Astrology's fate as an occult pursuit was further strengthened after 1687, by Newton's theory of universal gravitation. As Hutchison has described, this theory involved a re-definition of 'occult', in which *acceptable* occult causes were both universal and permitted one to 'detect [their] effects reliably'.[50] While this is far too large a subject to be explored here, it is permissable to speculate that in this way, Newton was able to draw on the conceptual resources of both alchemy (in his matter-theory) and astrology (in action-at-a-distance), while virtually devastating those subjects as known till then. Certainly he was able to use astrological ideas about comets – suitably transplanted and re-described – against astrologers.[51] At any rate, astrology was excluded from legitimate natural knowledge on both criteria. Gadbury and Goad had been unable to produce convincing empirical effects; and the concept of gravity undercut the attribution of different qualities or principles to each of the planets. By the 1690s, therefore, astrology stood more open than ever to the charge of pantheistic ('Conjurers') or deistic ('Aristotelian') naturalism – as opposed to a correctly Anglican scientism.

These considerations apply with even more force to the second group of reformers and their programme. That is evident from Partridge's and Whalley's still more marginal and extreme position, socially and intellectually, *vis-à-vis* their rivals. And such political extremism was matched by their astrology, which made no concessions to the contemporaneous intellectual climate, but taking its mandate from Ptolemy, relied on increasingly outmoded and illegitimate Aristotelian senses of 'natural' and 'rational'. Their failure is therefore largely explainable in the same terms as that of the first group. For if Latitudinarians and Whigs were sensitive to threats of seditious atheism, and unable to perceive astrology differently, how much more so were the High Church and Tories? There was simply no place left for astrology – reformed or otherwise – in either in Newtonian synthesis of the 1690s or the conservative Whig ascendancy beginning in 1714 – let alone in the intervening Tory rule.

Perhaps as importantly, astrology must have appeared as more of an encumbrance than otherwise even when in opposition. Only someone with as little to lose as Toland – we may speculate – could have found astrology

[50] K. Hutchison, 'What Happened to Occult Qualities in the Scientific Revolution?', *Isis* 73 (1982), 233–53, 251.
[51] See Simon Schaffer's paper in this volume.

attractive; there is no evidence he did, although a distant link could be perceived with his interest in Bruno's Hermeticism.[52] Resistance to Newton and Latitudinarianism from the 'right', on the other hand, was partly based precisely on objections (by men like Hutchinson and North) to the occult nature of gravitational 'attraction'. They perceived in this doctrine a pagan heresy, dressed in modern clothes. Less promising ground for an astrological revival is difficult to imagine.

When considering the decline of astrology in early modern England, then – including attempted reforms – all the reasons usually given still apply. That is: the hostility of the Protestant Church, the effects of new scientific discoveries and theories, urbanization, literacy, a new 'ideology of self-help', astrology's failure to adapt and/or become institutionalized and its radical associations. Insofar as the importance of the last has (in my view) been underestimated, and its close relationship with the other factors neglected, to these reasons must be added the consequences of astrology's entanglements in the intense political and religious divisions of Restoration, as well as Revolutionary, England.[53]

However, there is also a broader social context for the changes affecting astrology that I have discussed. That context is symbolized by the agreement about the undesirability of astrologers between two leading natural philosophers: Isaac Newton, an anti-trinitarian Arian and active Whig, and David Gregory, a high church Jacobite. The agreement across parties (and in this case extremes of parties) points to a deeper process, the formation of a new social mainstream. That process – which left astrology stranded on the 'wrong' side – consisted of a 'dissociation of polite and plebian cultures which has been identified as a fundamental feature of English society in the eighteenth century'. As Wrightson elaborates,

> a deep social cleavage of a new kind had opened up in English society. It was not simply between wealth and poverty, but between respectable and plebeian cultures, and it followed a line which divided not the gentry and the common people, but the 'better sort' and the mass of labouring poor.[54]

A key element in this development was the rise of the middle and professional classes, beginning around the end of the seventeenth century. It is hardly possible to discuss such a subject in this amount of space; I only want to note that these newly important classes basically adopted the ideology of their social superiors – just the people, as we have seen, with the

[52] M. C. Jacob, *The Newtonians and the English Revolution 1689–1720* (Hassocks, SX, 1976), chapter 6.

[53] The quote is from Thomas, 797; besides Thomas and Capp, see P. W. G. Wright, 'Astrology in Mid-Seventeenth Century England: A Sociological Analysis' (unpublished Ph.D. thesis, University of London, 1984).

[54] K. Wrightson, *English Society 1580–1680* (London, 1982), 227; drawing on E. P. Thompson, 'Patrician Society, Plebeian Culture', *Journal of Social History*, 7 (1974), 382–405.

least reason to like or trust astrology. (It took nearly a century before they began to express any real independence.) And their authentic voice was that of the literati: Swift, Addison and Steele.[55] This is the fundamental context within which to view Swift's attack on astrology. His attack was indeed 'rational' and 'critical', but it was not just that; it was also a Tory Bishop addressing a leading Whig populist and radical. Even more than that, however, Swift wrote as a representative of 'the wise and learned, who *alone* can judge whether there by any truth in this science . . .' which 'none but the ignorant vulgar' credit.[56] It is a sign of the power of this process, and its intellectual hegemony, that astrology was so thoroughly banished from respectable culture. Despite its survival, virtually intact, among rural labouring classes throughout the eighteenth century, and despite the appearance in the nineteenth of a middle-class urban form that is still very much with us, astrology's marginal status remains essentially unchanged today. On the other hand, its durability, in the face of such entrenched and habitual hostility, is also remarkable.

This paper has (I hope) revealed the superficiality and self-interestedness of much of the received wisdom about astrology: that it can be isolated from its historical context, because it 'is' (and therefore, essentially, always has been) subjective, irrational and/or superstitious – unlike science, purportedly its opposite in all respects; that scientific and educated criticism of astrology was purely that, in all the same respects, and had (in and of itself) devastating effects on astrology; or even that astrology's decline took place due to mysterious intellectual changes, whose coherence and grounding cannot be specified. Perhaps I may also be permitted to cast doubt on another view: that astrology's survival, too, is merely a mysterious aberration, with nothing interesting to tell us about ourselves and our present world.

[55] See T. Eagleton, *The Function of Criticism* (London, 1984). Addison and Steele continued Swift's onslaught in the *Tatler*.

[56] Swift, *The Accomplishment of the First of Mr Bickerstaff's predictions . . .* (London, 1708); my emphasis. Of course, Swift often had an ambivalent relationship to others within his class; that doesn't affect the point made here.

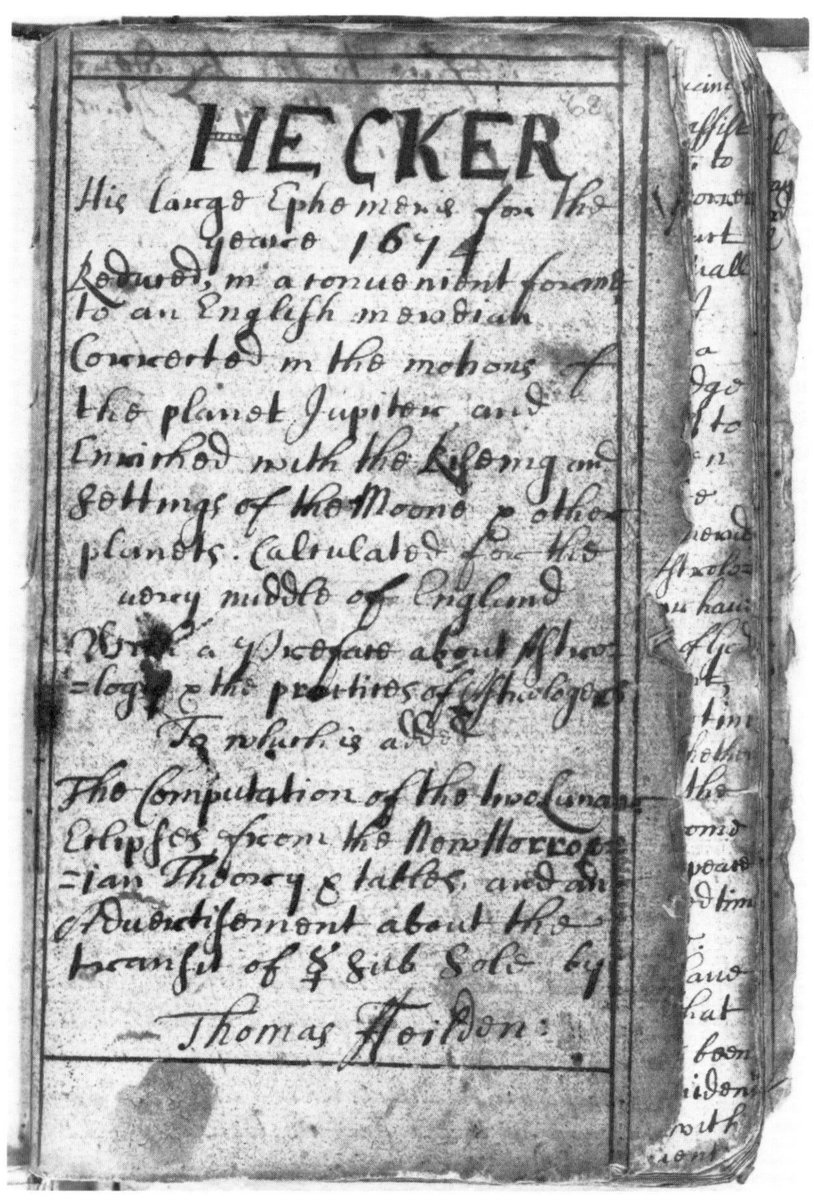

HECKER
His large Ephemeris for the
yeare 1674
Reduced, in a convenient forme
to an English meridian
Corrected in the motions of
the planet Jupiter and
Enriched with the Risings and
settings of the Moone & other
planets. Calculated for the
very middle of England
With a Preface about Astro=
=logy & the practices of Astrologers
To which is added
The Computation of the two Lunar
Eclipses from the New Horrox=
=ian Theory & tables, and an
Advertisement about the
transit of ☿ sub Sole by
Thomas Feilden:

Plate 1. The title-page of Flamsteed's 'Hecker', RGO MS 1/76, fol. 68.

SCIENCE AND ASTROLOGY IN SEVENTEENTH-CENTURY ENGLAND: AN UNPUBLISHED POLEMIC BY JOHN FLAMSTEED

Michael Hunter

In considering the decline of educated belief in astrology in late seventeenth- and early eighteenth-century England, more than one author has drawn attention to an apparent paradox. Though the new science was clearly linked to changing attitudes towards astrology, 'curiously enough, scarcely anyone attempted a serious refutation of astrology in the light of the new principles'. Apart from a couple of polemics translated into English from French, Keith Thomas and others who have chronicled astrology's decline have found only critical asides concerning astrology on the part of those associated with the new science, rather than outright assaults.[1]

For this, various reasons have been propounded. Bernard Capp has suggested that scientists' restraint regarding astrology was probably due, if not to a belief 'that its influence in educated society was crumbling, and that no frontal assault was needed', then to 'a lingering belief that the pseudo-science contained an element of truth'. Peter Wright, by contrast, has claimed that the absence of such assaults is connected with the fact that astrology was not really in competition with the new science, being a 'craft' and hence having a somewhat different rationale and sphere of activity.[2]

In fact, however, both the statement about the lack of such attacks and these explanations of it need to be modified. Though it is true that attacks on astrology by scientists are rare, one cogent and sustained critique of the art by a major scientist of the late seventeenth century does survive, though it

[1] Keith Thomas, *Religion and the Decline of Magic* (London, 1971), p. 352. Cf. Peter Wright, 'Astrology and Science in Seventeenth-century England', *Social Studies of Science*, 5 (1975), 404, 413–4; Bernard Capp, *Astrology and the Popular Press: English Almanacs, 1500–1800* (London, 1979), pp. 190, 278.

[2] Capp, Astrology, p. 278; Wright, 'Astrology', pp. 399–422 passim, esp. 410f.

The following abbreviations are used throughout these notes:

EL	Royal Society Early Letters
Oldenburg	A. R. and M. B. Hall, eds, *The Correspondence of Henry Oldenburg* (13 vols, Madison, Milwaukee and London, 1965–86)
Phil. Trans.	*Philosophical Transactions of the Royal Society*
RGO	Royal Greenwich Observatory
RS	Royal Society
Rigaud	S. J. Rigaud, ed., *Correspondence of Scientific Men of the Seventeenth Century* (2 vols, Oxford, 1841)

has hitherto been almost entirely overlooked.[3] This was compiled in the early 1670s by John Flamsteed, the future first Astronomer Royal, a scientist who, as we shall see, had a considerable knowledge of astrology and strong views on it.

That Flamsteed wrote such a work has long been familiar, since he explains in his autobiography how in 1673 he wrote an ephemeris 'wherein I showed the falsity of Astrology, and the ignorance of those who pretended to it: wherein I gave a table of the moon's risings and settings, carefully calculated; together with the eclipses and appulses of the moon and planets to fixed stars'.[4] In the past, this work has been presumed lost, evidently because it was not among the Flamsteed manuscripts at the Royal Greenwich Observatory when Francis Baily worked on them in the early nineteenth century, and was therefore not noted by him in his *Account of the Revd. John Flamsteed* (1835).[5] The manuscript in question – a small stitched paperbook written in Flamsteed's own hand – forms part of a group of miscellaneous sixteenth- and seventeenth-century items added to the main group of Flamsteed Papers by Sir George Biddell Airy later in the nineteenth century.[6] The object of the present paper is to provide a text of the relevant portion of Flamsteed's ephemeris, to comment on it, and to place it in the context of Flamsteed's life, thought and milieu.

The history of Flamsteed's ambivalent attitude towards astrology can be traced back to his formative years. Its background lies in the overlap which it is now commonplace to acknowledge existed between astrology and astronomy in this period. Astrological analysis required precise information on the positions and movements of the heavenly bodies, and, particularly in the 1650s and 1660s, a tradition of accurate computational astronomy emerged in England in the writings of such astrologers as Vincent Wing, Jeremy Shakerley and Thomas Streete, from which Flamsteed's own work on observational astronomy can be seen to emerge.[7] It is thus hardly surprising that the works of both Wing and Streete were to be found among the books that Flamsteed owned, together with sets of tables by such continental astrologers as Andreas Argoli and G. A. Magini.[8] Moreover Flamsteed repeatedly used and assessed the data of authors like Streete, which

[3] The only existing account is in Lesley Murdin, *Under Newton's Shadow: Astronomical Practices in the Seventeenth Century* (Bristol, 1985), pp. 136–7, where the work is, however, incorrectly described as 'a copy of Thomas Hecker's *Ephemeris*, or astronomical table, taken from his astrological almanac for 1674'.

[4] Francis Baily, *An Account of the Revd John Flamsteed* (London, 1835), p. 34.

[5] M. E. Bowden, 'The Scientific Revolution in Astrology: the English Reformers, 1558–1686' (Yale Ph.D., 1974), pp. 54, 60–1; Capp, *Astrology*, p. 425 n. 59.

[6] These are now shelved as RGO MSS 1/75–76. For a description of the 'Hecker', which forms section 5 of MS 1/76, see below, note 87.

[7] See J. T. Kelly, 'Practical Astronomy during the Seventeenth century: a Study of Almanac-Makers in America and England' (Harvard Ph.D., 1977), ch. 2 and pp. 240–1.

[8] E. G. Forbes, 'The Library of the Rev. John Flamsteed, F.R.S., First Astronomer Royal', *Notes and Records of the Royal Society*, 28 (1973), 119–43.

he saw as superior to that of some continental astronomers, though falling short of the precision to which he himself aspired and to which he was to devote his life's work. In the 1660s, he also corresponded with Wing.[9]

Information on the earliest phase of the development of Flamsteed's interests may be derived from the autobiography that he wrote partly at the time and partly later; this can be supplemented by what he says about his early acquaintance with astrology in the text below. From these sources it is clear that, as was the case with Newton and others, Flamsteed's initial encounter with the new science took place in an astrological context. The first astronomical book that he mentions his acquaintance with was the version of Shakerley's *Tabulae Britannicae* (1653) – which was based on the calculations of Jeremiah Horrocks and the approximation techniques of Ismael Boulliau – that was included by the astrologer, John Gadbury, in his *Genethlialogia; or, the Doctrine of Nativities* (1658); this was used by Flamsteed's friend, William Litchford, for calculating the positions of the planets.[10]

Having noted Litchford's possession of this work, Flamsteed claimed in a revealing aside that '(because I would not be seen with Mr Gadbury's book, lest I should be suspected astrological) I bought Mr Street's *Caroline Tables*', in other words Streete's *Astronomia Carolina* (1661). Though by a practicing astrologer, the latter was devoted entirely to computational astronomy without the astrological component which accompanied the tables in Gadbury's book, while it also improved on the Shakerley/Gadbury tables in accuracy although based on almost identical sources.[11]

What Flamsteed recorded about his acquaintance with Gadbury's work in his autobiography is, however, incomplete. This is shown by the fuller account given in the text below, where he refers to borrowing from 'a freind well versed in Astrology' (presumably Litchford) not only Gadbury's *Doctrine of Nativities* but also his *Collectio Geniturarum* (1662) – a collection of natal horoscopes of famous people with a commentary and conclusions – adding that he learnt the complex rules of astrology at this time since he was 'then much persuaded of its certeinty' (ff. 69v–70).

This early favourable attitude is borne out by a further passage in his autobiography in which he records how in 1665 'I also busied myself very much in calculating the nativities of several of my friends and acquaintance, which I have since corrected, and shall transcribe on a convenient paper'. In contrast to this tolerance, however, he subsequently reverted to a more reserved stance, explaining how in 1666 'I spent some part of my time in astrological studies, but so as my labours were rather astronomical'. After again noting how he cast nativities, and giving his views on the differing

[9] See *Oldenburg*, vi, 514–5, vii, 303–4, viii, 207–8, 268–9, 366, ix, 3, 177–8, 469, x, 157, 370; Rigaud, ii, 77, 81, 86, 88, 89, 96; and below, text, f. 80v. For Flamsteed's criticism of such men, see, for instance, Rigaud, ii, 80, 87, 89, 91, 93, 110. On Flamsteed's correspondence with Wing see ibid., ii, 92–3.

[10] Baily, *Account*, p. 11. On Newton, see F. E. Manuel, *A Portrait of Isaac Newton* (Cambridge, Mass., 1968), p. 81, and R. S. Westfall, *Never at Rest* (Cambridge, 1980), p. 88.

[11] See Kelly, 'Practical Astronomy', p. 51. It may be noted that in an overlapping passage of his autobiography Flamsteed adds that he obtained the copy of Streete in exchange for 'a piece of Astrology I found amongst my father's books' (Baily, *Account*, p. 26).

'arcs of directions' used for calculating the 'measure of time' – a topic on which further information is provided in the text below – he added equivocally: 'In fine, I found astrology to give generally strong conjectural hints, not perfect declarations'.[12]

This ambivalence is also in evidence in the manner in which Flamsteed first came into contact with London scientific circles in 1669, namely by sending papers on impending celestial phenomena to an astrological enthusiast and antiquary, John Stansby, a friend of the leading patron of astrologers of the day, Elias Ashmole, through whom they were presented to the Royal Society.[13]

Flamsteed's own recollection of this episode may be supplemented by a further retrospective and in detail clearly apocryphal account of his first encounter with London intellectuals, which evidently represents gossip in circulation in 1715, when it was recorded by the antiquary, Thomas Hearne. Hearne tells us that initially Flamsteed wrote to William Lilly, 'the famous Figure-Flinger, & took occasion to correct many of his Errors and Mistakes'. On this, Hearne continues, Lilly, Sir George Wharton, the royalist astrologer, and Sir Jonas Moore, Surveyor General of the Ordnance and Flamsteed's future patron, met Flamsteed halfway between Derby and London – a folklorish detail, perhaps – where they 'were so well satisfyed with Flamsteed's Skill in the art of Astronomy, that at their Return to London they recommended him to King Charles the IId, as a man of great Abilities in the foresaid profession'.[14] Contemporaries were thus conscious that Flamsteed had had significant links with astrologers in his early years.

Moreover there is evidence of a continuing prevarication regarding astrology on Flamsteed's part in periods subsequent to his formative years. One oddity is a horoscope which Flamsteed cast for the date of the foundation of the new Observatory at Greenwich in 1675, which faces a plan of the building in one of Flamsteed's early volumes of his observations there, but which has been endorsed by Flamsteed in pencil: 'Risum teneatis, amici'.[15] There are also notes that he made of the exact times of birth of Edmond Halley, Johann Hevelius, the Danzig astronomer, and others, presumably with astrological purposes in mind.[16]

In 1678 Flamsteed is to be found taking an interest in the research of John Goad on links between the weather and the aspects and positions of the planets. Goad had spent thirty years making observations on this topic, and Flamsteed sent a transcript of a selection of Goad's predictions to his correspondent, Richard Towneley, remarking how Goad's 'conjectures I find come much nearer truth then any I have hitherto met with'. 'You know I put

12 Baily, *Account*, pp. 12, 22. Cf. below, f. 70.

13 Baily, *Account*, p. 23; C. H. Josten, *Elias Ashmole, 1617–1692* (Oxford, 1966), iii, 1047–8 n. 5; W. H. Black, *A Catalogue of the Manuscripts bequeathed unto the University of Oxford by Elias Ashmole* (Oxford, 1845–66), index s.v. 'Stansby'.

14 Thomas Hearne, *Remarks and Collections*, 5 (1714–16), ed. D. W. Rannie (Oxford Historical Society, vol. 42, 1901), 130–1.

15 RGO MS 1/18, fol. 2v. This item is noted in Baily, *Account*, p. 34n., and reproduced in Murdin, *Under Newton's Shadow*, p. 137.

16 RGO MS 1/41 fol. 192. These notes apparently date from c.1680.

no Confidence in Astrology', he continued, concluding his letter: 'yet dare I not wholly deny the influences of the stars since they are too sensibly imprest on. Your affectionate servant, J[ohn] Flamsteed'. Moreover, even when citing annual almanacs to criticise them for their astronomical inaccuracy – as in his statement to Towneley the following April that 'Most of our Almanacks Tanners excepted tel us there will be no Eclipse visible this yeare in England' – Flamsteed shows that he was well-acquainted with the literature that he attacked.[17]

In the following decades, Flamsteed was to be found helping the astrologers John Wing and George Parker. In the case of Wing, Flamsteed assisted him with accurate data about planetary movements for a set of astronomical tables that he began to prepare in the late 1680s.[18] Similarly, in the 1690s Flamsteed provided astronomical information for insertion in Parker's almanac which, though more than usually accurate, was nevertheless quite unrepentantly astrological in its rationale. It may be symptomatic, however, that, though on two occasions actually naming Flamsteed, on others Parker showed a circumspection in this regard which could have been imposed on him by Flamsteed, referring to Flamsteed and his material by such formulae as 'the same hand that imparted those of former years'. That Flamsteed had objected to being mentioned in person is strongly suggested by Parker's reference to him in the issue for 1695, where, having alluded to 'the Correct Observations of the most celebrated Mathematician in our Nation', he continued: 'to publish his Name matters not, I have not his license for so doing'.[19]

Whatever his views on such publicity, Flamsteed was thus not so hostile to astrology that he refused to have any truck whatever with its practitioners, and the reason for this was surely the state of affairs already noted – that astrology had stimulated and continued to stimulate worthwhile astronomical work in the most proficient of its enthusiasts, which was of potential significance to all concerned with the heavenly bodies and their movements, regardless of their view on astrology. Clearly Flamsteed felt that it was good

[17] Flamsteed to Towneley, 4 July 1678, 5 April 1679, RS MS 243, nos 35, 38. For Tanner's almanac, see Capp, *Astrology*, p. 381.

[18] See George Parker, *A Double Ephemeris for 1703*, sigs B1, B3–5.

[19] George Parker, *Mercurius Anglicanus*, issues for 1692, sig. A3, 1698, sig. A1v (where Flamsteed is named), 1695, sig. A2, *An Ephemeris of the Coelestial Motions and Aspects*, 1696, sig. A3; see also *Mercurius Anglicanus*, 1691, sig. B5, 1694, sig. [A8]. For a contemporary comment on this collaboration and on Parker's quarrelling with one of his collaborators, see John Partridge, *Flagitiosus Mercurius Flagellatus* (London, 1697), p. '15' [i.e., 14]. The fact that Flamsteed's name was suppressed did not prevent contemporaries realising that the data was his (cf. Newton to Flamsteed, 26 January 1695, in J. F. Scott, ed., *The Correspondence of Isaac Newton*, iv (Cambridge, 1967), 74), while a further motive for withholding his name may have been a feeling that the data was lacking in precision, as revealed in this context (ibid., pp. 69, 71, 74). After the 1690s Flamsteed evidently ceased to assist Parker, and there are various critical comments concerning Parker's ephemerides in his correspondence: see Baily, *Acount*, e.g. pp. 265, 297, 305, 312, 313, 315, 328. In his 1717 almanac, Parker apparently attacked Flamsteed (see Sharp to Flamsteed, 28 May 1717, RGO MS 1/34, fol. 135, and evidently also Kennett to Flamsteed, 16 November 1716, RGO MS 1/37, fol. 68), but I have not managed to locate a copy of the almanac in question.

for correct data about the heavenly bodies to be disseminated, whatever the medium by which this was achieved. Indeed, one of the reasons why Flamsteed was to argue for the need for a 'reformed' almanac in the early 1670s was that the most accurate English ephemeris extant, that of Vincent Wing, only went up to 1671.[20]

Such an attitude was shared by others: thus Edmond Halley not only assisted Parker with data for his almanacs, like Flamsteed, but also supplied the first of these with an approving testimonial.[21] In addition, as has been pointed out elsewhere, an openmindedness about the possible efficacy of astrology such as that expressed by Flamsteed in his 1678 letter to Towneley was common among devotees of the new science. Halley professed at least an uncertainty about the validity of the art, while others, such as Robert Boyle, were more positive in their claims that astrological influences were worthy of fuller investigation, a topic which Boyle aired particularly in two tracts of the early 1670s; still others active in scientific circles, such as John Aubrey, were themselves practising astrologers.[22]

In contrast to these, such tolerance as Flamsteed showed towards astrology was balanced almost from the outset by an outspoken hostility, seen most explicitly in the text published here but also in evidence in critical comments on the subject throughout his career. In 1679, the year after the letter to Towneley in which he expressed his openminded or even approving attitude towards Goad's investigations, it was reported at a meeting of the Royal Society that Flamsteed 'had examined several of Dr Goad's predictions, but had not found one of them true'.[23] An overtly hostile stance is in evidence in the 1690s – during the years in which Flamsteed was providing data for Parker's almanacs – when Flamsteed was approached for advice on astronomical calculations by the Rye astrologer, Samuel Jeake. For, correcting an error which Jeake had made, Flamsteed went on to 'take Occasion from his Mistake to shew him how he may Imploy his time much better then in the Study of Astrology'.[24]

Reverting to Flamsteed's earliest years, his anxiety not to be 'suspected astrological' as early as 1664 has already been noted, while an openly critical attitude even at this early date is suggested by his compilation of an '*Almanac Burlesque*' for 1666, of which nothing is unfortunately known except that he

[20] Flamsteed to Brouncker et al, 24 November 1669, Rigaud, ii, 88–9.

[21] Parker, *Mercurius Anglicanus*, 1690, sig. [A2v]. The help of his 'friend' Halley (a formula not used to describe Flamsteed) is acknowledged by Parker in *Mercurius Anglicanus*, passim.

[22] See Michael Hunter, *John Aubrey and the Realm of Learning* (London, 1975), p. 140 and ch. 2, passim; Capp, *Astrology*, pp. 188–90; Bowden, 'Scientific Revolution', pp. 202–10 (on Boyle).

[23] Thomas Birch, The History of the Royal Society (London, 1756–7), iii, 454–55. It is perhaps worth noting here a manuscript evidently owned by Flamsteed, John Woodford's 'Cosmographiæ, et Partis eius præcipuæ Astronomiæ, definitio: & Divisio, & Judiciariæ partis Astrologiæ Reiectio' (1604), now RGO MS 1/67F: the section of this dealing with judicial astrology is almost wholly limited to a list of hostile authorities on f. 135, and there is no evidence that Flamsteed was influenced by it.

[24] Flamsteed to John Harris, 8 March 1698, RGO MS 1/33, fol. 202. Harris, who was Vicar of Winchelsea, wrote to Flamsteed on Jeake's behalf. In fact, Flamsteed's draft is incomplete and does not include the promised critique.On Jeake, see Michael Hunter and Annabel Gregory, An Astrological Diary of the Seventeenth Century (Oxford, forthcoming).

'never offered it to the press' – perhaps a significant remark in the light of subsequent developments, as we shall see.[25] Indeed, it might be argued that this enmity seems almost to predate Flamsteed's acquaintance with the new science, its roots perhaps lying in a tradition of the thought of the day which had long been associated with opposition to astrology, namely Puritanism, with its stress on God's omnipotence and its anxiety about ethical rectitude, both of them themes much in evidence in Flamsteed's autobiography.[26]

The extent to which Flamsteed's hostility to astrology merely represented an elaboration of a basic antipathy inherited from this tradition will be referred to again below in connection with the content of his anti-astrological tract. But here a further reason for Flamsteed's dislike may be recorded, which often appears in his correspondence and which reaches a climax in the text below, and this is his distaste for what might be called the 'populist' character of almanacs.

This had two components. In part, Flamsteed was hostile to almanac writers for their proneness to what he saw as irrelevant inclusions and extravagant interpretations of the significance of celestial phenomena. Indeed, in the latter connection he seems to have been concerned in part about the political partisanship which almanac writers had shown during and since the Civil War, sharing a general revulsion against this which (as others have noted) undoubtedly contributed to educated distaste for astrology in the late seventeenth century.[27] Here it is appropriate to scrutinise Flamsteed's ownership of religious books, for it is revealing that, though he possessed works by Calvin and the Elizabethan Puritan, William Perkins, these were outnumbered by Anglican and eirenic authors like the Dutch thinker, Hugo Grotius, and such mid-century divines as Henry Hammond, Richard Allestree and John Wilkins, the champion of the 'Latitudinarian' position which took shape in reaction to the bitterness and sectarianism of the Interregnum.[28]

Equally important, however, was Flamsteed's antagonism to almanac-writers for the inexactness and incompleteness of the astronomical data they gave. Indeed, contrary to Peter Wright's claim that astrology and the new science existed in separate spheres, the one practical and the other more theoretical, in fact Flamsteed's attitude reveals clearly the overlap in technique that existed between the two, and hence the potential for outright hostility towards what could be seen as incompetence in pursuit of a shared goal to which he, for one, was to give expression. Even popular almanacs used information about the movement of the celestial bodies of the kind that preoccupied Flamsteed, and the shortcomings of many almanac writers in this regard does much to explain the virtual crusade to reform the genre on Flamsteed's part which forms the context of the text printed here.

25 Baily, Account, pp. 11–12.

26 Baily, Account, pp. 7f., passim. On Puritan attitudes to astrology see Thomas, Religion and the Decline of Magic, ch. 12; Capp, Astrology, ch. 5, pt. i.

27 See ibid., p. 281, and N. H. Nelson, 'Astrology, Hudibras and the Puritans', Journal of the History of Ideas, 37 (1976), 521–36.

28 See Forbes, 'Library', passim.

Parenthetically, it may be noted that Flamsteed was not alone in this aim. Another contemporary who wanted to see popular almanacs reformed – mainly, it seems, from the point of view of introducing heliocentricity – was the virtuoso John Beale, who was anxious in the late 1660s to persuade the three almanac-writers whom he saw as the leaders in their field to purvey the new scientists' views on the planets and their motions, on the grounds that through this popular medium it might be possible to influence a social stratum normally immune to the arguments of intellectuals.[29] In addition, Flamsteed mentions in 1672 that, though obliged to turn it down on grounds of business and illness, he had had the idea of a reformed ephemeris proposed to him 'some while since' by Sir Robert Markham, a Lincolnshire landowner and Deputy Lieutenant, and a graduate of Wadham College, Oxford.[30]

As with such figures, a mission of intellectual enlightenment could be seen as Flamsteed's main incentive to produce a reformed almanac, but, in addition, it may be possible to discern a strange envy of the success which he saw almanac-writers as enjoying and a wish to share in it. For one thing, he seems to have hoped that he might gain financially from the venture, thus allowing him to supplement the meagre allowance from his father on which he survived at this time, though in fact all but the élite of almanac-writers of the day could have disabused him of any illusions of rapid success on this score.[31] Linked to this was a further possible – and arguably equally unrealistic – aspiration, to acclaim at a popular level of a kind which Flamsteed believed that almanac-makers enjoyed. In the text below he noted how 'the Vulgar have esteemed them as the very oracles of God' (f. 69), and certainly an author like Lilly achieved widespread fame through his almanacs.[32] It is perhaps in this light that one should read Flamsteed's retrospective comment on his 1669 almanac, which will be referred to shortly: 'that

[29] John Beale to John Evelyn, 2 January, 18 December 1669, Christ Church, Oxford, Evelyn Collection, Correspondence (bound volumes), nos 79, 93 (cited by permission of the Trustees of the Will of the late J. H. C. Evelyn). Though it is perhaps significant that Beale saw the death of Wing in 1668 as a setback to the cause of reform, he does not seem to have been very well in touch, since the other two of his three leading astologers were Phillippes, who had not produced an almanac since 1658, and Leybourne, who had not produced one of his own since 1651, though for his work on 'sorts' see Capp, *Astrology*, p. 318. For the context of Beale's concern with propaganda, see Michael Hunter, *Science and Society in Restoration England* (Cambridge, 1981), pp. 194–7.

[30] Flamsteed to Collins, 1 January 1672 (copy), EL F.1.81. On Markham see B. D. Henning, The House of Commons, 1660–1690 (London, 1983), i, 159. No other evidence apparently survives of scientific enthusiasm on Markham's part, whose time at Wadham postdated the Wardenship of Wilkins, though his son became an FRS in 1708. Sir Robert's memorandum book (British Library Additional MS 18721) is an interleaved copy of Gadbury's Ephemeris for 1681, but the notes in it are exclusively historical and personal.

[31] Flamsteed to Wilson, 3 July 1671, EL F.1.70. On Flamsteed's allowance see, e.g., Rigaud, ii, 167; *Oldenburg*, x, 157. On the rewards of almanac-writing, see Capp, *Astrology*, pp. 51f.

[32] See Derek Parker, *Familiar to All: William Lilly and Astrology in the Seventeenth Century* (London, 1975). Flamsteed's formula, quoted in the text, echoes the divine, Thomas Gataker's, attack on Lilly in 1651, quoted in Nelson, 'Astrology', p. 527, though there is no evidence that Flamsteed knew Gataker's book. It is interesting that Nelson postulates a similar resentment on Gataker's part.

rejection thrust me on ingenuous freinds, and acquaintance, which I shall ever valew more than all the applause could be gained by an Almanack'.[33]

Further evidence of a similar nature will be cited later in this paper, and it is perhaps not fanciful to see in this aspiration to acclaim a symptom of that self-esteem which Frank E. Manuel has perceptively discerned in Flamsteed in his relations with Newton later in life.[34] As we shall see, Flamsteed was to experience rejection at the hands of the almanac-publishing fraternity, and one might argue – in a vein similar to Manuel's – that this helps to explain the venom with which Flamsteed was to assault astrology, apart from the more intellectual rationale with which his attack was to be framed.

Flamsteed first attempted a serious reformed almanac in 1669. In that year, he presented predictions of celestial phenomena for the forthcoming year to the Royal Society, and he explains in his autobiography how this material had its origins as part of a reformed ephemeris, perhaps comparable to the surviving one for 1674: though he says nothing about any attack on astrology in it, he makes it clear that his reason for writing it was a dissatisfaction with existing almanacs, claiming that his was compiled 'not after the usual method, but much more accurately', including data on eclipses and appulses normally omitted from almanacs, which – when the work was rejected by the publishers – he excerpted and sent to Stansby, thus starting the contact with London virtuosi that has already been referred to.[35]

In a sense, the outcome of the 1669 episode was a development which made the publication of a reformed ephemeris less necessary, for the bulk of the data that Flamsteed sent to the Royal Society was inserted by the society's secretary, Henry Oldenburg, in the journal *Philosophical Transactions*: it was entitled, 'An Accompt of such of the more notable Celestial *Phaenomena* of the Year 1670, as will be conspicuous in the English Horizon'.[36] In subsequent years, Flamsteed was to follow this precedent. For 1671, he sent Oldenburg an elaborate Latin treatise on forthcoming celestial appearances, of which – perhaps to Flamsteed's disappointment – Oldenburg printed only a bare epitome, omitting Flamsteed's detailed calculations and commentary.[37] Thereafter, however, Flamsteed seems to have submitted his data 'so contracted that they may take up but a little roome in the transactions', and his contributions were printed verbatim in the journal, 'to which', as Flamsteed explained to Oldenburg when sending his data for 1672, 'since you have inserted them 2 yeares I hold them now a due'.[38] He continued the

33 Flamsteed to Collins, 1 January 1672, EL F.1.81.

34 Manuel, *Newton*, ch. 14.

35 Baily, *Account*, p. 23. See also ibid, p. 28, and Flamsteed to Brouncker et al., 24 November 1669, Rigaud, ii, 76f.

36 *Phil. Trans.*, 4 (1670), 1099–1112.

37 Royal Society MS Extra 5, item 5 (this hitherto neglected manuscript would merit proper study); *Phil. Trans.*, 5 (1670), 2029–34 (bis). Cf. Flamsteed to Brouncker et al., 24 November 1669, Rigaud, ii, 86–7, and Flamsteed to Oldenburg, 16 November, 5 December 1670, *Oldenburg*, vii, 265, 302. For the manuscripts of his subsequent contributions see EL F.1.78–, 78a, 88, 93 and 104. Royal Society MS Extra 5, item 4, comprises an apparently wholly unpublished item, Flamsteed's *Subsidium Selenographicum*, dated 2 February 1670.

38 Flamsteed to Oldenburg, 15 November 1671, *Oldenburg*, viii, 361; *Phil. Trans.*, 6 (1671), 2297–3001, 3061–3, 7 (1672), 5034–42, 5118–24, 8 (1673), 6162–66.

practice for two further years, terminating it in 1674 when these observations disappear from *Philosophical Transactions* because they were subsumed into the *Royal Almanack*, about which more will be said shortly.

Arguably, this was precisely the kind of function for which *Philosophical Transactions* was well-suited, disseminating accurate information to a widely scattered and discerning, but not particularly large, audience. Parenthetically, it is worth noting here that, when *Philosophical Transactions* was revived after a break in publication in 1683, Flamsteed was to return to publishing accurate data about planetary movements and eclipses there, as also an accurate tide-table for each successive year.[39] Indeed, an interesting sidelight on the journal's effectiveness in disseminating exact information is provided by evidence that Flamsteed distributed offprints of his articles in 1686.[40]

Despite finding this home for his material in *Philosophical Transactions*, however, in the summer of 1671 Flamsteed was again considering the idea of an almanac, communicating the notion to Oldenburg and to his other London correspondent, the government clerk and 'intelligencer', John Collins. The means by which he did so was indirect. Flamsteed wrote a letter on the subject to his chief contact with these two men, Thomas Wilson, a cousin of his and apparently an attorney, asking him to canvass Oldenburg and Collins about the idea, though this letter now survives in the archives of the Royal Society presumably because Wilson passed it on to Oldenburg.[41] That Flamsteed should have felt so uncertain of his plan as to want to have it tried out orally in this way is evidently itself symptomatic of scientists' unpredictability in their attitude to astrology in this transitional period, as already alluded to.

In this letter, which has been endorsed as 'exposing the folly of Astrological Predictions in Ephemerides and encouraging the predictions of Astronomical Appearances &c.', Flamsteed expressed some of the views that he was to air more fully in his 1674 ephemeris. 'I am vext to see our Ephemeridists spend the pages of their almanacks in Astrologicall whimseys tendeing onely to abuse the people & disturbe the publique with anxious & jealous prædictions, whilest the præmonition of Cælestiall appearances which ought to be theire onely charge are wholly contemned or neglected', he wrote. Moving on to his own plans, Flamsteed explained that, 'tho I might make good excuses yet I would (if I thought it might be printed & give me never so little for my paines toward my Astronomicall expenses,) write an Ephemeris in which it should be my businesse to explode the vanity of our vulgar Almanacks & to manage some small discourse concerneing anye Astronomicall doubt or question by which meanes my country men might

39 *Phil. Trans.*, 13 (1683), 10–15, 322–3, 404–15; 14 (1684), 458–62, 821–2; 15 (1685), 1215–30; 16 (1686), 196–206, 232–5, 428–32, 435–39. For an earlier tide-table by Flamsteed's assistant, Thomas Smith, see *Philosophical Collections*, no. 4 (1682), 102–3.

40 Flamsteed to Towneley, 15 March 1686, RS MS 243, no. 67.

41 Flamsteed to Wilson, 3 July 1671, EL F.1.70. Flamsteed refers to the letter in letters of 1 August 1671 to Collins (Rigaud, ii, 122) and Oldenburg (*Oldenburg*, viii, 179, which clearly alludes to the Wilson letter rather than to Flamsteed's own letter to Oldenburg of 4 July 1671, ibid., viii, 136–7). On Wilson, see ibid., vi, 515, vii, 302, 344; Baily, *Account*, p. 21; and evidently also Rigaud, ii, 92.

be informed & the sparkes of ingenuity blownes [sic] into fires & glorious flames of reall knowledge'. He then alluded to his 1669 attempt and its failure to get into print, adding somewhat ambiguously – presumably with reference to Oldenburg and Collins and his present ideas – 'which yet by meanes of them I hope they maye'. If encouraged, he continued, he would go ahead and 'aggresse the designe cheerefully & I question not the successe'.

How Oldenburg and Collins responded to this – if at all – is not known. Perhaps they simply ignored it, since on 15 November 1671, when forwarding his annual predictions for *Philosophical Transactions*, Flamsteed added a nonchalant – and perhaps slightly disingenuous – aside: 'I intended once to have composed all these calculations in an Ephemeris, but my other affaires & studies takeing up the best parte of my time, I had not leasure to performe what I intended. Yet I hope to get so much time as to performe it in parte the next yeare if providence permit mee health as hitherto'.[42]

Later in 1671, Flamsteed heard through Collins of a plan to compile a more accurate ephemeris like his by the Northampton surveyor and mathematical practitioner, Thomas Nunnes, and this episode stimulated Flamsteed to comments which are again worth quoting here by way of background to his 1673 views.[43] Nunnes was a friend of Vincent Wing's who had published a series of almanacs in the 1660s which had been strong on technical astronomical data and weak on astrological interpretation: Nunnes seems to have seen this virtually as an optional extra, though feeling obliged to include a commentary on the abnormal circumstances of 1666 as foreshadowed by the comets of 1664–5.[44] What he evidently now planned (perhaps building on comparable data in his earlier almanacs) was an 'Eclipsiography', as Flamsteed describes it, and we learn that he was encouraged in this by a person believed by Flamsteed to be Sir Robert Markham, who has already been referred to.

Writing to Collins on 1 January 1672, Flamsteed observed of Nunnes: 'I am glad to heare, that Astronomy hath such well willers abroad amongst us. And could only wish the labours of ingenious Artists had some other encouragements to draw them forth than what is afforded by the blockish stationer, who valews the saleable, but stupendious follies, of a Lilly, Gadbury, &c. before the ingenious endeavours of others, whose paines he esteemes not, because he understands not'. His further remarks are revealing. First, he seems to imply that the accurate prediction of eclipses by writers like Nunnes and himself should persuade 'the vulgar' 'which are to be esteemed Artists, & which not, & consequently which are most fitt to be licensed'. In particular, he looked forward to the solar eclipse in the following August as an event that would sort out true from false practitioners, itemising Lilly, Tanner, Swallow, 'Fly' and Dade among the latter, 'permitted, because of their names, to usurpe on the vulgar in annuall trifles,

42 *Oldenburg*, viii, 362.

43 Flamsteed to Collins, 1 January 1672, EL F.1.81. See also Flamsteed to Oldenburg, 23 December 1671, Oldenburg, viii, 427.

44 Thomas Nunnes, *An Almanack or Ephemerides*, 1661–2, 1664–6, esp. 1666, sig. C6. See also Capp, *Astrology*, pp. 174, 322.

or worse'. Speaking of Nunnes' scheme, he advised that he 'would not have it published, except, as he finds occasion in his annuall book; if hee expose it quickly, all our writers will flaunt it in his [sic] feathers', in contrast to the ignorance that they had been displaying since Vincent Wing's death in 1668. He then added, revealingly if not wholly consistently, that Nunnes 'should not print it, except the Stationers will give a very considerable rate for his paines, & such as may encourage an Artist'.[45]

Perhaps partly because he was waiting to see if anything would come of the Nunnes proposal, Flamsteed did nothing more in connection with his own projected ephemeris during 1672.[46] In any case, much of the time that he had available for study in that year was taken up by the assistance that he gave to the edition of the *Opera Posthuma* of Horrocks that Collins and others were preparing, the extent of Flamsteed's contribution to which is evident from the fact that he was the only living author referred to on its title-page.[47]

In the summer of 1673, however, Flamsteed again set to work on an almanac, evidently first sounding out Collins concerning the idea. Collins must have replied encouragingly in a letter now lost, and on 12 July Flamsteed wrote to him: 'Since yours of the 8th instant approves of my designe for an Ephemeris I shall take care of that taske & so to order it that however it bee it shall not looke like an Ephemeris'. (In the event, he evidently abandoned this notion, and the final version was condensed to 'the usuall compasse of an Almanack' (f. 69), presumably on the grounds that only thus was he likely to penetrate the normal almanac market.)

In addition, the letter deals with two other important topics. One was the timing of the project, which (as we shall see) was to prove crucial: 'pray let me know how long I may be ere I finish it for I have some other things under hand I would gladly dispatch at present & at what time at furthest it will be fit to print it, & if God spare mee life & health I will take care you shall have it by then'. Lastly, evidently replying to a point made by Collins in his lost letter, Flamsteed wrote, equally revealingly: 'as for peruseing of Authors I thinke it needlesse, since theire owne workes are too full of the Astrologers follies, & my owne experience has suggested some Arguments against them which will be enough to fill up one yeares worke' – the last phrase suggesting an intention to continue an annual series had the initial publication succeeded.[48]

Then, on 20 August, he wrote to Collins to report on his progress: 'I have more than half finished my Hecker, for so I call my Ephemeris, and, after the next week, hope to get time to conclude him, so that you may have him before Michaelmas'.[49]

The work as we have it is in fact entitled 'HECKER', and this alludes to the fact that the source of its data was a work by Johann Hecker (1625–1675),

45 EL F.1.81. On the almanacs referred to, see Capp, Astrology, pp. 357, 360–1, 368, 380–1.

46 Flamsteed refers to abortive attempts to establish contact with Nunnes in letters to Collins of 18 March and 17 April 1672: EL F.1.83; Rigaud, ii, 137.

47 See Jeremiae Horroccii Opera Posthuma (London, 1673), title-page and pp. 441f., passim; Baily, Account, pp. 30–1; and Rigaud, ii, 133–59, passim.

48 EL F.1.99.

49 Rigaud, ii, 167.

a Danzig intellectual and cousin of Hevelius, whose *Motuum Caelestium Ephemerides Ab Anno Aevi 1666 ad 1680* had appeared in 1662. This was a highly accurate work of computational astronomy, 'grounded upon the Tychonian observations and Keplerian Hypotheses, and the Rudolphin Tables, composed ad Meridianum Uraniburgecum' (in other words, the meridian of Tycho Brahe's observatory on the island of Hveen), as Henry Oldenburg had noted, paraphrasing the work's sub-title, when he had first heard of it in 1665.[50]

Flamsteed had evidently obtained a copy of Hecker's book in 1670 or 1671. Previously he had used Vincent Wing's *Ephemerides for the Coelestial Motions for XIII Years; Beginning Anno 1659* (1658), in conjunction with Thomas Streete's *Astronomia Carolina*, 'the exactest Tables, I know in being'.[51] As already noted, however, the last year covered by Wing was 1671, and Flamsteed must have been delighted to find this highly accurate replacement. It is apparent from letters to Collins that he had heard of Hecker's work by January 1670, attempting to get a copy during that year and succeeding at least by November 1671. Thereafter he used Hecker for all his calculations for the remainder of the 1670s, and a copy of the work was to be found in his library.[52]

Since Flamsteed had acquired his copy of the *Ephemerides*, Hecker had himself established contact with the Royal Society, writing in 1672 to enclose a copy of his tract, *Mercurius in Sole* (1672), which dealt with an impending transit of Mercury. This was delayed due to hostilities on the continent, but Flamsteed had finally received a copy by February 1673, when he gave his comments on the work in a letter to Oldenburg, comments which he echoed in the final, more technical section of the preface printed here.[53] In this section, Flamsteed also gave elaborate data on eclipses in the coming year, computed according to 'the New Horroxian Theory & tables' on which he had been working the previous year (ff. 68, 79), with diagrams to illustrate both these and his disquisition on Hecker's claims. This component of the work is similar in approach to the material that Flamsteed had been publishing in *Philosophical Transactions* for the last few years, and he inserted it here as an example of what ephemerides ought to contain in preference to the tendentious and erronious material whose rationale was undermined in the remainder of the preface.

This was further exemplified by the body of work, a sample opening of which is reproduced here (plate 2), from which astrological observations and comments were banished, and instead profuse and accurate information

[50] Oldenburg to Boyle, 18 September 1665, *Oldenburg*, ii, 512.

[51] *Phil. Trans.*, 4 (1670), 1100–1101. Cf. Flamsteed to Brouncker et al., 24 November 1669, Rigaud, ii, 77–8, 88.

[52] Flamsteed to Collins, 24 January 1670, 1 October 1670, ibid., ii, 94, 101; Flamsteed to Oldenburg, 15, 21 November 1671, *Oldenburg*, viii, 362, 366. See also Rigaud, ii, 127; *Oldenburg*, ix, 3, 177–8, 226, 326; Baily, *Account*, p. 32; Forbes, 'Library', p. 129. Flamsteed first refers to Hecker in *Phil. Trans.* in the issue for 19 August 1672 (7, 5040–1) and uses his data consistently thereafter; cf. Nicholas Stephenson, ed., *The Royal Almanack*, passim.

[53] See the letters between Hecker and Oldenburg, 1672–3, in *Oldenburg*, ix, 240–1, 284–6, 312–4, 502–4, and Flamsteed to Oldenburg, 17 February 1673, ibid., ix, 468–70.

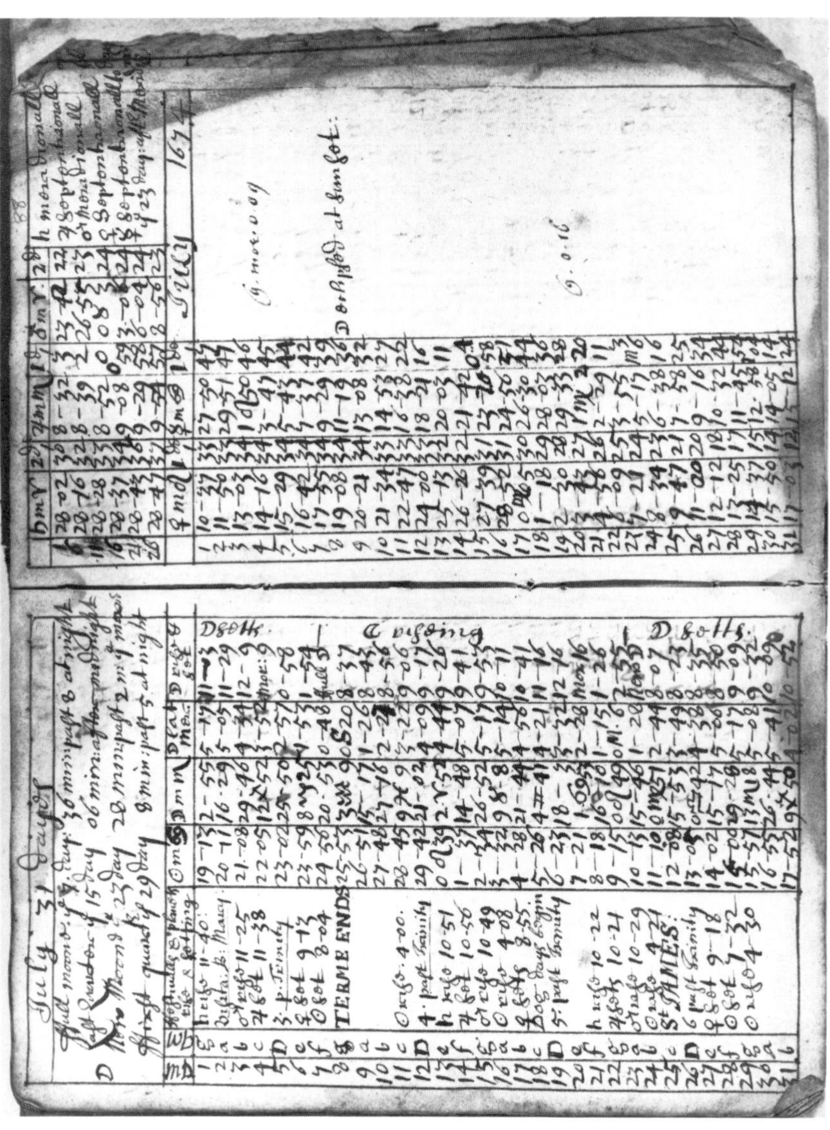

Plate 2. Flamsteed's ephemeris for July 1674, RGO MS 1/76, fols. 87v–88.

provided on the positions of the sun, moon and planets. The bulk of this came from Hecker's *Ephemerides*, slightly abbreviated and in some cases adapted to match the format normally found in English almanacs. In addition (as explained in the title) Flamsteed made various embellishments, correcting Hecker concerning 'the motions of the planet Jupiter', and including material on the 'Riseing and settings of the Moone & other planets' (f. 68).

All this data was, as the title states, 'Reduced, in a convenient forme, to an English meridian . . . Calculated for the very middle of England' (f. 68). The spot chosen was Derby, and this recalls Flamsteed's justification for basing his astronomical observations on Derby in his earlier writings for the Royal Society. Though claiming that he did this at first only for his own convenience, he subsequently justified his choice on the grounds that Derby was nearer to the middle of England than London, while its latitude also bissected the country, making it 'the most convenient place that can be elected, whereon to fix our Calculations'.[54]

On the title-page, Flamsteed gave as author an anagram of his name, 'Thomas Feilden'. This is perhaps surprising in view of the acclaim which he seems to have hoped to derive from publishing an almanac, but it is remniscent of the way in which he suppressed his identity under an anagram in his initial letter to the Royal Society of 1669.[55]

Exactly when Flamsteed finished the work is unclear. Another letter to Collins in which he claimed that he had advanced further with the project and hoped to finish it 'either this day or to-morrow' is frustratingly undated. 'If that my father's affairs had not pressed me extraordinarily, I had long since finished it', he explained, 'but hope now will be soon enough'.[56] Thereafter nothing is heard till 9 December, when Flamsteed wrote to Oldenburg: 'The weeke after I sent you the Ephemeris, I wrote a letter to Mr Collins about it, but have not since receaved any answer from him. which makes me suppose it comes rather too late for the presse'. This perhaps suggests that the date of despatch had been late rather than early in the autumn, but no further clues exist: though an allusion by Flamsteed in a letter to the Paris-based scientist G. D. Cassini of 5 September has been identified by the editors of Oldenburg's *Correspondence* as referring to this project, this is not conclusive.[57] Hence it is unknown exactly when Flamsteed forwarded the completed text to London, and thus impossible to be sure whether it is likely that it was rejected by potential publishers on grounds of lateness. The norm was for copy for the forthcoming year's almanacs to be submitted early enough for the books to be printed and available in good time for Christmas.[58]

After this, there is one further reference to the 'Hecker' in Flamsteed's letters to Collins, in a missive of 27 December 1673. From this it is apparent that by now the work had been turned down by potential publishers and

54 Flamsteed to Brouncker et al., 24 November 1669, Rigaud, ii, 78–80.

55 Flamsteed to Brouncker et al., 24 November 1669, Rigaud, ii, 90.

56 Rigaud, ii, 169.

57 *Oldenburg*, x, 191, 195 n.7, 382.

58 Cf. Capp, *Astrology*, p. 59.

Collins must therefore have suggested to Flamsteed that the Royal Society might take responsibility for it, perhaps similarly to the way in which the society *was* to sponsor an ephemeris in the 1680s. In reply, Flamsteed wrote: 'Yours informs me of the reasons why my Ephemeris is not printed, with which I am very well satisfied. But, if you have not already done it, I would not have you propose the printing it at the charge of the Royal Society, lest they lose by it, to whom I am so much already obliged for their respects, that I desire not to be made their further debtor'.[59] By now, Flamsteed was evidently under no illusions about the potential saleability of a reformed almanac like his, quite apart from its incendiary preface.

Thereafter, to complete the narrative, we hear little of the project. On 7 March 1674 Sir Jonas Moore wrote to Flamsteed saying: 'The Almanack Mr Collings had I onely saw . . . I pray yow that I may have the peruseing of it'; ironically, he added that 'if yow had sent it me I would have had it printed but I knew of it too late'. Flamsteed evidently then sent it, and on 23 March Moore replied: 'I am very much pleased with your sentiments of Astrology agreable in every thing to my thoughts'. Flamsteed refers in his autobiography to the fact that his almanac 'fell into the hands of Sir Jonas Moore', and that as a result 'I made a table of the moon's true southings for that year' at Moore's request, thereby assisting the latter's enterprise of devising a more accurate theory of the tides.[60]

Though one might have expected Flamsteed to repeat his experiment for the subsequent year, he did not in fact do so. Possibly this was because he was busy with other things, including a treatise on the diameters of the planets and their distances from one another, and an epistle to Moore 'concerning the Correction of the Motion of the planet Mars'.[61] But in fact both projects had already been in his mind before he carried out his work on the 'Hecker', and, in the same way that he was able to find time for that in 1673 because he evidently considered it worthwhile in its own right, so he could have repeated this later.

More significant was the fact that in 1674 Moore enrolled Flamsteed's help in an overlapping project which was intended to serve the utilitarian if not the polemical functions of the 'Hecker', namely the production of the so-called *Royal Almanack*, which provided a non-astrological account of forthcoming celestial phenomena. Published annually from 1675 to 1678 by the Stationers' Company, this had on its title-page the names both of Moore and of his assistant, Nicholas Stephenson, under whose supervision the work was evidently compiled, and it took over from *Philosophical Transactions* as a venue for Flamsteed's annual predictions of eclipses, appulses of the moon

59 Rigaud, ii, 170.

60 RGO MS 1/37, fols 73–74 (I am indebted to Frances Willmoth for these references); Baily, *Account*, p. 34.

61 Flamsteed to Oldenburg, 9 December 1673, *Oldenburg*, x, 381–2. For earlier references to these projects see ibid., ix, 327, x, 165; Rigaud, ii, 160; Flamsteed to Towneley, 25 January 1673, RS MS 243, no. 1; Baily, *Account*, pp. 30n., 32–3.

and the like.[62] The *Almanack*'s continuity in providing for a select and well-informed audience is indicated, however, by the fact that now, as earlier, this data was provided in Latin 'because it may also be of use to strangers abroad'.[63]

Flamsteed seems to have experienced some reservations about the *Royal Almanack*, complaining to his friend Richard Towneley in 1676 about errors that had crept into the forthcoming issue and how 'so many hands in a businesse spoyle it'; he also later confided to the Dublin virtuoso, William Molyneux, his resentment that minor shortcomings in his predictions were the subject of critical comment among 'our talkative townespeople'.[64] But he continued to contribute to it until it ceased to appear in 1679, after which there was a lacuna in Flamsteed's activity as an ephemeridist until, first, his new series of *Philosophical Transactions* articles from 1683 to 1686, and then his collaboration on the 'Royal Society' ephemeris: the latter was a further experiment with a reformed almanac which was put out under the editorship of Edmond Halley between 1686 and 1688.[65]

If the inauguration of the *Royal Almanack* evidently explains why the 'Hecker' enterprise was not repeated, what can be said about the failure of Flamsteed's book to get into print in 1673? As we have seen, Collins' suggestion that the Royal Society might assist with it was rejected by Flamsteed himself, though it may be noted that it is highly unlikely that the society would have been able to help in any case, for the institution was in a parlous state at this time due to a falling off of its members' support.[66]

What, however, of 'the reasons why my Ephemeris is not printed' which had led to the society's suggested involvement? It is possible that Flamsteed's

62 The first issue of the *Royal Almanack* had evidently been for 1674 although no copy of this survives: see Capp, *Astrology*, pp. 38–9 and 396 n. 96, and Moore to Flamsteed, 23 March 1674, RGO MS 1/37, fol. 74 (reference kindly supplied by Frances Willmoth). Flamsteed's collaboration seems to have begun with the 1675 issue. Cf. Baily, *Account*, p. 35. For a reference in a letter from Flamsteed to Boucher of 24 January 1678 to the almanac as being 'done by a Scholar of mine', see RGO 1/43, fol. 38: see also Flamsteed to Molyneux, 11 April 1682, Southampton Civic Record Office MS D/M 1/1.

63 Nicholas Stephenson, ed., *The Royal Almanack* (London, 1675), sig. D6. On this aspiration to an international rather than local audience, see also Flamsteed to Collins, 1 August 1671, Rigaud, ii, 119–120.

64 Flamsteed to Towneley, 11 December 1676, RS MS 243, no. 19; Flamsteed to Molyneux, 15 April 1682, Southampton Civic Record Office MS D/M 1/1.

65 On Flamsteed's contribution to this publication see esp. Flamsteed to Towneley, 4 November 1686, RS MS 243, no. 68. See also Flamsteed to Towneley, 15 March 1686, 12 January 1688, ibid., nos 67, 30 (and perhaps also 18 January 1687, no. 69, which refers either to this or to a further abortive proposal for an ephemeris of which nothing is otherwise known); Flamsteed to Molyneux, 12 February 1687, Southampton Civic Record Office MS D/M 1/1. At least some contemporaries thought the ephemeris was by Flamsteed: see the inscription on the title-page of Bodleian Library 8° F 131 Linc., item 11, a copy of the 1687 issue (cf. item 12, a copy of the 1688 issue). But the attribution has there been changed to Halley, to whom Aubrey's copy of the 1688 issue, now in Bodleian MS Wood 498, is also attributed. This venture is evidently to be seen in the context of the Royal Society's expansiveness in the mid-1680s: see Michael Hunter, *The Royal Society and its Fellows, 1660–1700* (Chalfont St Giles, 1982), pp. 41–2.

66 Ibid., pp. 36–39. The society's minutes contain no reference to any such approach by Collins at this time, though see Birch, History, iv, 4, for a later approach to the society's Fellows by Collins in connection with a publishing project.

copy had arrived too late for the publishers' deadline, despite the fact that he had sought Collins' advice as to when it was required. If this was the case, it might have been a victim of the business and ill-health which repeatedly interrupted Flamsteed's studies at this stage in his career: he is frequently to be found complaining how 'My distempers and affairs of late have been so intermutually urgent, that I have performed little'.[67] In addition, his apology for the residual deficiencies of the information in the ephemeris (f. 79) could suggest that – characteristically – Flamsteed had spent too long in trying to perfect the data before finally consigning it to the press.

On the other hand, here a revealing parallel may be sought in the different reasons for the non-publication of his 1669 almanac that Flamsteed gave in three separate accounts of the affair, in one claiming that 'lying too long in the hands of a friend, [it] was refused the press', in a second that it was 'rejected, as beyond the capacity of the vulgar' and in a third – perhaps a variation on the latter – that it was turned down 'because my pages had no prædictions'.[68] In 1673, as in 1669, it seems likely that the reluctance of the potential publishers had more to do with reasons of the latter kind than the former. Indeed, one might well wonder whether the Stationers' Company would have wanted to publish the book at all.

At the most straightforward level, they might have had reservations as to whether – contrary to Flamsteed's rather naive expectations – a reformed almanac like his was saleable. It is true that the Stationers' Company were to publish the *Royal Almanack* from 1675 to 1678, possibly due to pressure from people in high places exerted through Moore: but the relative rarity of this work in comparison with fully astrological almanacs like Lilly's or John Gadbury's is itself symptomatic of the limited sales which such austere volumes commanded, while the same is true of other reformed almanacs, such as the Royal Society ephemerides of the 1680s.

A realisation that demand for such an ephemeris was limited explains why in the late 1670s and 1680s even Flamsteed was reluctant to spend time on the preparation involved, and in this connection it is worth quoting a letter from Flamsteed to Towneley of July 1679, when the *Royal Almanack* had been defunct for a year. This illustrates how, though Flamsteed was by now more realistic about the potential appeal of a reformed almanac, he had still not entirely lost the wistful envy of the rewards which he believed to accrue to those who wrote the populist compilations of which he disapproved.

> There are so few that understand the uses of the Royall Almanack besides your selfe & a few particular friends & there are so few appulses observable by reason of our Cloudy weather that I find it scarce worth my time [to] undertake it. I have proposed it to the stationers who answer me so coldly as to point of reward that I find it would be much

[67] Flamsteed to Collins, 3 May 1671, Rigaud, ii, 113. Compare, e.g., ibid., ii, 109, 160; Oldenburg, vii, 266, ix, 158, 326; EL F.1.81, 83, 108.

[68] Flamsteed to Brouncker et al., 24 November 1669, Rigaud, ii, 78; Baily, Account, p. 23; Flamsteed to Wilson, 3 July 1671, EL F.1.70. In EL F.1.81, where he also mentions this episode, he gives no reason why the work 'could not be accepted'.

better to write a Poore Robin – Merlin propheticus [. . .] to fill up two sheetes of paper with pamphlet rubbish for them, then to under[take] the paines that are requisite to produce that creature which dies when the yeare is over. however I shall doe something for my owne use & if I publish it not I shall however send you a copy.[69]

Three and a half years later he echoed this (and his acknowledgment of the competence of the better astrological practitioners of the day) by noting of the *Compleat Ephemeris* that Thomas Streete published from 1682 to 1685 that it 'excuses me the unregarded labor of the Royall Almanack'.[70] In the late 1680s, there is evidence that the Royal Society ephemeris foundered because of waning enthusiasm for the work involved, while comparable problems almost led to the demise of George Parker's accurate almanac in the 1690s, from which, ironically, it was saved by the intervention of Flamsteed and Halley.[71] Even then, Parker found that it was hard to sell his unusually accurate and lengthy compilation, since the extra material included added to the cost and put it out of line with its more popular rivals, 'And the number of *Students* in this Science being few, did not take off an *Impression*'. From 1695, therefore, he experimented with a separate Ephemeris (which, significantly, is again a relatively rare book), while producing a briefer almanac which was almost regressive in its appeal to popular taste.[72]

Simple economics must thus have had something to do with the publishers' decision on the 'Hecker', but, beyond this, it would surely have been odd for the Stationers' Company to be very keen to accept for publication a work prefaced by an abrasive epistle questioning the rationale of the whole genre whose annual sales were so highly profitable to the company's members.[73] Arguably, it was the component of the 'Hecker' which is retrospectively chiefly of interest to us which made it unpublishable at the time. (It might be added that – prefixed as it was to a reformed almanac – Flamsteed was surely unrealistic in thinking that his preface would be read by those who purchased traditional almanacs in any case.)

Moreover, it may not only have been the Stationers Company who felt such reservations. Indeed, the fact that this unique assault on astrology by a scientist was never published may suggest a different reason for the lack of

69 Flamsteed to Towneley, 16 July 1679, RS MS 243, no. 41. A fragment has been torn away, obscuring the words in square brackets. There are illegible deletions before 'scarce worth' and 'if I publish' ('perhaps'?). On William Winstanley's satirical *Poor Robin*, see Capp, *Astrology*, esp. pp. 39–40. Flamsteed's reference suggests that he was as hostile to this as to standard almanacs.

70 Flamsteed to Towneley, 8 December 1682, RS MS 243, no. 61. Cf. Flamsteed to Molyneux, 11 April 1682, Southampton Civic Record Office MS D/M 1/1.

71 Flamsteed to Towneley, 4 November 1686, RS MS 243, no. 68, and perhaps also 18 January 1687, no. 69 (but see above, note 65); George Parker, *Mercurius Anglicanus*, 1692, sig. A3.

72 George Parker, *Mercurius Anglicanus*, 1695, sig. A1v, and 1696–7, passim; id., *An Ephemeris of the Coelestial Motions*, 1696, and see Capp, *Astrology*, p. 372, for other surviving issues, though he is wrong in implying that the *Ephemeris* was published before 1695.

73 See Cyprian Blagden, 'The Distribution of Almanacks in the Second Half of the Seventeenth Century', *Studies in Bibliography*, xi (1958), 107–16, and Capp, *Astrology*, ch. 2. For the company's ability to tolerate satire, however, see ibid., pp. 243–4.

evidence of open hostility towards astrology on the part of scientists, apart from the suggestions of Capp and Wright noted at the start of this paper, and this relates to the question of whether it was tactful to mount such an attack at all.

In other contexts, Flamsteed showed his awareness that he was prone to a bluntness which could verge on the counterproductive. Concerning his first letter to the Royal Society he apologised for 'a boldnesse that I feard would looke more like an impudence then an Ingenuitie', later reporting Oldenburg's advice on him not to 'advert too plainly on Hevelius, lest he should recede from his correspondency, and detain his observations from us, if he be disgusted'.[74] Possibly Flamsteed was influenced by comparable arguments in connection with his 'Hecker', though there is no evidence of this in our sources. One reason for this may have been the ambivalence of many scientists towards astrology, as already alluded to: after all, it was just at this time that Boyle was publishing tracts advocating open-mindedness concerning astrological phenomena. But in addition it is important to take into account a factor which is easily underrated, and this is the institutional weakness of the new science at this time, its lack of support and endowment and hence the danger of upsetting potential supporters.[75]

In view of the overlap between the priorities of astrological practitioners like Thomas Streete and George Parker and those of Flamsteed, it was clearly inadvisable to assault the astrological principles which were so important to such men, rather than silently ignoring these while encouraging the worthwhile work on computational astronomy that they carried out. It should also be remembered that, whatever Flamsteed's own position, his friend and patron Sir Jonas Moore was friendly with such astrologers as William Lilly and Elias Ashmole, and the 'established' position of astrology at this juncture is epitomised by the fact that Ashmole – with Lilly's assistance – was actually acting as a kind of court astrologer to Charles II and his advisers in these years.[76]

Moreover the distance that almanac-makers could go towards Flamsteed's aspiration to accuracy without abandoning the other components of their publications may be illustrated by a curious postscript. For it is probably not coincidental that, just at the time when Flamsteed was holding up Hecker as an example of the accuracy that was desirable in an ephemeris, one finds Hecker's work being used by Lilly, arguably the most popular of all almanac-writers, and one of the most extravagant in terms of the interpretative superstructure that Flamsteed so disliked. In 1674, Lilly cited Hecker concerning forthcoming eclipses, later using data from Hecker's *Ephemerides* as the basis of his almanac from 1678 until 1680, the last year to which

[74] Flamsteed to Oldenburg, 7 February 1670, *Oldenburg*, vi, 468; Flamsteed to Collins, 20 March 1671, Rigaud, ii, 109. See also ibid., ii, 91, *Oldenburg*, ix, 468–9.

[75] On this theme see Hunter, *Science and Society*, pp. 39–40, 83–4 and passim.

[76] Josten, *Elias Ashmole*, i, 188–90. On Moore's contacts with Lilly and Ashmole see ibid., i, 36, 118, 207, 235, ii, 397, iv, 1649. Moore was also a colleague at the Ordnance Office of Sir George Wharton: see above, p. 264.

Hecker's tables applied, perhaps having been alerted to the value of Hecker's work by the mutual friends he shared with Flamsteed.[77]

Clearly the point is that – contrary to anachronistic presumptions about its inexorable decline – the astrological lobby in the late seventeenth century was too strong to be easily assaulted, least of all on its home ground, by a new science that was still only marginally established. This was a lesson which Flamsteed painfully learned in 1673, and the non-publication of the work which now first sees the light of day is arguably one of the most significant things about it.

This fact, together with the evidence of ambivalence towards astrology on Flamsteed's part that has already been referred to, should be borne in mind in reading the text that follows. It is a statement of an extreme position about which Flamsteed himself probably had subsequent reservations, as is suggested by his failure to launch a similar attack at a later date. But it spells out with particular clarity a range of arguments against astrology which at the time seemed to Flamsteed likely to convince his putative audience.

Here again a caveat must be sounded, for – as has already been seen from Flamsteed's letter to Collins of 12 July 1673 – this was only intended as 'one yeares worke', and, both in passing in his preface and in a longer passage towards the end in which he visualised a barrage of criticism of his views in annual almanacs, Flamsteed implied that he had further material up his sleeve with which he could have extended his assault (ff. 72, 75v, 78). He may also have held back on one part of his argument – that concerning astrologers' errors in their predictions concerning current affairs – on the grounds that a book on the subject was forthcoming, though it seems never to have materialised (f. 72v). Not least for these reasons, one cannot be sure that the objections to astrology mentioned and the amount of space devoted to them necessarily reflect the relative weight of different factors in stimulating Flamsteed himself to reject astrology and wish to attack it. Nevertheless, they deserve careful scrutiny.

Some of the themes in Flamsteed's preface echo topics which have already been discussed. Thus he briefly brought up the objection to astrological prediction which, as already noted, may have predated any 'scientific' ones in his intellectual development – the rejection of astrology's claims to foreknowledge on religious grounds. Flamsteed refused to 'offend Christian eyes with a view & pretense of more foreknowledge then God ever indulged to his Prophets', seeing astrologers' predictions as being baffled by God's providence and later referring in passing to the art as 'opposite to reason & the divine law' (ff. 69, 78).

In addition, the preface echoes the letters already quoted in expressing Flamsteed's dislike of the populist character of astrology – the inclusion of sensationalist and irrelevant material in almanacs, the incompetence of many

[77] William Lilly, *Merlini Anglici Ephemeris*, 1674, sig. B4, 1678–80, sig. C1. (Hecker is also cited in *Poor Robin*, 1674, sig. C2v.) See also Capp, *Astrology*, pp. 195–6. In 1681, Flamsteed did in fact recommend Lilly's almanac to Towneley in lieu of anything better, 'which I thinke to be a transcript from Gadbury': RS MS 243, no. 51.

practitioners from the point of view of astronomical computation, and their provision of astrological advice in return for financial reward.

Taking these points in turn, Flamsteed attacked at the outset almanac-writers' 'vaine' and 'pernicious' predictions concerning the weather and affairs of state, alluding specifically to Lilly – 'Merlinus Anglicus', as his almanacs proclaimed him – in this connection (f. 69). Later, Flamsteed went on to emphasise briefly the ill consequences of astrological prediction for social stability by reference to 'the historys of all insurrections' and not least the Civil War (f. 72v).

It is worth noting, however, that Flamsteed's comments on this subject hardly suggest that it was simply political subversiveness that worried him about the extraneous material in almanacs. In fact, as just noted, he bracketed predictions concerning state affairs and the weather together as equally undesirable, while his remarks about the desiderata for an ephemeris show that he saw as irrelevant not only 'idle Aphorismes and judgments in Astrology' but also catalogues of fairs and roads: it is interesting, on the other hand, that he felt that it *was* appropriate to include useful notes on horticulture and agriculture (ff. 69, 78).

As for technical incompetence, Flamsteed gave an example from the 1673 almanac of John Gadbury, Flamsteed's particular bête noire, who was attacked at greater length than any other astrologer, perhaps because Gadbury claimed a commitment to the reform of the art in addition to being a highly popular practitioner of it. Flamsteed was no doubt highly incensed that in his almanac Gadbury offered 'As a Demonstration of the great perfection Astronomy is arrived unto in this our Age' the prediction of a conjunction which was in fact quite inaccurate.[78] An echo of this scorn for astrologers' failure to forecast celestial appearances correctly has already been noted in Flamsteed's letter to Towneley of 5 April 1679, quite apart from the disdain for astrologers' lack of technical skill that runs through his letters in the years preceding the compilation of the 'Hecker'. Clearly many popular astrologers offended against the astronomical exactitude which Flamsteed pursued throughout his career, and in his preface he makes particularly clear the contrast that he perceived between these vulgar practitioners and the true 'Astronomer', of whose role he had a high view. Indeed, in a revealing passage which looks forward to the high valuation of the natural philosopher in the Enlightenment, Flamsteed apologised that the task of compiling even a correct almanac might seem 'too triviall & beneath the paines of an Astronomer'.[79]

[78] Below, f. 75; John Gadbury, *Ephemeris: or A Diary . . . for the Year of our Lord 1673*, sigs A5v–6. On Gadbury and reform see Hunter, *John Aubrey*, pp. 119–20, 144–5; Capp, *Astrology*, pp. 184–6.

[79] Below, f. 68v. Compare the opinion of William Neile concerning the 'dignity' of a natural philosopher quoted in Michael Hunter and Paul B. Wood, 'Towards Solomon's House: Rival Strategies for Reforming the Early Royal Society', *History of Science*, xxiv (1986), 78. On the Enlightenment see Roger Hahn, *The Anatomy of a Scientific Institution* (Berkeley and Los Angeles, 1971), ch. 2. Cf. Manuel, *Newton*, pp. 296–7.

Linked to this are a number of critical asides implying that astrologers were mercenary creatures, too dependent on their art for their livelihood to question its rules: Flamsteed does not seem to have conceived of the possibility of a disinterested astrological reformer. In his view, an astrologer would respond favourably to any approach, however inadequate the data proffered, if it were made 'not forgetting to pay him handsomely'; elsewhere he accused the astrologer of renouncing his reason 'for the profit hee receaves by fooleing simple people' (ff. 71v, 75v, 77, 78). Here one perhaps sees the obverse of that wistful envy on Flamsteed's part for the acclaim that astrologers enjoyed, which, as already noted, may help to account for the zeal with which he attacked the art.

On the other hand, the preface also offers something which is barely in evidence in Flamsteed's letters, and this is an exposition of the rationale of his rejection of astrology, the arguments which in his view made the influence of the heavenly bodies on sublunary affairs dubious, especially as currently interpreted by astrologers.

A substantial part of the preface is taken up with a critique of astrology, for which the familiarity with the art that Flamsteed had acquired in his youth had well prepared him. To some extent Flamsteed raised issues of which any intelligent layman who looked into the subject might have been aware: inconsistencies among astrologers even about the basic techniques and applicability of their art (ff. 73, 76); the endless elaboration of causal factors in which they could indulge (f. 71); the self-confirming and circular quality of their analysis (ff. 71, 74v–5); and their ignorance of factors which, by their own admission, might be crucial (f. 76). But in addition, Flamsteed's astrological expertise allowed him to speak with authority on technical issues like the calculation of the arc of direction and its problems (f. 70v), while his familiarity with the actual rules of judicial astrology made it possible for him to give a devastating critique of one of the more questionable cases included in John Gadbury's *Collectio Geniturarum* (1662) (ff. 73–4). This was the natal horoscope of Bishop Joseph Hall – Flamsteed's interest in whom is evident from his possession of more than one of Hall's works – and Flamsteed drew attention to a discrepancy between the symbolism of the horoscope and Hall's actual career which Gadbury had simply ignored.[80]

Here one sees an empirical streak in Flamsteed's case against astrology, which is also in evidence in the experiments on the rectification of nativities that he refers to making in the 1660s, citing his difficulties in getting any technique to work satisfactorily, despite attempting to deploy them 'with greater curiosity then is usuall' (f. 70).

On the other hand, it might be argued that there were limits to the open-mindedness with which Flamsteed applied his empiricism, and that an a priori element entered into his presumption that the problems he had with established techniques necessitated the complete rejection of astrology rather than a search for superior methods to those employed hitherto. His response

[80] John Gadbury, *Collectio Geniturarum* (London, 1662), p. 91. For Flamsteed's copies of Hall, see Forbes, 'Library', p. 128.

contrasted with that of others at the time who, though accepting the limitations of traditional astrology, saw this as an argument for the need to reform the art rather than to abandon it.[81] There is a distinctly Luddite quality about Flamsteed's criticism of astrology as practiced in his time without allowing for any possibility of improvement. For instance, in his views on weather prediction (ff. 72v–3), Flamsteed was rhetorically dismissive of astrologers to an extent that might have been applicable to the more casual of almanac-makers, but was hardly fair to the painstaking efforts of a man like John Goad, Flamsteed's respect for which has already been cited. Moreover, though Flamsteed refers to his reading of astrological authorities, it is revealing that he seems to have presumed that the older, Arabian writers were likely to be correct, and that modern deviations from them must be due to error (ff. 70v–1).

Indeed, much of Flamsteed's appeal to 'experience' was rather generalised and rhetorical, differing little from the rationalistic critiques of astrology that had been published since the time of Pico della Mirandola in the late fifteenth century, often by people who, like Flamsteed, had dabbled in the art before rejecting it.[82] That Flamsteed's animus against astrology was served rather than caused by empiricism is illustrated by a telling passage in which he made it clear that he already *knew* he was right: arguing that there was no rationale for the influence which astrologers claimed that planets had over specific mundane phenomena, he assured his readers that they would be unable to prove the contrary before Doomsday (f. 72v).

What, however, of the new science? Did this add anything to a case against astrology which had long been current? Certainly, Flamsteed was aggressively confident in the new philosophy, dismissing Ptolemy's 'Epicyclar Hypothesis' as fallacious, glorying in the discoveries of the telescope concerning the nature of the heavenly bodies, and discussing the theories of Borelli and others on the likelihood of a plurality of worlds (ff. 74, 76v–77).

Moreover he states clearly what it was 'which wrought most with me to desist the Study of this pretended Art' – even though he claims that this came only after his early empirical trials – and this was that astrology was inconsistent with his conception of the natural world. It was according to this criterion that he found the art capricious and unconvincing: 'I found there was no reason in nature', he wrote, 'why the Planets should have any influence upon our actions, or our thoughts which moderate them' (ff. 71v–2). In the mechanistic universe that he accepted, he could visualise no 'Mechanicall operations' by which the planets could have the power attributed to them by astrologers, dismissing on these grounds any parallel with the effects of the sun on vegetation or the moon on the tides, and revealingly using an analogy with the rays of light cast by a candle or a fire to argue against any influence on thought or action by the heavenly bodies (ff. 72, 77).

[81] See Bowden, 'Scientific Revolution', passim; Capp, *Astrology*, ch. 6, pt. i; Hunter, *John Aubrey*, pp. 117f.

[82] See particularly D. C. Allen, *The Star-crossed Renaissance. The Quarrel about Astrology and its Influence in England* (1941; reprinted London, 1966).

Indeed, what comes across strongly is a wish to distance the heavens from sublunary affairs. To the astrologers' argument that the heavens were there to influence events on earth, he answered that this trivialised them and that they were self-contained and awesome in their own right, the discoveries of the new science being seen to militate against an anthropocentric view of the universe in favour of 'a greater and truer use of the Cælestiall bodies', as being there for the glory of God (ff. 76v–78). The implications of this attitude were spelled out by Flamsteed in a letter to Richard Towneley of 11 May 1677 which may be quoted here, in which he referred to comets and his hypothesis that they returned in a regular cycle, which

> will wholly overthrow the conjectures & fearefull praedictions of our Astrologers who conceat [?] all actions succeding comets to be influenced by them. for if theire returnes be stated why are not theire effects the same [:] but nothing of ill hapned to England after either the 3 Comets of 1618. or the great one of 1653[.] & therefore the evills succedeing the comet of 1665 were not to be attributed to its influences for then the like Mischeifes had succeeded the former, but to other causes. but this I suppose is enough to have mentioned to you who are not taken with this superstition of the vulgar.[83]

In many ways, Flamsteed illustrates exactly the kind of objections to astrology which one might have expected a scientist of his generation to feel, but which have hitherto otherwise been frustratingly undocumented. Clearly the new science was a significant component in his alienation from astrology, largely since the general view of the universe that it imbued made astrology seem peculiarly implausible in Flamsteed's eyes. On the other hand, it would be wrong to confuse this with empirical disproof, since Flamsteed's critique of the art – though cogent – added relatively little to the attacks on astrology which had been made since the Renaissance. Equally familiar and a priori was his distaste for the populism and ignorance of astrological practitioners, while his religious objection of the principle of foreknowledge also had a long history. Moreover, as we have seen, though relatively briefly adumbrated in the preface, the two latter themes may have had more significance in forming Flamsteed's views than the intellectual arguments to which more space was there devoted.

Indeed, before this text is acclaimed too quickly as signalling astrology's deathblow – 'Science rebuts astrology', in the words of its only commentator hitherto[84] – it is worth reiterating various points. First, perhaps the most crucial fact about it is that it was never published. The possible reasons for this non-publication have already been canvassed, but here its significance is as a warning against overestimating the influence of the new science and underestimating the power of astrology in this transitional period. Ideas like

[83] Flamsteed to Towneley, 11 May 1677, RS MS 243, no. 26. For a comparable remark by Flamsteed a few years later – this time concerning the conjunction of Saturn and Jupiter – see *Phil. Trans.*, 12 (1683), 244–5.

[84] Murdin, *Under Newton's Shadow*, p. 137.

Flamsteed's may have played a part in the rejection of astrology by the educated, but they did so not through aggressive statements by scientists – whose direct influence in this matter can easily be anachronistically overrated – but less directly, by being mediated through spokesmen for educated opinion as a whole.[85]

In addition, it is worth recalling the ambivalence towards astrology among scientists, of which Flamsteed himself partook to some extent. It is thus revealing, for instance, that Robert Boyle took a diametrically opposite view to Flamsteed's on astrology's anthropocentric view of the heavens: for while Flamsteed strongly disapproved of this, Boyle saw in it a valuable defence against a godlessly mechanical universe.[86]

Flamsteed's case is in some respects the most curious of all, an ambivalence towards the art coexisting with overt hostility in a manner that we may find puzzling. Moreover the exploration of the background to Flamsteed's opposition to astrology presented here has been revealing in itself: for the degree of documentation available in this instance is probably unique, making this a crucial case-study in the rejection of astrology in the period and illustrating how a range of factors seems to have contributed to Flamsteed's antipathy – religious, social, professional and intellectual – none of which can be proved predominant. This should be taken as a caveat against too simplistic an interpretation of the text now first presented to the public. However cogent its arguments, it can only be properly understood against the background outlined here.

[85] See Capp, *Astrology*, pp. 280–1. I hope to develop this point further elsewhere.

[86] Bowden, 'Scientific Revolution', pp. 209–10.

THE TEXT[87]

f. 68/

HECKER

His large Ephemeris for the yeare 1674
Reduced, in a convenient forme, to an English meridian
Corrected in the motions of the planet Jupiter, and Enriched with the Riseing and settings of the Moone & other planets. Calculated for the very middle of England
With a Preface about Astrology & the practices of Astrologers
To which is added
The Computation of the two Lunar Eclipses, from the New Horroxian Theory & tables, and an Advertisement about the transit of [Mercury] sub Sole by
Thomas Feilden:

f. 68v/ A Preface to the Readers Concerneing the Vanity of Astrology, & the practices of Astrologers:[88]

Some moneths agone, meeting with an ingenuous Mathematicall acquaintance,[89] amongst other discourses hee informed mee, that by by [sic] the helpe of some tables I had formerly given him, hee had calculated the times of the Moones riseings, & settings, for the yeare 1674; to which hee intended to adde the rise and settings of the other planets: Hee affirmed this taske to be but badly performed in our vulgar Almanacks, & therefore requested me to make use of his paines to fill up an Ephemeris for that yeare: I had beene

[87] The manuscript, RGO MS 1/76, item 5, is a paperbook of twenty-six leaves, stitched at the central fold. Each leaf is 15.3 cm tall and between 9.5 cm and 9.8 cm wide. The preface occupies fols 68v–81. Fols 81v–94 comprise a series of double-page spreads giving an ephemeris for each month in 1674: that for July is reproduced here as plate 2. The verso of fol. 94 is blank except for the lines, in Flamsteed's hand:

'Nullum Numen abest, si sit pruden[tia]/ sed te/
Nos facimus Fortuna deam,/ caeloque locamus. Juv: Sat:'

This is a quotation of the concluding lines of Juvenal's tenth satire, which may be translated: 'You would have no divinity, Fortune, if we had wisdom; but we make a goddess of you, and place you in the heavens'. In addition, fol. 94v has on it a printed label 'Royal Observatory/ Manuscripts' on which the figure '8' has been entered in a nineteenth-century hand.

The manuscript is slightly worn, and a few letters are lost at the edges of the pages. In the transcript, abbreviations and contractions have been slightly expanded and 'u' and 'i' modernized to 'v' and 'j' where appropriate. Deletions are recorded in the footnotes, but insertions are ignored, as are Flamsteed's alterations to parts of words made in the course of composition. It should be noted that the fewness of corrections suggests that the manuscript as we have it may well be a fair copy.

[88] This title is added in what appears to be a later hand.

[89] The friend who encouraged Flamsteed to carry out the project is perhaps likeliest to have been Immanuel Halton (cf. Baily, *Account*, pp. 21, 23, 26, 27, and *Oldenburg*, vi, 515, x, 156, 381). Other possibilities, however, are Richard Towneley and Sir Jonas Moore.

often urged to this taske by others of my ingenuous freinds, but still declind it, as too triviall & beneath the paines of an Astronomer; but at last considering with my selfe, what some had urg'd That the proper & primitive use of the Ephemeris (which was onely to shew the common Festivalls with the planets places, riseings, & settings) was either wholly lost, or perverted; our usuall & common Almanacks, besides the Calendar, conteining nothing lesse then these, I yeelded for once to undergoe the trouble of Composeing a Reformed Ephemeris, and to expose it to the publick View./f. 69

I resolved therefore to leave the idle or vaineglorious formes in use, and that I might assist the studies of the ingenuous Astronomer, to reduce the learned Heckers Ephemeris corrected, to an English Meridian, & so to contract it that it might be conteined in the usuall compasse of an Almanack. And (that I might not offend Christian eyes with a view & pretense of more foreknowledge then God ever indulged to his Prophets) to reject & omit those always vaine, often pernicious prædictions of the Weather, & State affaires, with which our miscalld Ephemerides have swelld of late yeares; and the Astrologer has so boasted of, that the Vulgar have esteemed them as the very oracles of God and thought that Artist very ignorant, that booke worth nothing, that has not impudently undertaken to determine, Whether the French should proceed victorious, the Dutch be humbled, the Imperialists come downe, or England enjoy her usuall peace & plenty the next yeare, with a faire seedtime & a seasonable harvest.[90]

These thinges the superstious [sic] Vulgar have Expected in every Ephemeris, & hee that has foretold most boldly of them, has been esteemd the very MERLIN: but Providence of late has so baffled these prædictions with /f. 69v/ events of actions, much different from those they pretend to have derived from theire schemes of the heavens, at the Suns Ingresses into the Cardinall pointes of the Eclipick, Eclipses, new and full moones præ & postventionall to them, and the Nativitys of illustrious persons[,] that tho, by theyre pretensions of things foreseene which either they durst not disclose, or had not liberty to print, or something afterwards found out, which then they thought not of, they make shift to hold up theire credit with the vulgar; Yet all knowing persons see theire evasions so poore, that they have long since questioned, and now can by experience prove theire arte to be so little infallible, that it may be justly doubted whether theire [sic] be any thing of truth in it. What my owne experience hath taught me concerneing it[91] I shall here relate, and I hope it will so far satisfie my reader, as that hee will not thinke it worth his while to gaine the like, with the expence of so much precious time & labor, as I wasted about it.

Tis now above seven yeares, since at the instigation of a freind well versed in Astrology, & then much persuaded of its certeinty, I began to study it: Hee lent me Gadbury's Doctrine, and afterwards his Collection /f. 70/ of Nativity's, on perusall of which I was veryly induced to beleive that this Art, for its many conveniencys, was worthy my greatest paines & study; and that

[90] 'the next yeare' deleted after 'harvest'.

[91] 'it' replaces an illegible deletion: 'this arte'?

nothing could be more certeine, or usefull, if my Memory could but retaine its many rules. I quicly learnt the fundamentalls, my freind discourseing them often to mee; after which, being curious of my owne fate, I resolved to experiment my skill, both in the genitures of my selfe, & some freinds, I easily learnt the estimate times of our births & not thinkeing my selfe scrupulouse enough, and that I might mistake much in correcting them by the *Trutine* of Hermes, or the *Animodar* of Ptolemy; I diligently enquired the times of severall notable accidents which had hapned in our younger yeares, that so I might correct the æstimate times by their much applauded methods of directions, which I did with greater curiosity then is usuall; & thus sometimes I could make a scheme represent two or three remarkable accidents, but then it would either misse in the discription of the person, (as I found it did egregiously in that of [an] ingenuous freinds which was noted very neare by his father) or that it would erre in timeing other accidents more remote: At first I thought the fault might lie in /f. 70v/ the Arkes of direction, which I found different in various Authors; Ptolemy numbring a yeare by a degree of Ascention, Kepler counteing the difference of the Suns Right ascentions at the Noone of the day of birth & the Noone following for one yeare, of the next day at noone for two, the third for three &c. Some others reckning the difference betwixt the Suns place at the noone of the birth day, & the noone following, & so of 2, 3, 4, 5, 6 days after, for 1, 2, 3, 4, 5, 6, yeare &c. And others mightily commending the sagacious choice & skill of Valentine Naboyd[92] who allowed always the Suns Meane diurnall motion for a yeares measure: I tried all these very carefully, & found in halfe a score nativity's, which I tried, that they would never answer the accidents but nearely, what way soever I tooke, or however I altered the parts of the figure.

This made me doubt sometimes the æstimate time of the nativity mistaken, at other times, & oftner, my owne skill; rarely at first, the Art, or Artists; I therefore set my selfe againe to the perusall of their Authors, wherein I found some differences betwixt the Arabian, and our Astrologers mentioned, & in Maginus Ephemerides,[93] that the tables of essentiall dignitys, the fundamentall of theire /f. 71/ pretended skill, were very much different from those used by the Arabian, the Eastern professors of it: This made mee cautious how I yeelded my assent to either of them, & with some further considerations, gave me occasion to take notice, how, & what Artifices the Astrologers used to defend the truth[94] of their Art, & cover its falsenesse: Now I soone observed, that when they could not find any direction, in a corrected nativity, to shew some notable accident hapned to the native, it shall be said the effect of that yeares Revolution; or, if that serve not theire turne, of the præcedents yet operateing. And something, in one of them at least, shall be found that may seeme to intimate the Event tho

92 I.e. Valentine Naibode or Nabod, 1527–1593, Professor of Mathematics at Cologne.

93 I.e. Giovanni Antonio Magini, 1555–1617, Professor of Mathematics at Bologna and an important astrological author.

94 'truth' is inserted, replacing 'higth', deleted. In the following line, 'I' has been repeated and the first deleted.

perhaps many better arguments might have beene drawne from them for the Contrary: And that it was usuall & common with them, when they come to shew the truth & exactnesses of their prædictions, to dissemble whatever aspects, directions, or transits shew accidents, contrary to those which have hapned; And to proclaime that (tho never so slight) which may have intimated them,[95] with so much noise, as may drown the least voice of the Contrary.

But if it be proved from the Revolution, that no such thinge could be by it foreshewne, my Artist /f. 71v/ will tell you presently it was the effect of the Profection, and if that cannot be proved rather fairely confesse the present defect or confusion of his memory, & desire further time to search his bookes, & consult his Authors, then acknowledge the least error in the art hee professes & lives by; And it shall be a miracle, if hee find not something, which as hee will interpræt it, may foreshew the accident; for hee pretends, so many rules, and has so many Aspects, Transits, Directions, Revolutions, Profections with the effects of Eclipses, Comets, & aspects to fixed stars, & theire positions (tho never made use of but at a dead list) to consider, and those judgd of by so many severall wayes, & methods, that it is impossible he should not meet with some thinge, which he may affirme to have influenced the event; tho it is hard to conceive, how the contrary aspects came to be overpowred, & lose theire operation; Nay if all faile, & you baffle him never so much, yet has he a rule, tho not in his booke, to tell you of, that shall both save his credit for Religion, and Art, & magnifie his ingenuity, tho hee may thanke the Poet for it,

Astra Regunt homines, sed regit Astra Deus.

But that which wrought most with me to desist the Study of this pretended Art, & reject it as false /f. 72/ after the experience of these failures, was, that I found there was no reason in nature,[96] why the Planets should have any influence upon our actions, or our thoughts which moderate them; And that it was impossible to conceive, how theire raies meeting each other at our earth with Trine, quartile, or sextile aspects; should be either beneficiall or nocent to us; or why the sun strong and well placed in a Nativity, should make the native capable of Command, & fortunate him with it; whereas there can not be any cause assigned, why hee should be more strong in one parte of the heavens then another; nor doth constant experience evince; that those attaine to higher preferment in whose nativitys hee was either essentially or accidentally dignified, before others, at whose births, accordeing to their rules, hee was ill placed, & debilitated, as I thinke I can sufficiently prove, & shall not forebeare to shew, if occasion serve. In the meane time my reader may take notice of theire sophistry if you demand how they can prove the stars to have any influences upon humane bodys and actions, they seeme to wonder at or pitty your ignorance, & to informe you

[95] 'it' has been deleted after 'intimated'.
[96] 'for it, or' is deleted after 'nature'. At the end of f. 71v, the catchword is '& vaine', although these words do not appear at the start of f. 72.

better they will readily evince, that by the influences of the sun the plants receave their growth /f. 72v/ and that the tydes are governd by the Moon from which effects of the luminaries upon our earth (whereof no Junior Sophister but thinkes himselfe able to give a reason, from the known frame of Nature, & the vulgar principles of Philosophy) they leave you to inferre the Influences of the other planets, & stars upon humane thoughts, & actions; as if there was no difference betwixt them, & these Mechanicall operations; which til they can prove, & shew us how the planets obteind theire Regiment over & came to divide Nations, Countrys, townes, trades[,] beasts, birds, herbes, & flowers amongst them & after what manner they direct theire operations; til then, (which will be till Doomsday,) they must permit us to esteeme them but the impudent canters, of an equally false & pernicious Juggle.

Of what ill consequences theire prædictions have beene, & how made use of in all commotions of the people against lawfull & established sovereignty, the historys of all insurrections, & our owne sad experience, in the late Wars, will abundantly shew the considerate; how they have erred in their Judgments, the same experience will informe us; & the pen of a learned person lately deceased, has, as I am informed, made evident to the World, & saved mee something of my intended paines:[97] Their prædictions of /f. 73/ the Weather, being wholly conjecturall (for I veryly beleive they are not so much fooles, as to trouble themselves to consult their rules or authors for it) are no lesse ridiculous; the aspects from which they pretend to derive it, serveing as well for *Egipt*, or *America* as *England*; Nay so small is the verity of their Art, that, if you examine their severall Authors, you will find them at a Contest wherein it consists: Mr Ramsey contends for the truth of the Elections of times, & Mr Sanders, if I mistake not, accords with him; Mr Lilly is wholly for Horary questions (he has raisd his estate by them;) but ingenuous Mr Gadbury laughs at both, thinkes the election of times a vanity, and, where the Nativity is not to be had, the judgment from an horary scheme but uncerteine.[98] I am of his opinion, which I beleive hee wants not proofe for, both against Mr Ramsey, & Lilly; & to prove that the Genethliacall part of Astrology, which hee professes, is as uncerteine as either of the other two; I shall here produce him the Nativity of the learned Bishop Hall, which I tooke from his owne Collection, pag. 91;[99] that wee may see

[97] This author has unfortunately not been identified; his work was apparently never published.

[98] Flamsteed here refers to William Ramesey, *Astrologia Restaurata* (London, 1653), esp. Book 3; Richard Saunders, *Palmistry, the Secrets thereof Disclosed . . . As also that Most Useful Piece of Astrology (long since promised) concerning Elections for every particular Occasion* (London, 1663); William Lilly, *Christian Astrology* (London, 1647), esp. Book 2 (on Lilly's practice, see Thomas, *Religion and the Decline of Magic*, pp. 305–22; Parker, *Familiar to All*, ch. 4); and John Gadbury, *Genethlialogia; or, the Doctrine of Nativities* (London, 1658). The latter work ignores elections altogether: although it has a section on horaries (ii, 233f.), it makes it clear that the interpretation of these should be dependent on a knowledge of nativities (ii, 236), to which the bulk of the work was devoted. Cf. also Gadbury's *Collectio Geniturarum* (London, 1662), sigs biv–2, in which he claimed nativities as his special study.

[99] The reference is to Gadbury's *Collectio Geniturarum*. Flamsteed has made a few very slight alterations to the notations on the horoscope, and he has omitted the figures for the right ascension of the mid-heaven and the oblique ascension of the ascendent given by Gadbury on the top and left-hand margins of the chart.

whether the heavens have any influence on the braine, or shew that ingenuity & learning, for which some persons have beene famous in their days./f. 73v

Let now Mr Gadbury shew us what signes here are, or promittors of that egregious learneing, eloquence, & piety, for which his workes have made this father deservd'ly famous; what significators of a Divine, or of preferment to be obteined by the study of Divinity: The [Sun], lord of the Ascendent, is not any good aspect either of the Ninth or its Lord: the Moone applyes to an opposition of [Jupiter] Lord of the Ninth, the significators of Divines, & their profession; which should shew, according to theire præcepts, an aversion in the Native to the study of it: further [Mercury] theire patron of learneing, & ingenuity is /f. 74/ both peregrine, & cadent, and in no Aspect either to the Ascendent, its Lord, the Moone or her dispositer, so that had this Geniture beene proposed to any skilfull pretender to Astrology with the name & partes of the Native conceald hee would imediately have judgd it rather of an Ideot then an eloquent & learned Divine.

But wee may not omit (least hee glory in noteing it for us) that here are some significators of the Natives præferment: the [Sun] who is Lord of the Ascendent being in [trine] of [Mars] Lord of the 10th house: [Saturn] and [Venus] tho in [opposition] to each other, yet if [sic] one in [sextile], the other in [trine], to the ascendent and Midheaven: But I dare affirme, that some persons with as greate signes have never risen to such high places; & others without them have acquired greater honors. which are therefore to be thought, onely to accord[100] with these positions by chance, and no more to commend Astrology then the accidentall concurrence of Ptolemys Numbers with an odde observation, may confirme his Epicyclar Hypothesis; which every one that is but very meanely versed in Astronomy knows to be false.

I know the Astrologer will yet urge, that the admirable concurrence of effects with directions, in all the other genitures, may very /f. 74v/ well excuse a faileing in one; and that in all the rest, they not onely shew the accidents hapned to the natives, but time them exactly too; Yes I acknowledge Mr Gadbury has so stated the times of his genitures that hee makes a shift to find directions for all the actions he mentions, or something to intimate them in the Revolution. But infinite other actions, hapning to his Natives, when there was nothing found to foreshew them, are passed over in silence; as may be noted in the late reverend Bishops geniture, in which, hee produces but one accident, with a direction to it, altho if hee could have found directions to suite them, hee might easily have collected the times of severall more remarkable occurrences, in those specialtys of his life, written by his owne hand, from which he tooke the time of his birth.[101]

But the Astrologer has his excuse ready for it; that it would have beene more labor to collect them, then hee had time to spare for, or that the time of the geniture was not very well noted or correct. Indeed Mr Gadbury is the most ingenuous person that can be in dissembling his faults and palliateing them by confessing some little inadvertencys; by which I suppose hee will endeavor /f. 75/ to excuse himselfe for haveing told us in his pittifull volume of Ephemerides that the Planet [Mars] shall cover a good part of the planet Jupiter on November the 5th. in the Morneing 1673. a little before Sun rise. when, by his owne tables, they are at that time above 27′ minutes distant, and both their diameters taken togeather make not above 50″ seconds.[102] This Astrologers cunning in covering the faults of his Art, & ignorance of those events, hee pretends to derive from it, is very superlative. Most of his genitures are chosen of deceased persons, wherin tis very hard to controvert the moment with him, in which haveing so stated the time that hee might find directions for the Most Notable accidents, he exposes them with no little pompe, & thence mightily prædicates the truth of Astrology, as if hee had foreseene each particular occurrence in the Natives life, & confuted all gainesayers: in the meane while concealeing, how much the estimate time of birth differs from what his correction makes it; and not heeding, that it is impossible to be certeine of the exact moment of a birth without carefull observation of it, since the Astrologer may erre in his judgment, & mistake an accident /f. 75v/ to be the effect of one direction when it may be the operation of the Revolution, onely, or some other direction or not

[100] 'with' has been accidentally repeated after 'accord' and the first deleted. On f. 74v, there is a tiny illegible deletion before 'hapning'.

[101] Hall wrote an autobiographical tract called 'Observations of some Specialties of Divine Providence in the Life of Joseph Hall, Bishop of Norwich'.

[102] See above, n. 78; for Gadbury's tables, see his *Genethlialogia*, part ii.

apprehended cause: for so many operators & influencers are there taught to be in Astrology, that should any one propose the time of his nativity two or three yeares false, & yet give an estimate time with the times of severall occurrences, not forgetting to pay him handsomely, hee shall find him a promiseing scheme, directions for all the accidents, a probable description of person, with severall, perhabs [sic] wholesome cautions; which if they deny, I thinke I can make good by Experiment.

But in the Genitures of persons liveing my author has beene more advisedly spareing of his prædictions, least the Event should not answer them; and not without some sence of a necessity as I veryly beleive. for where hee has more boldly adventured, it has beene with such successe, as I am apt to thinke will make him carefull hereafter, not to assert the truth of Astrology from the Genitures of any but deceased persons, in which hee may have some advantage in case of controversy; for from the Geniture of the present King /f. 76/ of Sweden Charles Gustave the 2d hee predicts the danger of death in the yeare 1661. *but*, says hee, *in the yeare 1663 at what time hee hath the [Moon] directed to the [conjunction] of [Mars], & the Ascendent to [conjunction] [Saturn] sine latitud[ine] hee may be in great danger of death should hee this yeare escape.* which he proves substantially enough in his Collection pag 42:[103] And to the prince of Orange pag 54. he foretells no lesse then certeine death in the Words of *Origanus*[104] in the 11th yeare of his age 1660. whereas all the World knows they are both no lesse alive then his Royal[105] highnesse the Duke of Yorke, to whom, from its evill directions & Revolution, hee prædicted the year 1667 to be fatall; Whereas hee has outliv'd it now six yeares, & wee hope, God will continue him yet many more, for the good of the Nation & service of his Majesty:

Suppose wee grant the planets some such influences as the Astrologers require; yet may wee aske him, how hee can be confident of his judgments, since there is not onely a controversy about the *rationalis modus* of erecting a figure yet undecided,[106] but himselfe is also confessedly ignorant of the influences of the fixed stars. That they have influences all confesse, & some pretend to use them when every /f. 76v/ thinge else fayles; but the aspects of planets to them, tho they confesse that they may have effects, they never consider. I demand therefore if it be not possible, that what is promised by the aspects of the planets, or directions of the Hylegiacall or other parts to

[103] Flamsteed has slightly altered the quotation from p. 42 of Gadbury's book, adding '[conjunction]' before 'Saturn', and omitting the additional phrase 'and the [Sun] nearly to the [square] of [Saturn], all by Direction' before 'he may'.

[104] I.e. David Origanus or Dost of Glatz, 1558–1628, Professor of Mathematics at Frankfurt on the Oder, whose *Astrologia Naturalis* was published in 1645.

[105] 'Royal' replaces 'Illustrious', deleted. The Duke of York does not appear in Gadbury's *Collectio Geniturarum*, but in Gadbury's *The Nativity Of the late King Charls Astrologically and Faithfully performed* (London, 1659), p. 111, the year 1667 is seen as 'most perillous and dangerous' for the Duke of York.

[106] Flamsteed here refers to the debate among astrologers as to where the cusps between the houses in the horoscope should be placed. The 'rational mode' was that of Regiomontanus (1436–76).

them, in a nativity, may not be contradicted by their aspects and directions to fixed stars, and how, on this account, wee may be ascertein'd of the truth of theire prædictions.

But what use, sayes the Astrologer, of the stars, if you admit not theire influences upon the bodys and actions of sublunary beings? Shine they for nothing but to direct by night the wandring saylour? or to excuse the absence of the sun, to the benighted traveller at land, worse then would some small candle? no, with the poet

> I'le ne're beleive that each particular stone,
> Has some peculiar vertue of its owne,
> And that the glorious stars of heaven have none:[107]

I know no one will force him, as for my selfe I shal ever say to him, *Jubeo stultum esse libenter*: But if Astrology fall, wee need not feare the heavens tumbling on our heads togeather with it, or the stars and planets becomeing uselesse. The Telescope, that best invention of this age, has some while since informed us; That the Moone is a solid body, full of inæqualities, vales, & prominencys, /f. 77/ & longer ones of late yeares, that the superficies of the other planets are alike unequall, & solid: yea that some of them are 1000 times bigger then our earths globe, & none lesse then halfe as bigge, and that all of them receave theire light as our earth from the sun: so that as the ingenuous Borellus[108] thinkes, it is very probable, they are the habitations of creatures, who are as wee, created to prayse theire maker. And for the fixed stars, since by probable conjectures, from cælestiall observations, some of them are found as much greater then our sun, as hee is bigger then the earth, it may accordeing to the Cartesian philosophy, be very reasonably thought, That they are placed immoveable in the centers of other *Vortices* where they may illuminate, as so many suns[,] systemes of planets, like our earth inhabitable & fild with creatures, perhaps more obedient to the lawes of theire maker, then its inhabitants; which planets yet, by reason of theire small remotions from the centers, or suns of theire severall systemes, and theire vast distances from our earth, become to us invisible.

And tell mee Reader, nay let the ingenuous Astrologer himselfe speake, if hee has not renounc'd his reason, & sworne to defend the Art, for the profit hee receaves by fooleing simple people with it; how it is possible the /f. 77v/ planets, reflecting, onely as the Moone, some small part of the suns light upon our earth, should have more influence on us then a good fire, a torche, or a Candle; from which, tho wee receive no small accommodations to our lives by theire heate, or light (neither of which from the planets much assist's us) yet we find no such influences on our thoughts or actions as the

[107] The sources of this quotation and of the Latin quotations here and on f. 71 have unfortunately not been identified. In the following sentence, 'need' has accidentally been repeated and the first one deleted.

[108] I.e. Giovanni Alphonso Borelli (1608–79), the Italian natural philosopher, whose *Theoricae Mediceorum Planetarum ex Causis Physicis Deductae* (of which Flamsteed owned a copy: Forbes, 'Library', p. 123) was published in 1666.

Astrologe[r] pretends to proceed from the stars: Nay let him say, if the Cælestiall bodies are not far more excellently usefull, whilest they may, & probably doe receave and susteine creatures to give theire maker perpetual praises; then hee would willingly suppose them, whilest he argues that they operate, hee knows not how, upon our globe; and influence, not onely the great, but triviall affaires, thefts, rapines, & debaucheries of it.

Since therefore Astrology finds no ground to sustaine it in nature, Since experience convinces us of its equall vanity and falsehood and observation has taught us a greater and truer use of the Cælestiall bodies, then the Astrologer[109] supposses them made for, I hope my ingenuous Reader will withdraw his credit, if formerly hee gave any, to this imposture. As for the fam'd professors of this science, I have no hopes to /f. 78/ reclaime them, because the profession – howmuchsoever foolish, and opposite to reason & the divine law, is yet no little gainefull, which is the onely reason why they so much endeaver to sustaine its credit by cloakeing those failures of which they cannot but be conscious. And that I expect no other reward for this plaine dealeing then the *titles* of *ignorant* & *peevish*, and to have theire Almanacks raile at mee perhaps a yeare or two togeather; for which I intreate them provide theire most substantiall Arguments, and I doubt not but theire owne workes and Authors will afforde mee, easy and ample answers to them:

Hee that will write a warrantable and usefull Ephemeris, ought in my opinion, togeather with the usuall account of the feasts moveable, and fixed, to give the places and latitudes of the planets, with theire riseings and settings, and if hee think it convenient, theire transits by fixed stars and each other, whereby hee may give the ingenuous occasion to exercise theire witts, and skill, in contemplateing the magnificent workes of an Almighty Wisdome, & give him praise and honor for and from them to which if any one shall adde any ingenuous /f. 78v/ notes concerneing the ordering of Orchards[,] gardens, or Agriculture, hee may doe well: but to fill up a sheet or two with idle Aphorismes and judgments in Astrology, or catalogues of the faires and Roades, in my mind shew no lesse, then an emptinesse of matter, conjoyned with an itch of scribling in the Exposer; with which most of our yearlinge writers are questionlesse infected, otherwise would they never trouble us with them, since they cannot be ignorant of theire error, & vanity; & that alltho they passe among the more ignorant vulgar, yet the more sober and ingenuous make them but theire scorne, and laughter.

To give a good example to such as are mindeinge to alter and reforme theire methods; I have adventured to transcribe an Ephemeris of the planets motions from the learned Heckers volume. How I have contracted it, & what is conteined in every Columne my reader will find by inspection, & the inscription on the top of it. The planets riseings and settings are computed for the latitude of 53 degrees, and theire places reduced to a meridian passing 5 minutes to the West from London, the parallell & Meridian intersecting each other, as nearely as can be estimated by our Maps exactly on the Navel

109 'the' has been adapted from 'they' and 'Astrologer' inserted.

of this Kingdome, to which place it being fitted, may very well and without sensible error serve for the whole: onely because /f. 79/ I find [Jupiter']s place in the heavens hath this last yeare beene observed some 14 minutes in consequence of Heckers numbers I have every where added them to his place therein. in Mars[110] Keplers numbers agree much better with the heavens, then theires who have most pretended to correct them[.] in the other planets I find there is some need of alteration, but I have beene forced to let them passe as I found them in the Ephemeris, it being a taske too large for my shreds of time, to give them at present that correction which I desire & intend hereafter.

But because that Keplers numbers, have not onely beene convinced of much falsity, but corrected also by the ingenuous Mr Horrox[111] in the motions of the luminaries; I leave the Ephemeris, to give you an account of the two Eclipses hapning this yeare, from the Horroccian Theory, and numbers accomodated to it; which I know by experience, to have sometimes agreed with the heavens, when most others would not come neare them within 15 minutes: The Visible Eclipses of this yeare to us are onely two, both of the Moone, whereof the first happens January the 12/22 in the morneing before sun rise: the apparent time of the Middle at Derby is 5h 15′ 02″ in the morneing at which time accordeing to the tables fitted to that Theory I calculate. /f. 79v/

The Suns true place	[Aquarius] 2° 40′ 12″
The Moones meane motion	4 04 08 59
The meane place of her Apog[eon]	10 20 37 18
The Argumentum Annuum	11 12 03 04
The Equation of the Apog[eon] sub.	5 52 53
The true place of the Apogeon	10 14 44 25
The excentricity of the orbe	64648
The moones meane Anomaly	5 19 24 34
The æquation sub.	0 1 28 48
The moones place in her orbite [Leo]	2 40 12
The Argument of latitude	5 23 38 29
distance betwix[t] the centers of the [Moon] & shade	33 18
the horizontall parallax of the [Moon] 61 58 [Sun] 15 [=]	62 13
the suns semidiameter	16 21
the semidiameter of the earths shadow	45 52
the moones semidiameter	16 33
therefore the partes obscured	29 07
motion of semiduration	52 47
time of semiduration	1h 28 41
of: reduction	0 2 26

Therefore.

110 An illegible deletion follows 'Mars': 'I esayd of'? It is presumably Hecker whom Flamsteed is implicitly criticising concerning Kepler's numbers. In the previous sentence, 'it' is deleted before 'may'; 'in consequence of' is a technical astonomical term denoting movement from east to west.
111 After 'Horrox', 'I leave the E . . .' has been deleted. For the 'Horroccian theory', see Horrocks, *Opera Posthuma*, pp. 465f.

The begining of the Eclipse at	3h	45′	21″	in the morn
Eclipticall Conjunction	5	11	36	at Derby
Middle or Greatest obscur[ation]	5	14	02	
The end	6	42	43	
duration	2	57	22	

digits eclipsed 10 dig. 33½′

The latter eclipse hapneth July the 7th in the evening: its middle at Derby 8h 35′ 02″ p:m: at which time by the aforesaid tables I find /f. 80/

The Suns true place	[Cancer] 25° 18′ 23″
The moones mean motion	9 21 42 05
The meane place of her Apog[eon]	11 10 18 03
The Argumentum Annuum	4 15 00 09
The excentricity of the orbe	55238
The Equation of the[112] Apogeon Sub.	11 32 50
The true place of the Apog[eon]	10 28 45 13
The moones meane Anom[aly]	10 22 54 15
The æquation add	3 36 28
The Moones place in her orbite [Capricorn]	25 18 23
The argument of latitude	11 25 37 56
distance betwixt the [moon] & shades centers	22 55
the horizontall parallax of the [moon] 55 32 [sun] 15 [=]	55 47
the Suns semidiameter. sub.	15 51
semidiameter of the earths shadow	39 56
the moones semidiameter	14 50
therefore the moone totally eclipsed & the darknesse extends beyond her limbe	2 11
the motion of halfe duration	49 44
of halfe duration in tenebris.	10 14
time of halfe duration hor	1 45 04
of halfe duration in tenebris	21 37
Therefore The begining of the Eclipse	6h 49′ 56″
Immersion: begining of totall darknes	8 13 25
true [opposition][113]	8 33 54
Middle of the Eclipse	8 35 02
Emersion: End of totall Darknesse	8 56 39
Finall end of the Eclipse.	10 20 06

The Sun sets 3 minutes after 8: at which time wee may behold the moone riseing Eclipsed 10 digits & presently after covered with the totall darkenesse. /f. 80v/

Last yeare came to my hands a little tractate of the learned Heckers intituled *Mercurius in Sole, seu Admonitio ad Astronomos Geographos rerumque cælestium curaosos de incursu [Mercur]ii in discum solis observando, Anno Christi 1674 April: 26:* in this hee determines from the Rudolphin tables, the time of

112 'Equation of the' is written over something else, which is not now legible.

113 The word 'conjunction' has been deleted and replaced by the symbol for opposition.

The begining of this transit at 11h 29′32″ before noon
the Middle 1 41 59: lat[itudine] [Mercurii]: 12′20″ A
the end 3 54 26 p m
the duration 4 24 54: at braniburge so that at London the Apparent time of these phases ought to be 48′ sooner.

Tho I justly might yet I shall not question the goodnesse of his Calculation but with his leave I will assert that *this Eclipse with us will scarce be visible* for the Rudolphin tables never were correct by any such observation nor could bee, & Mr Streets, corrected by two or more of these transits, make at Derby April 26

the begining: 11h 00′ 34″ lat[itudine] [Mercurii] 12′ 29″ A
Middle 12 49 34 cent[ral] dist[ance] 14 01
Emersion 14 38 34 lat[itudine] [Mercurii]. 15 05: So that this Appearance happens wholly in the night: & to us invisible:

Yet by reason Mr Street erres something in the place of the [upper node] of [Mercury],[114] I beleive his latitude will be lesse & the duration something longer; which those who are resident with our Antipodes may observe. In the /f. 81/ meane time I here present my reader with the types of these appearances & bid him farewell.

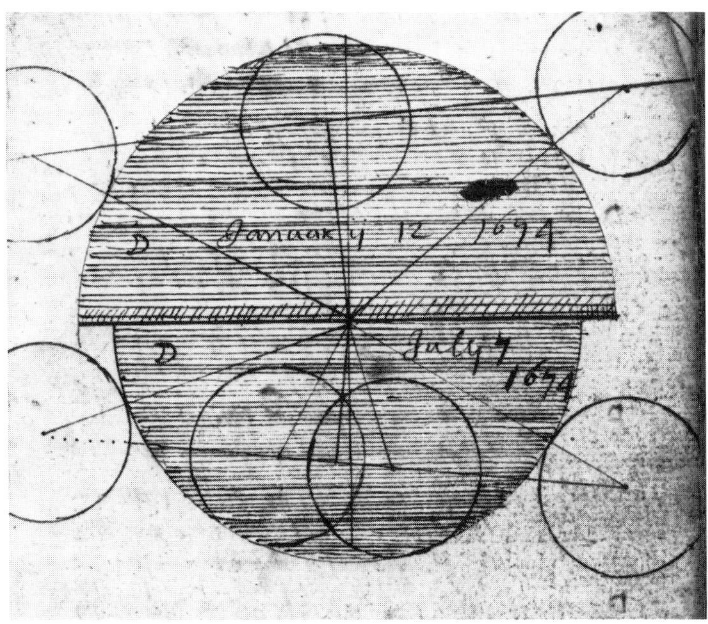

[114] Cf. Thomas Streete, *Astronomia Carolina* (London, 1661).

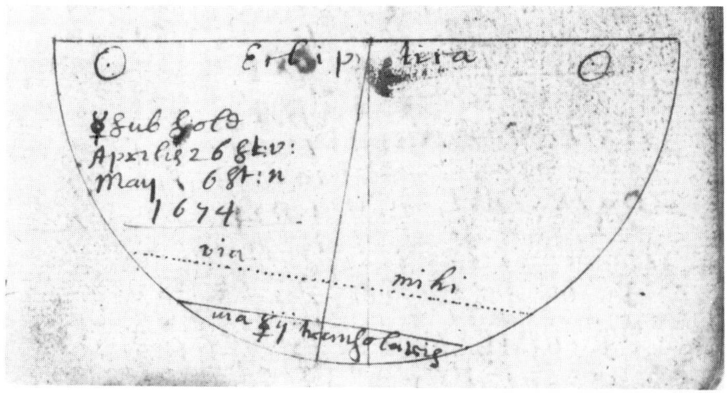

ACKNOWLEDGMENTS: The research for this article has been assisted by a grant from the Royal Society under their history of science research grant scheme. Material from RGO MS 1/76 has here been reproduced by permission of the Royal Greenwich Observatory. I am also grateful to the following for their help: Jenny Bray, Harold Brooks, Geoffrey Cornelius, Patrick Curry, Janet Dudley, Sally Grover, John Henry, Fred Lock, Giles Mandelbrote, Anthony Turner, Keith Walker, Mari Williams, Frances Willmoth and Paul Wood.

INDEX